普通高等教育园林景观类"十二五"规划教材

风景园林专业英语（第二版）

（景观设计）

主编 王欣 方薇

中国水利水电出版社
www.waterpub.com.cn

内 容 提 要

本教材是一本适用于风景园林（景观设计）专业的英语教材，从专业的角度针对提高该专业学生的英语能力编写。

本教材包括两部分，第一部分精选了西方现代风景园林（景观设计）思想的文章，主要内容包括理论历史、宪章法规、城市设计、公园设计、历史文化保护、生态设计、种植设计和设计过程等；第二部分精选的文章介绍12位对西方现代风景园林发展有重要贡献的著名风景园林师，以及他们的作品及设计理念。本教材针对该专业教学特点，对提高相关专业师生的英语水平有很大的帮助。同时本教材作者还精心收录了大量与专业相关的背景知识和网络信息，对该专业师生大有裨益。

本教材第二版在之前重印修订的基础上进行了改版，在原有基础上增加了词汇表，增加了部分难句和长句的译文，重新整理了全部译文，使之更符合中文阅读习惯。本教材还配套提供了相关教学辅助材料，可在 http://www.waterpub.com.cn/softdown 下载。

本教材可作为高等院校景观设计、风景园林、环境艺术设计、园林等专业的专业英语教材，也可供相关专业人员参考。

图书在版编目（CIP）数据

风景园林（景观设计）专业英语 / 王欣，方薇主编
. -- 2版. -- 北京：中国水利水电出版社，2013.3(2022.1重印)
普通高等教育园林景观类"十二五"规划教材
ISBN 978-7-5170-0672-5

Ⅰ．①风… Ⅱ．①王… ②方… Ⅲ．①园林设计－英语－高等学校－教材 Ⅳ．①H31

中国版本图书馆CIP数据核字(2013)第036601号

书　名	普通高等教育园林景观类"十二五"规划教材 **风景园林（景观设计）专业英语（第二版）**
作　者	主编　王欣　方薇
出版发行	中国水利水电出版社 （北京市海淀区玉渊潭南路1号D座　100038） 网址：www.waterpub.com.cn E-mail：sales@waterpub.com.cn 电话：(010) 68367658（营销中心）
经　售	北京科水图书销售中心（零售） 电话：(010) 88383994、63202643、68545874 全国各地新华书店和相关出版物销售网点
排　版	中国水利水电出版社微机排版中心
印　刷	天津嘉恒印务有限公司
规　格	210mm×285mm　16开本　22.5印张　846千字
版　次	2008年9月第1版　2008年9月第1次印刷 2013年3月第2版　2022年1月第5次印刷
印　数	12001—14500册
定　价	49.00元

凡购买我社图书，如有缺页、倒页、脱页的，本社营销中心负责调换

版权所有·侵权必究

编 委 会

主 编

王　欣（浙江农林大学园林学院）
方　薇（浙江农林大学外国语学院）

副主编

王洪涛（山东农业大学林学院）
其其格（北华大学林学院）
宋钰红（西南林学院园林学院）
赵彩君（中国．城市建设研究院）

编 委

王洪义（黑龙江八一农垦大学植物科技学院）
王　瑛（浙江农林大学园林学院）
纪　鹏（黑龙江八一农垦大学植物科技学院）
吴晓华（浙江农林大学园林学院）
唐世斌（广西大学林学院）
柳　丹（浙江农林大学环境与资源学院）
傅　凡（北方工业大学建筑工程学院）

如 何 使 用 本 书

写给读者

本教材适用于具有大学英语四级或相当水平的景观设计、风景园林及相近专业的专业人员。旨在使读者了解西方现代风景园林发展历史、常见理论、著名设计师和经典作品，并进一步提高他们专业英语阅读和表达的能力。

本教材第一部分材料大多选自风景园林名著，使读者最大限度地"原真性"地了解西方现代风景园林经典思想和理论。第二部分选择了在西方现代风景园林发展史上具有重要影响的 12 位设计师，介绍了他们的设计理念和作品，图文并茂，能较好地引发读者的阅读兴趣。

本教材英文难度总体适中，但由于作者个人写作风格及写作内容，各篇难度略有差异，读者可根据自身英文水平选择阅读。本书对难词难句均作了注释，为了让读者对相关西方历史文化、西方现代风景园林等背景知识有更深的了解，以便更好地理解课文，我们在文后编写了课文背景，可供参考的扩展阅读信息及 Tips——知识背景、查询技巧和表达技巧等，希望对读者有所帮助。

对学有余力的读者，本教材可作为阅读外文专著的入门读物。文中列举了很多扩展阅读的信息和途径，也通过课文的形式对西方现代风景园林名著做了一个概要的纵览。

兴趣是最好的老师，渴望了解西方现代风景园林，有志于推动中国风景园林建设，这才是学习本课的最大动力。

写给教师

对大多数教师而言，专业英语教学是一项颇具难度的课程。该课程时常面临一种尴尬：在有限的学时内，教学目标到底重在提高学生的英语水平，还是重在通过英文材料介绍专业内容。教学目标的模糊定位，严重影响了教学效果。

我们认为，专业英语教学重点不应该是大学英语式的语言教学，也不应该简单地用英文重新"解读"其他专业课程内容。专业英语应该成为风景园林相关专业人才培养环节中的有机组成部分——把英语作为一种工具，通过阅读英文专业文献拓展专业视野，掌握西方现代风景园林常见理论和经典作品，了解西方现代风景园林学科发展，而这正是国内一般高校相关专业教学内容中的空白。正因为此，此教材也适用于西方现代景观设计课程的双语教学。

本教材第 1 部分的内容基本是名著名篇，每个单元的第一课建议作为精读，后三课作为泛读；第 2 部分可作为学生扩展阅读材料。全书第 1 部分内容包括 9 个单元，加期终复习和考核，可分 10 个单元时间进行教学。建议教师根据人才培养方案、学生水平和专业差异等实际情况选择和安排教学内容。

本书对难词难句均作了注解，对课文及作者背景也进行了简要的介绍。教师还可根据 Sources of Additional Information 所列线索在互联网上找到更多相关材料。在使用教材中有任何疑问或建议请与主编单位联系，我们愿意免费提供本书相关音像资料、教案等与您交流，共同改进和完善本课程教学。

第二版前言

2011年，风景园林学科正式成为我国110个一级学科之一。风景园林行业迅速发展，日益走出中国国门成为世界风景园林大家庭中的重要一员，专业英语越来越成为新一代风景园林师的必备技能。自2008年出版《景观设计（风景园林）专业英语》以来，我们收到了全国各地许多热心师生的邮件，他们对本教材给予了充分的肯定，也提出了很多很好的建议。感谢广大读者对本教材的支持，你们的建议是本次改版的主要依据。

本教材是在2008年初版后重印修订了几次的基础上进行改版的，在基本保持原来内容不变的情况下，主要做了以下三方面的改进：

（1）增加了词汇表。把每个Unit中出现的专业词汇进行了汇总列表，方便读者学习查找。

（2）增加了部分难句和长句的译文。补充翻译了一些原来没有中文翻译的难句和长句，降低了英文原文的阅读难度。

（3）重新整理了全部译文，使之更准确更能符合中文阅读习惯；订正了错漏部分和因时代变化不再合适的内容。

在我们编写和修订教材过程中，始终得到中国水利水电出版社编辑的悉心指导，浙江农林大学有关部门的关心和帮助，各参编高校和教材使用单位师生发来各种反馈意见，在此一并表示真诚的感谢。

由于编写人员水平有限，时间仓促，书中缺点和错误难免，望读者批评指正，以便今后进一步修改补充。

本教材还配套提供了相关教学辅助材料，以便于读者学习和参考，可以在http：//www.waterpub.com.cn/softdown下载。

编 者

2013年1月

第一版前言

随着国际交流的增多和我国高等院校双语化教学的趋势，为景观设计、风景园林等专业师生编写一本题材对路、语言难度适中、能反映西方现代风景园林理论和实践概况的专业英语教材，已经成为一种需要。

本教材主要供景观设计（包括风景园林、环境艺术设计和园林设计等）相关专业本科生和研究生使用，同时也适用于具有大学英语四级水平或相当水平的专业人员。本教材旨在进一步提高他们对西方现代风景园林基本情况、常见理论和经典作品的了解；提高英语应用水平，特别是查找、阅读和理解风景园林专业英文文献，以及用英文交流专业内容的实际能力。

本书的编写主要遵循了以下几个原则：

（1）**注重选择经典材料**。本教材选材范围涉及西方现代风景园林发展有重大影响的名著名篇，重要英文教材、著名设计师和著名作品、权威机构发布的宣言和公告，在其中选择语言难度适中，符合我国高等教育实际情况的篇章。

（2）**突出展现西方现代风景园林概况**。本教材全面介绍西方风景园林学科的发展历史、主要理论、代表人物和经典作品。编者主要选择了20世纪70年代初到90年代末的内容，也选择了部分21世纪以来的思想和作品，以及少量能反映学科和职业发展历史的60年代以前的经典篇章。本教材除作为专业英语教材之外，也适于用作西方现代景观设计双语教学教材，以及西方园林史和园林规划设计等课程的教学参考书。

（3）**力求从实际出发，注重教材的实用性和针对性**。考虑到我国高等院校风景园林专业英语教学的一般状况，对有一定难度的语言点作了注解，增加了背景情况介绍、获取扩展阅读信息的途径等。文后问题和作业注重和我国风景园林规划设计实际问题的结合，以及从英文查询到英文专业论文写作的循序渐进训练方法。

景观设计（风景园林）专业英语是一个拥有双重目的的课程，既需要介绍西方现代风景园林规划设计基本知识，又是对专业英语运用的一项综合训练。本教材在七校专业英语教学探索的基础上编写而成，编委基本是具有丰富教学经验的教授、副教授和风景园林规划设计博士，突出教材的专业性和针对性。另外，本书的编写还邀请了英语语言文学专业教师加盟，并有 J. O. Simonds, Julius Fabos 等学者的排疑解惑，使我们更透彻地了解了英文材料和英文学习规律。在此对全体编委的通力协作和各位朋友的大力帮助表示感谢。

本教材编写分工如下：

王　欣（浙江农林大学园林学院：总体设计，材料收集，第 1、2、8 章）

方　薇（浙江农林大学外国语学院：英文审核）

王洪涛（山东农业大学林学院：第 3、9 章）

其其格（北华大学林学院：第 5、6 章，风景园林师 Peter Walker, Martha Schwarzt）

宋钰红（西南林学院园林学院：第 4 章）

赵彩君（中国·城市建设研究院：第 7 章，风景园林师 Julius Fabos）

王洪义（黑龙江八一农垦大学植物科技学院：风景园林师 Burle Marx）

王　瑛（浙江农林大学园林学院：风景园林师 John Ormsbee Simonds）

纪　鹏（黑龙江八一农垦大学植物科技学院：风景园林师 George Hargeaves）

吴晓华（浙江农林大学园林学院：风景园林师 Frederick Law Olmsted）

唐世斌（广西大学林学院：风景园林师 Geoffrey Jellicoe, Lawrence Halprin）

柳　丹（浙江农林大学环境与资源学院：风景园林师 Thomas D. Church）

傅　凡（北方工业大学建筑工程学院：风景园林师 Ian McHarg, Dan Kiley）

本教材的编写得到了浙江农林大学园林学院政策和经费上的大力支持，以及同行朋友的具体关心和协助，浙江农林大学刘桂玲、蒋健等学生为本教材承担了大量文字校核工作。在参编七校风景园林专业英语教学和教材编写过程中，广大学生给予了积极的配合和帮助，使我们的教学和教材越趋成熟。事实上，对本课给予热情支持的广大师生也是本教材的编写者。在此，对关心和帮助我们的师生领导表示衷心感谢。

鉴于作者学识所限，在文章选材和编写等方面存在不足之处，恳请广大读者和使用者批评指正。

编　者

2008 年 1 月

CONTENTS

如何使用本书
第二版前言
第一版前言

PART 1　Theory

Unit 1　Landscape Architecture & Landscape Architect ·················· 3
　　　　　风景园林师职业发展历史，工作范畴，工作程序
　Text　Harmony with the Living Earth ·················· 4
　　Further Reading A　Definition of the Profession of Landscape Architect ·················· 11
　　Further Reading B　What is Landscape Architecture ·················· 13
　　Further Reading C　Occupational Outlook: Landscape Architects ·················· 18
　Exercises ·················· 24
　Tips 欧美风景园林学相关称谓 ·················· 25

Unit 2　History & Theories ·················· 28
　　　　　现代主义风景园林设计理论，后现代主义风景园林设计理论，演变历史
　Text　Contemporary Meanings in the Landscape ·················· 29
　　Further Reading A　Post Modernism Looks beyond Itself ·················· 35
　　Further Reading B　Post-Postmodernism ·················· 41
　　Further Reading C　Landscape For Living ·················· 46
　Exercises ·················· 51
　Tips 风景园林主要英文网站 ·················· 52

Unit 3　Charters & Policy ·················· 54
　　　　　风景园林设计相关宪章，相关宣言，相关法规
　Text　The Venice Charter 1964 ·················· 55
　　Further Reading A　The Convention on Biological Diversity ·················· 60
　　Further Reading B　Beijing Charter: Towards an Integral Architecture ·················· 65
　　Further Reading C　Federal & State Regulations ·················· 70
　Exercises ·················· 74
　Tips 检索和获取风景园林英文资料方法 ·················· 75

Unit 4　Urban Landscapes Planning & Design ·················· 78
　　　　　城市规划设计，城市绿地设计
　Text　The Granite Garden ·················· 79

  Further Reading A The Growth of Suburbs ··· 84
  Further Reading B Landscape Design in the Urban Environment ················· 88
  Further Reading C The 100 Mile City ·· 95
 Exercises ··· 99
 Tips 欧美风景园林教学常见课堂活动形式 ·· 100

Unit 5 Parks & Recreation 102
   公园规划设计，风景游览地规划设计
  Text Urban Parks and Recreation ··· 103
  Further Reading A People's Parks: Design and Designers ······················ 110
  Further Reading B Parkways and Their Offspring ································· 115
  Further Reading C National Parks & National Forests ·························· 121
 Exercises ·· 125
 Tips 城市公园的发展历程 ·· 126

Unit 6 Historical & Cultural Landscape 129
   风景园林历史文化保护原则，保护方法，实例
  Text The Conservation Policy ··· 130
  Further Reading A The Heritage of Garden Art ··································· 138
  Further Reading B New Birth for Gettysburg? ····································· 142
  Further Reading C Place & Project: Moody Historical Gardens Design ········ 148
 Exercises ·· 152
 Tips 风景园林中的历史文化保护 ·· 153

Unit 7 Ecological Planning & Design 155
   风景园林生态设计理论，设计方法，设计实例
  Text Silent Spring: The Obligation to Endure ··· 156
  Further Reading A Bioregional Planning and Ecosystem Protection ············ 160
  Further Reading B Translating Environmental Values into Landscape Design ··· 166
  Further Reading C Sustainable Stormwater Management Program ············· 171
 Exercises ·· 176
 Tips 风景园林生态规划与生态设计 ·· 177

Unit 8 Planting Design 178
   风景园林种植设计
  Text Planting Design Through the Ages ··· 179
  Further Reading A Plants Dictate Garden Forms ·································· 187
  Further Reading B The Aesthetics of Planting Design ···························· 190
  Further Reading C Plant Strata, Size and Spatial Issues ························ 194
 Exercises ·· 198
 Tips 植物拉丁学名简介 ··· 199

Unit 9 The Process of Landscape Design 201
   风景园林设计过程解析，用地分析，方案设计，详细设计，方案解说
  Text The Design Process ·· 202
  Further Reading A The Site Planning Process ····································· 210
  Further Reading B Site Analysis ·· 214
  Further Reading C Five Rules for Explaining a Project ························· 218

Exercises 223
Tips 风景园林专业英语论文写作 224

PART 2 Major Figures & Their Works

Unit 10 229
  Frederick Law Olmsted 230
  John Thomas D. Church 238
  Geoffrey Jellicoe 245
  Roberto Burle Marx 255

Unit 11 262
  John Ormsbee Simonds 263
  Lawrence Halprin 269
  Dan Kiley 276
  Ian Lennox McHarg 284

Unit 12 290
  Julius Gy. Fabos 291
  Peter Walker 301
  Martha Schwartz 309
  George Hargeaves 317

附录 I   Bibliography 324

附录 II   Illustration Credits 329

附录 III   Glossary 333

PART 1
Theory

该篇共设九个专题，内容可分三个部分：第一部分是西方现代风景园林概述，包括风景园林与风景园林师，风景园林历史与理论，以及相关宣言法规；第二部分是风景园林职业范畴，包括城市设计，公园与风景游览地，风景园林历史文化保护，以及生态规划设计和种植设计；第三部分是风景园林设计过程。这九个专题基本涵盖西方现代风景园林的各个方面，以期让读者熟悉常用专业词汇和句法，学会阅读和理解风景园林专业英文文献。课文材料基本来自经典文献，力图让读者了解西方现代风景园林职业和学科的发展，掌握常见理论。

从19世纪中期到20世纪20年代初，是西方现代风景园林的探索时期。随着机器工业的发展和资本社会的成熟，人们对物质环境的要求也大大提高，思想、文化、审美情趣等发生了巨大的变化，由此启发并引导了西方风景园林师们对现代主义的追求。得益于弗雷德里克·劳·奥姆斯特德（Frederick Law Olmsted）等人的开拓性工作，一个新兴的行业——继承传统而又有长足发展的风景园林行业（Landscape Architecture）随之诞生。1938～1941年间，哈佛大学设计研究生院学生詹姆士·罗斯（Jame Rose）、丹·凯利（Dan Kiley）及盖瑞特·埃克博（Garrett Eckbo）发表了一系列文章，提出郊区和市区园林的新思想。这些文章及其中的革命性观点动摇并最终导致哈佛大学风景园林系"巴黎美术学院派"教条的解体和现代设计思想的建立，并推动美国的风景园林规划设计行业朝向符合时代精神的"现代主义"方向发展，这就是著名的"哈佛革命"（Harvard Revolution）。第二次世界大战以后到20世纪70年代，西方社会和经济状况的稳定和繁荣使现代风景园林蓬勃发展，设计领域不断扩展。20世纪70年代以后，西方风景园林进入"后现代主义"时期。这是一个对现代主义进行反思和重新认识的时期，现代风景园林在原有的基础上不断地进行调整、修正、补充和更新。传统园林的价值重新得到尊重，其他学科的介入使其知识领域更为广阔，西方现代风景园林进入了多元化发展的时期。

从1900年以来，西方现代风景园林经历了从产生、发展到壮大的过程，但是它并没有表现为一种单一的模式。不同的风景园林设计师们结合当地的特点和各自的美学认识形成了多样化的集团，产生了不同流派和风格，他们丰富多彩的理论和实践不断地为现代风景园林添加新的内容。虽然具体形式不尽相同，但西方现代风景园林各流派面对着相似的社会文化生态问题，普遍产生了"现代主义"或"后现代主义"思想，总体上呈现出有规律的思想理论发展。

西方现代风景园林也许是最难描述的行业，其工作范畴涉及从区域规划到庭院设计的广大范围。其中城市公园和风景游览地是风景园林师最传统的职业领域之一。从奥姆斯特德开始，一代代美国风景园林师锐意进取，美国城市公园已经从游憩园、休闲设施发展到开放空间系统，以国家公园体系为代表的风景游览地建设也硕果累累。西方风景园林师在城市建设的另一个领域也大放光彩：风景园林师介入城市规划设计在西方本来就有悠久的历史，而20世纪60年代以来约翰·奥姆斯比·西蒙兹（John Ormsbee Simonds）等美国风景园林师主导规划设计了大量新镇，更是大大拓宽了风景园林师的职业领域。

文化保护和生态保护是西方现代风景园林运动中的两大主题。20世纪70年代，人们对自身生存环境和人类文化价值的危机感日益加重，在经历了现代主义初期对环境和历史的忽略之后，传统价值观重新回到社会，环境保护和历史保护成为普遍的意识。从那时起，不仅仅保护传统意义上的"历史景观"，也出现了一些保护现代工业景观的设计，说明70年代以后人们对社会发展的每一个脚印的珍惜。伊恩·伦诺克斯·麦克哈格（Ian Lennox McHarg）、克莱尔·瑞尼格（Clair Reiniger）等人的生态规划设计方法是整个西方社会环境保护运动在风景园林规划设计中的具体折射，美国六七十年代以《寂静的春天》为号角的"环境保护运动"，对西方现代风景园林的发展产生了深远的影响。科学，而不仅仅是艺术，成为西方现代风景园林的又一主题。

种植设计是风景园林设计中一项具有悠久历史的工作，从古罗马花园到英国自然风景园到今天的自然生态式种植，其设计美学和设计方法具有很多相似的成分。而罗伯特·布雷·马科斯（Roberto Burle Marx）等现代风景园林师以其独特的个人风格，为传统的种植设计增添了新的内涵。

本篇的最后一章介绍风景园林设计过程，选择了从用地调查分析、场地规划、细部设计，到方案汇报的一系列文章，课文材料基本来自英文风景园林优秀教材。该篇内容集中介绍常用风景园林规划设计专业词汇和表述方法，旨在切实提高读者专业英语应用能力。

Unit 1

Landscape Architecture & Landscape Architect

- 风景园林师职业发展历史
- 工作范畴
- 工作程序

It's believed that the lifetime goal and work of the landscape architect is to help bring people, the things they build, their communities, their cities—and thus their lives—into harmony with the living earth.

By John Ormsbee Simonds

Text

Harmony with the Living Earth

By John Ormsbee Simonds

> 本文从人的自然秉性，人类营建居住环境的成功和失败经验，论述 LA 职业的社会责任：协调人与自然的关系，创造理想化的人居环境，全文可分以下四个部分。
> - 人类成为构筑环境的牺牲品
> - 不恰当的规划
> - 创造健康的生活环境
> - 重新发现自然，建设美好家园

We are the victims of our own building. We are trapped, body and soul, in the mechanistic surroundings we have constructed about ourselves (Figure 1.1). <u>Somewhere in the complex process of evolving our living spaces, cities, and roadways, we have become so absorbed in the power of machines,</u> so absorbed in the pursuit of new techniques of building, so absorbed with new materials that we have neglected our human needs. Our own deepest instincts are violated. Our basic human desires remain unsatisfied. <u>Divorced from our natural habitat, we have almost forgotten the glow and exuberance of being healthy animals and feeling fully alive.</u>

在生活空间、城市、道路逐步进化发展的复杂过程中，我们太依赖机械的力量。

离开了自然生境的我们几乎已经忘记身为一个健康的动物并感受生命意义时所拥有的活力与激情。

Figure 1.1 We are trapped in the fuming working of our own machinery

Many contemporary ailments—our hypertensions and neuroses—are no more than the physical evidence of rebellion against our physical surroundings and frustration at the widening gap between the environment we yearn for and the stifling, artificial one we planners have so far contrived.

Life itself is dictated by our moment-by-moment adjustment to our environment. Just as the bacterial culture in the petri[1] dish must have its scientifically compounded medium for optimum development and the potted geranium[2] cutting its proper and controlled conditions of growth to produce a thriving plant, so we—as complicated, hypersensitive[3] human organisms—must have for our optimum development a highly specialized milieu[4]. It is baffling that the nature of this ecological framework has been so little explored. Volumes have been written on the conditions under which rare types of orchids may best be grown; numerous manuals can be found on the proper raising and care of guinea pigs, white rats, goldfish, and parakeets[5], but little has been written about the nature of the physical environment best suited to human culture. Here is a challenging field of research.

The naturalist tells us that if a fox or a rabbit is snared in a field and then kept in a cage, the animal's clear eyes will soon become dull, its coat will lose its luster, and its spirit will flag[6]. So it is with humans too long or too far removed from nature. For we are, first of all, animals. We are creatures of the meadow, the forest, the sea, and the plain. We are born with a love of fresh air in our lungs, dry paths under our feet, and the penetrating heat of the sun on our skin. We are born with a love for the feel and smell of rich, warm earth, the taste and sparkle of clear water, the refreshing coolness of foliage[7] overhead, and the spacious blue dome of the sky. Deep down inside we have for these things—a longing, a desire sometimes compelling, sometimes quiescent but always it is there.

It has been proposed by many sages[8] that, other things

---

[1] petri：培养皿（实验室用于培养细菌等的有盖小玻璃盆）
[2] geranium：天竺葵
[3] hypersensitive：非常敏感的
[4] milieu：环境
[5] parakeet：[鸟] 长尾小鹦鹉
[6] flag：枯萎，衰退
[7] foliage：树叶
[8] sage：贤人，圣人

being equal, the happiest person is one who lives in closest, fullest harmony with nature. It might then be reasoned: Why not restore humans to the woods? Let them have their water and earth and sky, and plenty of it. But is the primeval forest-preserved, untouched, or simulated—our ideal environment? Hardly. For the story of the human race is the story of an unending struggle to ameliorate the forces of nature. Gradually, laboriously[1], we have improved our shelters, secured a more sustained and varied supply of food, and extended control over the elements to improve our way of living.

在这个说法之下，继而可以质问：为什么不使人类重返森林？让他们拥有丰富的纯自然的水、土地和天空呢？但是这种原始森林——保存完好的，未被染指的森林——就是我们的理想环境吗？很难说是。因为人类的历史是一场永不停息的和自然斗争并改善自然的过程。

What alternatives, then, are left? Is it possible that we can devise a wholly artificial environment in which to better fulfill our potential and more happily work out our destiny? This prospect seems extremely doubtful. A perceptive analysis of our most successful ventures in planning would reveal that we have effected the greatest improvements not by striving to subjugate nature wholly, not by ignoring the natural condition or by the thoughtless replacement of the natural features, contours, and covers with our constructions, but rather by consciously seeking a harmonious integration. This can be achieved by modulating ground and structural forms with those of nature, by bringing hills, ravines[2], sunlight, water, plants, and air into our areas of planning concentration, and by thoughtfully and sympathetically spacing our structures among the hills, along the rivers and valleys, and out into the landscape.

通过深入分析那些最成功的规划，可以看出：我们的成功之处在于有意识地寻求一种和谐的整体而不是粗暴地去征服自然，不是忽略自然条件，也不是武断地用人工建筑掩盖自然特征和外观。

We are perhaps unique among the animals in our yearning for order and beauty. It is doubtful whether any other animal enjoys a "view", contemplates the magnificence of a venerable oak, or delights in tracing the undulations of a shoreline. We instinctively seek harmony; we are repelled by disorder, friction, ugliness, and the illogical. Can we be content while our towns and cities are still oriented to crowded streets rather than to open parks? While highways slice through our communities? While freight trucks[3] rumble past our churches and our homes? Can we be satisfied while our children on their way to school must cross and recross murderous[4] traffic ways? While traffic itself must jam in and out of the city,

是否有其他动物能欣赏"景观"，会对古老的橡木带来的壮观景致产生思考，或者喜欢追寻海岸线的波动，这是值得怀疑的。

---

[1] laboriously: 费劲地
[2] ravines: 沟壑，峡谷
[3] freight trucks: 货车
[4] murderous: 危险的

morning and evening, through clogged and noisy valley floors, although these valley routes should, by all rights, be green free-flowing parkways leading into spacious settlements and the open countryside beyond (Figure 1.2).

Figure 1.2 The visual clutter of strip roadside development

We of contemporary times must face this disturbing fact: our urban, suburban, and rural diagrams are for the most part ill-conceived. Our community and highway patterns bear little logical relationship to one another and to our topographical[1], climatological, physiological[2], and ecological base. We have grown, and often continue to grow, piecemeal[3], haphazardly[4], without reason. We are dissatisfied and puzzled. We are frustrated. Somewhere in the planning process we have failed.

Sound planning, we can learn from observation, is not achieved problem by problem or site by site. Masterful planning examines each project in the light of an inspired and inspiring vision, solves each problem as a part of a total and compelling concept which, upon consideration, should be self-evident. Stated simply, a central objective of all physical planning is to create a more salubrious[5] living environment—a more secure, effective, pleasant, rewarding way of life. Clearly, if we are the products of environment as well as of heredity, the nature of this environment must be a vital concern. Ideally it will be one in which tensions and frictions have been in the main eliminated, where we can achieve our full potential,

---

〔1〕 topographical：地形学的
〔2〕 physiological：生理学的
〔3〕 piecemeal：一个个地
〔4〕 haphazardly：无规则地
〔5〕 salubrious：有益健康的

and where, as the planners of old Peking envisioned[1], man can live and grow and develop "in harmony with nature, God, and with his fellow man."

Such an environment can never be created whole; once created, it could never be maintained in static form. By its very definition it must be dynamic and expanding, changing as our requirements change. It will never, in all probability, be achieved. <u>But striving toward the creation of this ideal environment must be, in all landscape design, at once the major problem, the science, and the goal.</u>

> 但是在风景园林中，创造这样一个理想环境，必定是一个主要的问题，是一种科学，也是一个目标。

All planning must, by reason, meet the measure of our physical dimensions. It must meet the test of our senses: sight, taste, hearing, scent, and touch. It must also consider our habits, responses, and impulses. Yet it is not enough to satisfy the instincts of the physical animal alone. One must satisfy also the broader requirements of the complete being.

As planners, we deal not only with areas, spaces, and materials, not only with instincts and feeling, but also with ideas, the stuff of the mind. Our designs must appeal to the intellect. They must fulfill hopes and yearnings. <u>By empathetic planning, one may be brought to one's knees in an attitude of prayer, or urged to march, or even elevated to a high plane of idealism.</u> It is not enough to accommodate. Good design must delight and inspire.

> 带有感情色彩的规划可以使人如祈祷般地虔诚，或者充满了前进的斗志，或者甚至被提升到了理想主义的境界。

Aristotle, in teaching the art and science of persuasion, held that to appeal to any person an orator[2] must first understand and know that person. He described in detail the characteristics of men and women of various ages, stations, and circumstances and proposed that not only each person but also the characteristics of each person be considered and addressed. A planner must also know and understand. <u>Planning in all ages has been an attempt to improve the human condition. It has not only mirrored but actively shaped our thinking and civilization.</u>

> 在任何时期，规划的目的都是为了改善人类的生存条件，这不仅反映了我们的思想和文明，而且积极地推动了它们的发展。

With our prodigious[3] store of knowledge, we have it within our power to create on this earth a veritable garden paradise. But we are failing. And we will fail as long as our

---

[1] envisioned: 构想
[2] orator: 演讲者，雄辩家
[3] prodigious: 巨大的

plans are conceived in heavy-handed violation of nature and nature's principles. The most significant feature of our current society is not the scale of our developments but rather our utter disdain[1] of nature and our seeming contempt for topography, topsoil, air currents, water sheds, and our forests and vegetal mantle[2]. We think with our bulldozers[3], plan with our 30-yard carryalls. Thousands upon thousands of acres of well-watered, wooded, rolling ground are being blithely plowed under and leveled for roads, homesites, shopping centers, and factories. <u>Small wonder that so many of our cities are (climatologically speaking) barren deserts of asphalt, masonry, glass, and steel.</u>

难怪这么多的城市已经成为（从气候学角度看）由沥青、石块、玻璃和钢铁等组成的荒漠。

For the moment, it seems, we have lost touch. Perhaps, before we can progress, we must look back. We must regain the old instincts, relearn the old truths. We must return to the fundamental wisdom of the gopher[4] building a home and village and the beaver engineering a dam. We must apply the planning approach of the farmer working from day to day in the fields, fully aware of nature's forces, forms, and features, respecting and responding to them, adapting them to a purpose. We must develop a deeper understanding of our physical and spiritual ties to the earth. We must rediscover nature.

Notes

1. Extracts from

Text

Simonds, John Ormsbee. *Landscape Architecture: A Manual of Site Planning and Design* [M]. New York: McGraw-Hill Publishing Company, 1997: 4~9.

Pictures

(1) Figure 1.1: John Ormsbee Simonds. *Landscape Architecture: A Manual of Site Planning and Design* [M]. New York: McGraw-Hill Publishing Company, 1997: 6.

(2) Figure 1.2: John Ormsbee Simonds. *Landscape Architecture: A Manual of Site Planning and Design* [M]. New York: McGraw-Hill Publishing Company, 1997: 7.

2. Background Information

(1) John Ormsbee Simonds 约翰·奥姆斯比·西蒙兹（1913～2005 年）。美国现代风景园林先驱，著名的风景园林师、规划师、生态学家。曾任美国风景园林师协会

---

[1] disdain：蔑视
[2] mantle：覆盖物
[3] bulldozers：推土机
[4] gopher：囊地鼠

（ASLA）主席等职。一生在著作、设计实践、教育三方面均取得令人羡慕的成就。

（2）*Landscape Architecture：A Manual of Site Planning and Design*《风景园林学：场地规划设计指南》。

1961 年出版，初版名为 *Landscape Architecture：The Shaping of Man's Natural Environment*《风景园林学：人类自然环境的形成》，自第二版后副标题改为 *A Manual of Site Planning and Design*（场地规划与设计指南）。第三修订版于 1997 年出版。

该书由作者在卡内基——梅隆大学授课内容基础上编写而成，是北美 60 多所大学风景园林专业学生必读教材。全书以风景园林规划设计工作步骤为次序；从基本原则基地选择到各组成要素的具体设计，从私家花园设计到区域规划，形成一个完整有序的体系。在 1999 年 ASLA 成立 100 周年世纪回顾活动中，被评为"十大最具影响的风景园林书籍"之一。

本文选自该书绪论部分，有删节。

Further Reading A

Definition of the Profession of Landscape Architect

By IFLA

Landscape Architects conduct research and advise on planning, design and stewardship of the outdoor environment and spaces, both within and beyond the built environment, and its conservation and sustainability of development. For the profession of landscape architect, a degree in landscape architecture is required. Tasks include:

(1) Developing new or improved theories, policy and methods for landscape planning, design and management at local, regional, national and multinational levels.

(2) Developing policy, plans, and implementing and monitoring proposals as well as developing new or improved theories and methods for national parks and other conservation and recreation areas.

(3) Developing new or improved theories and methods to promote environmental awareness, and undertaking planning, design, restoration, management and maintenance of cultural and/or historic landscapes, parks, sites and gardens.

(4) Planning, design, management, maintenance and monitoring functional and aesthetic layouts[1] of built environment in urban, suburban, and rural areas including private and public open spaces, parks, gardens, streetscapes, plazas, housing developments, burial grounds, memorials; tourist, commercial, industrial and educational complexes; sports grounds, zoos, botanic gardens, recreation areas and farms.

(5) Contributing to the planning, aesthetic and functional design, location, management and maintenance of infrastructure such as roads, dams, energy and major development projects.

(6) Undertaking landscape assessments including environmental and visual impact assessments[2] with view to developing policy or undertaking projects.

(7) Inspecting sites, analysing factors such as climate, soil, flora[3], fauna[4], surface and subsurface water and drainage; and consulting with clients and making recommendations regarding methods of work and sequences of operations for projects related to the landscape and built environment.

(8) Identifying and developing appropriate solutions regarding the quality and use of the built environment in urban, suburban an rural areas and making designs, plans and working drawings, specifications of work, cost estimates and time schedules.

---

[1] layout：设计，布局
[2] assessments：评估
[3] flora：植物群
[4] fauna：动物群

(9) Monitoring the realisation and supervising the construction of proposals to ensure compliance[1] with plans, specifications of work, cost estimates and time schedules.

(10) Conducting research, preparing scientific papers and technical reports, developing policy, teaching, and advising on aspects regarding landscape architecture such as the application of geographic information systems, remote sensing, law, landscape communication, interpretation and landscape ecology.

(11) Managing landscape planning and design projects.

(12) Performing related tasks.

(13) Supervising other workers.

Notes

1. Extracts from

Text

International Federation of Landscape Architecture. Definition of the Profession of Landscape Architect for the International Standard Classification of Occupations [EB/OL]. http://www.iflaonline.org/resources/policy/pdf/ifla_definition.pdf, 2007-04-29.

2. Background Information

IFLA (International Federation of Landscape Architecture) 国际风景园林师联合会。

IFLA 于 1948 年在英国剑桥成立，是一个风景园林的国际组织。在世界范围内分设三区，包括中区——非洲中部和欧洲，东区——中国等，以及西区——美国和加拿大等。IFLA 秘书办公室于 2007 年在比利时布鲁塞尔成立。

IFLA 网站：http://www.iflaonline.org.

IFLA 在线期刊：www.iflajournal.org.

---

[1] compliance：依从；和……一致

Further Reading B

What is Landscape Architecture

By ASLA

A Profession in Demand

From city council rooms to corporate boardrooms, there is increasing demand today for the professional services of landscape architects. This trend reflects the public's desire for better housing, recreational and commercial facilities, and its expanded concern for environmental protection. Residential and commercial real estate developers, federal and state agencies, city planning commissions, and individual property owners are all among the thousands of people and organizations in America and Canada that will retain the services of a landscape architect this year.

More than any of the other major environmental design professions, landscape architecture is a profession on the move. It is comprehensive by definition-no less than the art and science of analysis, planning design, management, preservation and rehabilitation[1] of the land. In providing well-managed design and development plans, landscape architects offer an essential array of services and expertise that reduces costs and adds long-term value to a project.

Clear differences do exists between landscape architecture and the other design professions. Architects primarily design buildings and structures with specific uses, such as homes, offices, schools and factories. Civil engineers apply scientific principles to the design of city infrastructure such as roads, bridges, and public utilities. Urban planners develop a broad overview of development for entire cities and regions. Landscape architects touch on all the above mentioned design professions, integrating elements from each of them. While having a working knowledge of architecture, civil engineering and urban planning. landscape architects take elements from each of these fields to design aesthetic and practical relationships with the land.

A Diverse Profession

Landscape architecture is one of the most diversified of the design professions.

Landscape architects design the built environment of neighborhoods, towns and cities while also protecting and managing the natural environment, from its forests and fields to rivers and coasts. Members of the profession have a special commitment to improving the quality of life through the best design of places for people and other living things.

In fact, the work of landscape architects surrounds us. Members of the profession are involved in the planning of such sites as office plazas, public squares and thoroughfares[2]. The attractiveness of

---

[1] rehabilitation: 修复
[2] thoroughfare: 大街

parks, highways, housing developments, urban plazas, zoos and campuses reflects the skill of landscape architects in planning and designing the construction of useful and pleasing projects.

Tracing the Profession's Roots

The origin of today's profession of landscape architecture can be traced to the early treatments of outdoor space by successive ancient cultures, from Persia and Egypt through Greece and Rome. During the Renaissance[1], this interest in outdoor space, which had waned during the Middle Ages, was revived with splendid results in Italy and gave rise to ornate villas, gardens, and great outdoor piazzas.

These precedents, in turn, greatly influenced the chateaux[2] and urban gardens of 17th-century France, where landscape architecture and design reached new heights of sophistication and formality. The designers became well-known, with Andre le Notre[3], who designed the gardens at Versailles and Vaux-le-Vicomte[4], among the most famous of the early forerunners of today's landscape architects.

In the 18th Century, most English "landscape gardeners", such as Lancelot "Capability" Brown[5], who remodeled the grounds of Blenheim Palace[6], rejected the geometric emphasis of the French in favor of imitating the forms of nature.

One important exception was Sir Humphrey Repton[7]. He reintroduced formal structure into landscape design with the creation of the first great public parks Victoria Park in London (1845) and Birkenhead Park in Liverpool (1847). In turn, these two parks would greatly influence the development of landscape architecture in the United States and Canada.

Frederick Law Olmsted: "Father of American Landscape Architecture"

The history of the profession in North America begins with Frederick Law Olmsted, who rejected the name "landscape gardener" in favor of the title of "landscape architect", which he felt better reflected the scope of the profession.

In 1863, official use of the designation "landscape architect" by New York's park commissioners marked the symbolic genesis of landscape architecture as a modern design profession. Olmsted became a pioneer and visionary for the profession. His projects illustrate his high professional standards, including the design of Central Park in New York with Calvert Vaux[8] in the late 1850's and the U.S. Capitol Grounds in the 1870's. Olmsted and the Brookline, Mass., firm he founded advanced the concept of parks as well-designed, functional public green spaces amid the grayness of the urban areas through the well practiced principles of landscape architecture and city planning.

---

[1] Renaissance：文艺复兴时期

[2] chateaux：城堡

[3] Andre le Notre：勒·诺特尔（1603~1700年），17世纪法国古典主义造园大师

[4] the gardens at Versailles and Vaux-le-Vicomte：凡尔赛宫花园和沃—勒—维贡特符邸花园，两个著名的法国17世纪古典主义园林

[5] "Capability" Brown："可为"布朗（1716~1783年），英国著名自然风景园设计师。由于他对任何立地条件下的园林建设都表现得很有把握，并有一句口头禅"大有可为"（It had great capabilities），因此人们称他为"可为布朗"

[6] Blenheim Palace：布伦海姆宫（邱吉尔庄园），位于英国牛津郡伍德斯托克镇附近，修建于1705~1722年 园林部分由"可为"布朗设计，是英国自然风景园代表作品

[7] Humphrey Repton：汉弗莱·莱普顿（1752~1818年），英国著名园林设计师

[8] Calvert Vaux：卡尔弗特·沃克斯，建筑师，奥姆斯特德的长期合作伙伴

Early Developments: Late 1800's

In the ensuing years, the profession of landscape architecture broadened. It played a major role in fulfilling the growing national need for well-planned and well-designed urban environments.

Urban parks, metropolitan park systems, planned suburban residential enclaves and college campuses were planned and developed in large numbers, climaxing with the City Beautiful movement[1] at the turn of the century.

Although the profession itself grew slowly, its early practitioners, including Olmsted, Vaux and Horace Cleveland, were among the first to take part in the town planning movement and to awaken interest in civic design. Olmsted also joined other early landscape architects in working on projects in other urban settings, such as at Yosemite Valley and Niagara Falls.

In 1899, the American Society of Landscape Architects was founded by 11 people in New York, most of them associated with Olmsted. The Society continued to represent landscape architects throughout the United States. In 1900, Olmsted's son, Frederick Law Olmsted Jr., organized and taught at Harvard University's first course in landscape architecture.

Broadening and Diversifying: The 20th Century

Landscape architecture continued to influence the city beautification and planning movement well into the 20th century, as growing cities used the services of professionally- trained landscape architects.

The L'Enfant Plan[2] for the nation's capital was revived and expanded by the McMillan commission[3] of 1901. Chicago, Cleveland and other cities also used landscape architects to lay out comprehensive development plans.

By the 1920's, urban planning separated from architecture and landscape architecture as a separate profession with its own degree programs and organizations. Yet, landscape architecture continued to remain a major force in urban planning and urban design.

During and after the Depression[4], opportunities to design national and state parks, towns, parkways and new urban park systems broadened the profession. The orientation of American landscape architecture returned to its roots in public projects—a trend which has continued throughout the mid-20th century to today.

The Profession of the Future

The years ahead promise new developments and challenges to the ever-broadening profession. With environmental concerns becoming increasingly important, landscape architects are being called upon to bring their expertise to the table to help solve complex problems. Rural concerns are attracting landscape architects to farmland preservation, small town revitalization, landscape

---

[1] City Beautiful movement：城市美化运动。19世纪90年代到20世纪初在北美兴起的城市景观建设运动

[2] L'Enfant Plan：郎方规划（1791年）。美国华盛顿特区最初城市规划，由法国人郎方完成

[3] McMillan commission：麦克米伦委员会，即1901年成立以参议员詹姆士·麦克米伦为主席的"华盛顿特区公园改造委员会"（Park Improvement Commission of the District of Columbia），成员包括小奥姆斯特德等

[4] Depression：美国经济大萧条（1929~1933年）

preservation, and energy resource development and conservation. Advances in computer technology have opened the field of computerized design, and land reclamation has become a major area of work for members of the profession.

Landscape architects have even begun to use their skill within indoor environments (e.g. atriums) and enclosed pedestrian[1] spaces have been incorporated into commercial development projects. From southern California to the Maine coast, the names of landscape architecture firms appear on signs heralding future developments, as more people seek the expertise and services of the profession.

Furthermore, the future also promises increase cooperation among landscape architects and other design professionals. As interest in the profession continues to grow, students are studying of the profession in increasing numbers nearly 60 universities and colleges in the United States and Canada now offer accredited baccalaureate and post-graduate programs in landscape architecture. Forty-five states license landscape architects. Today, headquartered in Washington, D.C., the American Society of Landscape Architects has grown to nearly 12,000 members in 47 chapters.

During the past decades, landscape architects have responded to the increased demand and professional responsibilities with new skills and expertise. More and more businesses appreciate the profession and the value that it brings to a project. The public praises the balance achieved between the built and natural environments. According to landscape architectural educator, author and ASLA Fellow, Lane Marshall:

"The future of the profession is bright. We are growing in size and stature each day. The profession is expanding its borders constantly and stands at the cutting edges of exciting new practice areas. There are landscape architects who are mortgage bankers, developers, business managers, architects, engineers, and lawyers. Since 1899, the profession has grown steadily and now stands at the threshold of a new period of growth."

The profession of landscape architecture continues to evolve as it meets the challenges of a society interested in improving the quality of life and the wisdom with which mankind uses the land in many ways, landscape architects are shaping the future.

Notes

1. Extracts from

Text

American society of landscape architects, What is Landscape Architecture [EB/OL]. http://www.asla.org/nonmembers/what_is_asla.cfm, 2007-04-29.

2. Background Information

ASLA (American Society of Landscape Architects) 美国风景园林师协会。

1899年，在弗雷德里克·劳·奥姆斯特德等人的发起下，11个风景园林师在美国纽约成立该协会，如今已发展成为一个国际性组织。

3. Sources of Additional Information

沃—勒—维贡别墅（Vaux-le-Vicomte）

---

[1] pedestrian：行人

http://www.vaux-le-vicomte.com

布伦海姆宫（Blenheim Palace）

http://www.blenheimpalace.com

美国华盛顿特区（Washington DC）规划建设历史

http://dcpages.com/History

奥姆斯特德

http://www.olmstedsociety.org

ASLA 网站

www.asla.org

Further Reading C

Occupational Outlook: Landscape Architects

By Bureau of Labor Statistics, U. S. Department of Labor

Nature of the Work

Everyone enjoys attractively designed residential areas, public parks and playgrounds, college campuses, shopping centers, golf courses, parkways, and industrial parks. Landscape architects design these areas so that they are not only functional, but also beautiful, and compatible with the natural environment. They plan the location of buildings, roads, and walkways, and the arrangement of flowers, shrubs, and trees (Figure 1.3).

Figure 1.3 Landscape architect

Landscape architects work for many types of organizations—from real estate development firms starting new projects to municipalities constructing airports or parks—and they often are involved with the development of a site from its conception. Working with architects, surveyors, and engineers, landscape architects help determine the best arrangement of roads and buildings. They also collaborate with environmental scientists, foresters, and other professionals to find the best way to conserve or restore natural resources. Once these decisions are made, landscape architects create detailed plans indicating new topography[1], vegetation, walkways, and other landscaping details, such as fountains and decorative features.

In planning a site, landscape architects first consider the nature and purpose of the project and the funds available. They analyze the natural elements of the site, such as the climate, soil, slope of the land, drainage, and vegetation; observe where sunlight falls on the site at different times of the day and examine the site from various angles; and assess the effect of existing buildings, roads, walkways, and utilities on the project.

After studying and analyzing the site, landscape architects prepare a preliminary[2] design. To account for the needs of the client as well as the conditions at the site, they frequently make changes before a final design is approved. They also take into account any local, State, or Federal regulations, such as those protecting wetlands or historic resources. In preparing designs, computer-aided design (CAD) has become an essential tool for most landscape architects. Many landscape architects also use

---

[1] topography: 地形学
[2] preliminary: 初步的

video simulation to help clients envision the proposed ideas and plans. For larger scale site planning, landscape architects also use geographic information systems technology, a computer mapping system.

Throughout all phases of the planning and design, landscape architects consult with other professionals, such as civil engineers, hydrologists[1], or architects, involved in the project. Once the design is complete, they prepare a proposal for the client. They produce detailed plans of the site, including written reports, sketches, models, photographs, land-use studies, and cost estimates, and submit them for approval by the client and by regulatory agencies. When the plans are approved, landscape architects prepare working drawings showing all existing and proposed features. They also outline in detail the methods of construction and draw up a list of necessary materials. Landscape architects then mainly monitor the implementation[2] of their design, with general contractors or landscape contractors usually directing the actual construction of the site and installation of plantings.

Some landscape architects work on a variety of projects. Others specialize in a particular area, such as residential development, street and highway beautification, waterfront improvement projects, parks and playgrounds, or shopping centers. Still others work in regional planning and resource management; feasibility, environmental impact, and cost studies; or site construction. Increasingly, landscape architects are becoming involved with projects in environmental remediation[3], such as preservation and restoration of wetlands or abatement of stormwater run-off in new developments. Historic landscape preservation and restoration is another important area where landscape architects are increasingly playing an important role.

Most landscape architects do at least some residential work, but relatively few limit their practice to individual homeowners. Residential landscape design projects usually are too small to provide suitable income compared with larger commercial or multiunit residential projects. Some nurseries offer residential landscape design services, but these services often are performed by design professionals with fewer formal credentials such as landscape designers, or by others with training and experience in related areas.

Landscape architects who work for government agencies do site and landscape design for government buildings, parks, and other public lands, as well as park and recreation planning in national parks and forests (Figure 1.3). In addition, they prepare environmental impact statements and studies on environmental issues such as public land-use planning. Some restore degraded[4] land, such as mines or landfills[5]. Other landscape architects use their skills in traffic-calming, the "art" of slowing traffic down through use of traffic design, enhancement of the physical environment, and greater attention to aesthetics[6].

Working Conditions

Landscape architects spend most of their time in offices creating plans and designs, preparing

---

[1] hydrologist: 水文学者
[2] implementation: 执行
[3] remediation: 纠正
[4] degraded: 退化的
[5] landfill: 垃圾填埋场
[6] aesthetics: 美学

models and cost estimates, doing research, or attending meetings with clients and other professionals involved in a design or planning project. The remainder of their time is spent at the site. During the design and planning stage, landscape architects visit and analyze the site to verify that the design can be incorporated into the landscape. After the plans and specifications are completed, they may spend additional time at the site observing or supervising the construction. Those who work in large national or regional firms may spend considerably more time out of the office traveling to sites away from the local area.

Salaried employees in both government and landscape architectural firms usually work regular hours; however, they may work overtime to meet a project deadline. Hours of self-employed landscape architects vary depending on the demands of the projects on which they are working.

Training, Other Qualifications, and Advancement

A bachelor's or master's degree in landscape architecture usually is necessary for entry into the profession. A bachelor's degree in landscape architecture takes 4 or 5 years to complete. There also are two types of accredited master's degree programs. The most common type of master's degree is a 3-year first professional degree program designed for students with an undergraduate degree in another discipline. The second type of master's degree is a 2-year second professional degree program for students who have a bachelor's degree in landscape architecture and who wish to teach or specialize in some aspect of landscape architecture, such as regional planning or golf course design.

In 2004, 59 colleges and universities offered 77 undergraduate and graduate programs in landscape architecture that were accredited by the Landscape Architecture Accreditation Board of the American Society of Landscape Architects. College courses required in these programs usually include technical subjects such as surveying, landscape design and construction, landscape ecology, site design, and urban and regional planning. Other courses include history of landscape architecture, plant and soil science, geology, professional practice, and general management. The design studio is another important aspect of many landscape architecture curriculums[1]. Whenever possible, students are assigned real projects, providing them with valuable hands-on experience. While working on these projects, students become more proficient[2] in the use of computer-aided design, geographic information systems, and video simulation.

In 2004, 47 States required landscape architects to be licensed or registered. Licensing is based on the Landscape Architect Registration Examination (L.A.R.E.), sponsored by the Council of Landscape Architectural Registration Boards and administered in two portions, graphic and multiple choice. Each portion of the testing is conducted over two days. Admission to the exam usually requires a degree from an accredited school plus 1 to 4 years of work experience under the supervision of a registered landscape architect, although standards vary from State to State. Currently, 14 States require that a State examination be passed in addition to the L.A.R.E. to satisfy registration requirements. State examinations, which usually are 1 hour in length and completed at the end of the L.A.R.E., focus on laws, environmental regulations, plants, soils, climate, and any other characteristics unique to the State.

---

[1] curriculums: 课程

[2] proficient: 精通

Persons planning a career in landscape architecture should appreciate nature, enjoy working with their hands, and possess strong analytical skills. Creative vision and artistic talent also are desirable qualities. Good oral communication skills are essential; landscape architects must be able to convey their ideas to other professionals and clients, and to make presentations before large groups. Strong writing skills also are valuable, as is knowledge of computer applications of all kinds, including word processing, desktop publishing, and spreadsheets. Landscape architects use these tools to develop presentations, proposals, reports, and land impact studies for clients, colleagues, and superiors. The ability to draft and design using CAD software is essential. Many employers recommend that prospective landscape architects complete at least one summer internship with a landscape architecture firm in order to gain an understanding of the day-to-day operations of a small business, including how to win clients, generate fees, and work within a budget.

Many landscape architects are self-employed because start-up costs, after an initial investment in CAD software, are relatively low. Self-discipline, business acumen[1], and good marketing skills are important qualities for those who choose to open their own business. Even with these qualities, however, some may struggle while building a client base.

Those with landscape architecture training also qualify for jobs closely related to landscape architecture, and may, after gaining some experience, become construction supervisors, land or environmental planners, or landscape consultants.

Employment

Landscape architects held about 25,000 jobs in 2004. Almost 6 out of 10 workers were employed in firms that provide architectural, landscape architectural, engineering, and landscaping services. State and local governments were the next largest employers. About 1 out of 4 landscape architects was self-employed.

Employment of landscape architects is concentrated in urban and suburban areas throughout the country; some landscape architects work in rural areas, particularly those employed by the Federal Government to plan and design parks and recreation areas.

Job Outlook

Employment of landscape architects is expected to increase faster than the average for all occupations through the year 2014. In addition to growth, the need to replace landscape architects who retire or leave the labor force will produce some additional job openings. Employment will grow because the expertise of landscape architects will be highly sought after in the planning and development of new residential, commercial, and other types of construction to meet the needs of a growing population. With land costs rising and the public desiring more beautiful spaces, the importance of good site planning and landscape design is growing. In addition, new demands to manage stormwater run-off in both existing and new landscapes, combined with the growing need to manage water resources in the Western States, should cause increased demand for this occupation's services.

New construction also is increasingly contingent upon compliance with environmental regulations,

---

[1] acumen: 敏锐

zoning laws, and water restrictions, which will spur demand for landscape architects to help plan sites that meet these requirements and integrate new structures with the natural environment in the least disruptive way. Landscape architects also will be increasingly involved in preserving and restoring wetlands and other environmentally sensitive sites.

Continuation of the Transportation Equity Act for the Twenty-First Century also is expected to spur employment for landscape architects, particularly through State and local governments. This Act, known as TEA-21, provides funds for surface transportation and transit programs, such as interstate highway construction and maintenance, and environment-friendly pedestrian and bicycle trails.

In addition to the work related to new development-and construction, landscape architects are expected to be involved in historic preservation, land reclamation, and refurbishment[1] of existing sites. They are also doing more residential design work as households spend more on landscaping than in the past. Because landscape architects can work on many different types of projects, they may have an easier time than other design professionals finding employment when traditional construction slows down. Opportunities will vary from year to year, and by geographic region, depending on local economic conditions. During a recession[2], when real estate sales and construction slow down, landscape architects may face greater competition for jobs and sometimes layoffs[3].

Earnings

In May 2004, median annual earnings for landscape architects were $53,120. The middle 50 percent earned between $40,930 and $70,400. The lowest 10 percent earned less than $32,390 and the highest 10 percent earned over $90,850. Architectural, engineering, and related services employed more landscape architects than any other group of industries, and there the median annual earnings were $51,670 in May 2004.

In 2005, the average annual salary for all landscape architects in the Federal Government in nonsupervisory[4], supervisory, and managerial positions was $74,508.

Because many landscape architects work for small firms or are self-employed, benefits tend to be less generous than those provided to workers in large organizations.

Notes

1. Extracts from
Text
Bureau of Labor of U. S. Department of Labor, Occupational Outlook Handbook (2006 - 07 Edition): Landscape Architects [EB/OL]. http://www.bls.gov/oco/ocos039.htm, 2007 - 04 - 28. (Unit 1/Further Reading C)
Picture
Figure 1.3: Bureau of Labor of U. S. Department of Labor. Occupational Outlook Handbook (2006 -

---

[1] refurbishment: 整修
[2] recession: 经济不景气
[3] layoff: 失业
[4] nonsupervisory: 非管理的

07)：Landscape Architects [EB/OL]. http://www.bls.gov/oco/ocos039.htm, 2007 - 04 - 28.

2. Background Information

本文节选自美国劳工部劳工统计署《2006～2007年职业展望手册·风景园林师》，有部分删节。该手册是选择职业的官方指导书，相关统计数据截止到2006年8月4日。"风景园林师"部分全面描述了美国风景园林职业的基本状况，并对该职业的前景做出预测。全文分以下9个部分：

(1) Significant Points（重要提示）。

(2) Nature of the Work（工作特征描述）❶。

(3) Working Conditions（工作条件）。

(4) Training, Other Qualifications, and Advancement（技能训练、资格和提高）。

(5) Employment（求职状况）。

(6) Job Outlook（职业展望）。

(7) Earnings（收入）。

(8) Related Occupations（相关职业）❷。

(9) Sources of Additional Information（更多信息）❸。

---

❶❷❸ 本书略。

Exercises

1. Questions（阅读思考）

认真阅读本单元文章，并讨论以下问题：
（1）风景园林学的根本宗旨和目标是什么。
（2）谈谈风景园林师的社会责任。
（3）谈谈风景园林师的工作范畴。
（4）联系我国情况，谈谈风景园林学科和职业将面临的机遇和挑战。

2. Skill Training（技能训练）

（1）查找并阅读某国风景园林协会官方网站英文网页，记录其内容构成、特色和资源。
（2）简要介绍某国风景园林协会英文网站主要内容，重点介绍其特色，并以"面向 21 世纪的世界各国风景园林学科和职业"为主题，进行交流和讨论。

Tips

欧美风景园林学相关称谓

风景园林学，一般译为 Landscape Architecture。由于国情和发展状况不同，本学科在欧美各国有一些不同称谓，反映着研究和实践的侧重点。讨论各种称谓的含义及出现背景，对我们了解欧美风景园林学科发展历史，深入理解相关英文资料是很有帮助的。

在英文资料中出现的风景园林学相关称谓主要有 Gardening，Landscape Gardening，Landscape Architecture 和 Landscaping 等。

1. Gardening

几乎每个国家的相关协会和组织都把 Gardening（庭园学）作为现代风景园林学的一部分。在美国风景园林师联合会（ASLA）对 LA 的解释中，把法国、意大利花园，乃至罗马、希腊、波斯对"户外空间"的关注和营造看成"LA 职业的源头"，把英国造园家兰斯洛特·布朗（Lancelot Brown）和法国造园家安德鲁·勒·诺特尔（Andre le Notre）看成是奥姆斯特德之前该职业伟大的先行者（参看本节 Further Reading B）。直至今天，欧洲一些国家仍然喜欢使用 Gardening、Garden designer 等称谓。

在英国，Garden Designers（society of garden designers，http://www.sgd.org.uk）和 Landscape Architects（The Landscape Institute，http://www.landscapeinstitute.org）各有协会，但实际上工作范畴大多重叠，名称也经常混用（如 Garden Designers 和 Landscape Designers），并无特别明显的界线。

美国一些文献提道：Gardening 和 LA "都关注户外空间的设计（the design of outdoor space）"，并且都用五种要素来进行设计：植物（Vegetation），地形（Landform），水体（Water），铺地（Paving），建构筑物（Structures）。正因为此，Gardening 和 LA 在今天并非那么泾渭分明，名称的不同只是工作的侧重点不同罢了。

2. Landscape Gardening

Landscape Gardening 指 18 至 19 世纪英国盛行的自然风景园理论和实践。18 世纪早期英国庭园设计的理论家 J. 爱迪逊（J. Addition）、A. 波贝（A. Pope）等都直接或间接地将绘画作为庭园设计的范本。18 世纪中叶，诗人与庭园理论家 W. 申斯通（W. Shenstone）用 Landscape 和 garden 的组合词 Landscape garden 表示这种按自然风景画构图方式创造的庭园，以示和欧洲几何式庭园的不同。杰出的造园学家 C. 布里奇曼（C. Bridgeman）、W. 肯特（W. Kent）、布朗和 H. 雷普顿（H. Repton）等人把英国自然风景园推向高潮，产生了一系列理论和重要作品（如 Humphrey Repton，*Sketches and Hints on Landscape Gardening*，1795），Landscape Gardening 一词也逐渐流行，成为自然风景园学的称谓。

Landscape Gardening 和 Landscape Architecture 关系非常密切。英国自然风景园传入美国，安德鲁·杰克逊·唐宁（Andrew Jackson Downing）把雷普顿等人的设计思想应用到美国乡村庄园园林设计之中；奥姆斯特德曾仔细研究过英国著名风景画家和园林设计师的作品，从他的纽约中央公园等作品中不难发现英国自然风景园的影子。无论是称为 landscape gardener 的唐宁，还是称为 Landscape architect 的奥姆斯特德及其后继者，都深受 Landscape Gardening——英国自然风景园学的影响。

3. Landscape Architecture

美国学者 P. 普雷基尔与 N. 沃克曼（P. Pregill & N. Volkmao）在《园林史》中认为 Landscape Architecture 一词最早出现于 1828 年吉尔伯特·凉·梅森（Gilbert Laing Meason）写的一本讨论意大利绘画的专著中（*Landscape Architecture of the Great Painters of Italy*），特指意大利风景画中的风景建筑。英国造园师约翰·克劳迪斯·罗顿（John Claudius Loudon，1783～1843 年）在花园杂志上发表文章时把该词借用在英国自然风景园中，罗顿的美国崇拜者唐宁沿用了这一名称，用 Landscape architecture 特指"风景建筑"。奥姆斯特德是唐宁的好友和崇拜者，他使用了 Landscape architecture 一词但修改了该词的含义重点：特指一种在建筑或建筑群之中的风景——美国纽约中央公园就是最早的例子。

而职业称谓 Landscape architect 的出现也与奥姆斯特德及其合作建筑师考福特·弗克斯（Calvert Vaux）有关。1863 年，奥姆斯特德和弗克斯在给纽约公园委员会的信中落款使用"Landscape Architects"，此后成为他们两人共同落款的惯例（见本书 Frederick Law Olmsted 课文）。据说奥姆斯特德本人对该称谓并不十分满意。

随着 1899 年美国风景园林师协会（ASLA）成立，1900 年哈佛大学率先开设 Landscape Architecture 专业方向，尤其是 1948 年国际风景园林师联盟（IFLA）成立，Landscape Architecture 和 landscape architect 逐渐为业内人士所接受。今天，Landscape Architecture 已经成为影响最广的风景园林学称谓。

4. Landscaping 和其他

尽管 Landscape Architecture 和 landscape architect 已基本得到业内认同，但仍然存在许多反对声音。英国于 1929 年成立风景园林师协会，英文名为 The Institute of Landscape Architects，但后来改名为 The Landscape Institute，据说是为了和"建筑师"相区别——确实，无论从字面意思还是历史渊源，Landscape Architecture 容易给人造成是建筑学的一部分或分支学科的错觉。与奥姆斯特德同时代的很多设计师仍然坚持用 Landscape Gardening 和 Landscape gardener，今天也有一些设计师喜欢称呼自己为 Landscape planner 或 Landscape designer，把学科称为 Landscaping，认为这样更加贴切。

20 世纪 30 年代以来，尤其是 70 年代"环境运动"大潮兴起之后，环境保护观念日益深入人心，风景园林逐渐从私人宅园、公司园区、城市公园中走出来，走向更为广阔的自然生态保护、人居环境建设等。风景园林学科的内涵更加丰富，外延更加广阔，由此出现了 Environmental planning，Earth-cape planning 等称谓，称谓的发展生动地说明了工作重心的延伸和学科的发展。

5. 结语

如上所述，现代风景园林学正是基于协调人与自然的关系，改善人居环境的目标，伴随着城乡建设中不断涌现的新问题而发展——由于文化背景、经济发展速度、社会制度等众多因素差异，世界各国风景园林学发展水平和侧重点都会有所差别，从而产生了各种称谓。值得注意的是，在学习西方文化的时候，我们最需要做的或许是针对国内人居环境建设中的实际问题，从学科发展角度出发，借鉴西方发达国家经验，为改善人居环境提高人们生活质量而架构、发展和完善学科与职业体系。

Landscape Architecture 该译成什么？有译为"景观建筑学"，有译为"景观设计学"，也有译为"风景园林学"的，国内学术界争论日久，莫衷一是。从某种程度上说，称谓的争执没有多大意义，学科发展的动力并非来自"接轨"某国，或者"赶上"某某世界潮流。无视解决本国实际问题的根本目标，纠缠名称的更新，并把某一国某一时的做法引为圭臬，这不是学科发展的正确途径。

附：1899～1999 年影响美国风景园林发展进程的十本经典书籍（选自 1999 年 ASLA 成立 100 周年纪念活动中 *Landscape Architecture* 期刊文章）

(1) *Town Planning in Practice*（Raymond Unwin, 1909）.
(2) *Landscape Design*（Henry Hubbard & Theodora Kimball, 1917）.
(3) *Gardens in the Modern Landscape*（Christopher Tunnard, 1938）.
(4) *Landscape for Living*（Garrett Eckbo, 1950）❶.
(5) *Gardens are for People*（Thomas Church, 1955）❷.
(6) *Landscape Architecture*（John Simonds, 1961）❸.
(7) *Design with Nature*（Ian McHarg, 1969）❹.
(8) *Design on the Land*（Norman Newton, 1971）❺.
(9) *Handbook of Landscape Architectural Construction*（Jot Carpenter, 1976）.
(10) *Social Life of Small Urban Spaces*（William H. Whyte, 1980）.

---

❶❷❸❹❺ 本书中有摘选文章。

Unit 2

History & Theories

- 现代主义风景园林设计理论
- 后现代主义风景园林设计理论
- 演变历史

Our new profession is essential to the future of civilization in any part of the world, for it is the only one that can unite these two aspects. It can explore the meaning and purpose of life and express its findings on a scale never before conceived; and in doing so, it can lift the spirit from the body, like the dream of the Chinese philosopher of old.

By Sir Geoffrey Jellicoe

Text

Contemporary Meanings in the Landscape

By Sir Geoffrey Jellicoe

> 本文从风景园林的艺术性入手，强调了其特殊的使命及现阶段的涵义，并且指出风景园林师的最终目标应该是创造出与自然相和谐的人居环境。全文可分以下三个部分。
> - 风景园林的现实意义
> - 人类发展经历了四个阶段，而风景园林的涵义也相应地经历了四个阶段
> - 风景园林师的终极目标

I think there is little doubt that landscape design is on the threshold[1] of becoming the most comprehensive of the arts. It can never be the most pure, for the abstract quality that creates music, poetry, painting and sculpture is too compromised by the realities of life. But it has one quality possessed by no other art that is not ephemeral, which relates it uniquely to the way of thought of the modern world: the sense of constant change.

它永远不可能是最纯粹的艺术，因为音乐、诗歌、绘画及雕塑这些的抽象的艺术形式太受现实的影响了。但它具备了其他永恒的艺术形式所不具备的特质：不断的变化性。正是这一特质和现代社会的思维方式密切相关。

In reviewing the vast sweep of this Congress, the tremendous achievement of its members since the war, and its potential future, I think there is one aspect that does not come easily to practitioners involved in day-to-day anxieties but is to my mind outstandingly the most important for all who are concerned with the created environment. Today, the repression of instinct[2] is the cause of quite unnecessary unhappiness. I believe it is a prime purpose of our profession to release the instincts, and that only by probing into the different layers of the mind can we acquire the knowledge to do so.

纵观该艺术对人类的强大冲击，它在战后取得的辉煌成就，以及它璀璨的前景，我认为设计师们平常容易忽略一个问题，而在我看来，这个问题正是最为重要的，所有关心现有环境的人都应该对此高度重视。

There seem to be four recognizable phases of the subconscious: that of the sub-tropical forester, when our physiology[3] was perfected; the hunter; the settler; and

人类的潜意识似乎分为四个阶段：亚热带丛林居民意识，那时人类的生理机能是最完美的；狩猎者意识；开拓者意识；以及今天的旅

---

[1] threshold：开始，开端，极限
[2] instinct：本能
[3] physiology：生理学

today the voyager (as he can truthfully be described). In what way, to quote Siegfried Giedion[1], are these phases The Eternal Present[2]? To answer this we must begin with the first and deepest, and work upwards.

In 1877 a professor of mental and moral philosophy, Grant Allen, published an analysis of aesthetics called *Physiological Aesthetics*. Today this little classic is virtually forgotten, but if a copy can be found, I commend it to all. The aesthetic conclusions were geared to the emotions of the day and to us now seem horribly sentimental, but the scientific method is unanswerable. It explains the hostility[3] of modern man to what he feels is a soul-less, machine-made environment, and is a basic guide to landscape. From this primitive treasure house my own first choice is the power of stereoscopic[4] vision to create feeling as well as sight in environment, how we can and should design for it, and how a felt foreground will connect our psyche to middle and far distance.

But of course the perceptions[5] are only a technical part in the long manufacture of the emotions in that marvellous[6] green and restless habitat. In a period described as the golden age of the mammals[7], we were vegetarian, peace-loving and monarchs[8]. No wonder the irrational[9] parts of us hanker for[10] that carefree if cruel existence, which created our basic sense of beauty and subconsciously inspires most private gardens today.

Then, as *Homo erectus*[11], ambitious and adaptable, man leaves the protection of the forests for the dangerous life of carnivore[12] and hunter in the open savannah[13]—the only animal to have within him the dual personality of peace-lover

行者意识（这是对现代人类恰如其分的描述）。

那些美学方面的结论符合当时人们的情感取向，对今天的我们而言也许过于多愁善感，但其中的科学方法却是无可辩驳的。

从这个人类最初的宝库中，我首先想选的是三维想象力，让人们不仅能看见而且能感觉到风景，让我们思考如何能够以及应该如何为此进行设计，这一想象力还会让我们思考感知的风景如何使我们的心灵与远方的景致进行交汇。

无怪乎深藏于我们心底的情感向往着那种原始但却自由自在的生活经历。这份向往形成了我们对美的基本感觉，并且作用于我们的潜意识，从而产生了今天的大部分私家园林。

接下来，雄心勃勃、适应力强的直立人就离开了树林来到大草原，过起了猎人的生活，以食肉为主——这就使得人类具有和平性与侵略性双重特征。

[1] Siegfried Giedion：即 Sigfried Giedion。基提恩（1888～1968年），瑞士建筑评论家，历史学家
[2] The Eternal Present：源自基提恩所著的 *The Eternal Present*，该书主要探讨永恒与变更的关系
[3] hostility：敌意，恶意，对抗，反对
[4] stereoscopic：有立体感的
[5] perceptions：有理解的
[6] marvelous：绝妙的，了不起的
[7] mammals：哺乳动物
[8] monarchs：君主政体，君主政治，君主国
[9] irrational：无理性的
[10] hanker for：渴望，追求
[11] Homo erectus：直立人
[12] carnivore：食肉动物
[13] savannah：大草原

and aggressor. Society changes from the individual to the collective, for hunting calls for organized teamwork. Man now spreads over the globe, adapting himself to climate and geography and thus creating diversity of species. In Europe he endures the incredible hardship of ice age and mountain terrors—a challenge that called forth the first works of art as an expression of inner reality and probably implanted in technocratic man an occasional urge to return to, and experience, the sublime in landscape. He so venerates[1] the special power of his monster prey that he hopes the spirit will pass into him when he eats the flesh.

Ten thousand years or so later the hunter society in the west had already been superseded[2] by the settler but seems, in essence, to have re-appeared in China. The abstract evidence of this is to be found in a generic similarity in painting, the factual in the early hunting parks. China is a genial[3] land and once the monsters had been domesticated and lost their mystique, veneration passed from them to the natural environment as a whole. So the aggressive hunter emerged into the peace-loving agriculturalist. Taoists believed that all man's works should be gentle and submissive to the earthmother, a very different concept from that of the western settler, who saw himself as master of all and centre of the universe. Today the west takes a more modest view and, as usual, the first evidence of this was seen in the arts rather than the sciences.

　　中国这片土地气候温和宜人，野兽一旦被驯服，就失去了神秘感，而人们也就把对它们的崇敬转向对整个自然界的崇敬。

The third phase, the western settler, was concerned with the beauty of mathematical proportion, and environment was solely the province[4] of the architect.

　　在第三阶段，西方开拓者考虑的是根据数学比例而产生的美，改造环境则完全是建筑师的职责。

Phase four began when Galileo and others broke out of the secure classical box that had taken 10,000 years to come to maturity. The search for the infinite beyond the finite had begun. The effect upon landscape design was revolutionary. Although Vignola in the Villa Lante[5] had already shown that the garden could be greater than the house, the search meant a totally different conception of space. Landscape was freed from architecture and opened up perspectives of imaginative space and ideas denied to its ancient parent. For me, the two symbols

　　风景学从建筑学独立出来，并且展示出建筑学所没有的虚构的空间观和理念。

[1]　venerate：崇敬
[2]　supersede：代替，取代，接替
[3]　genial：亲切的
[4]　province：研究范围，管辖范围
[5]　Villa Lante：朗特庄园

of this liberation are Bernini's Piazza[1] of St. Peter's, which created the sense of infinity beyond its columns and is still the greatest collective urban space in the world; and the contemporary Villa Gamberaia[2] near Florence, which for the first time in history set out to record in space the contrasting facets of the individual mind.

The liberation become absolute in the English 18th century park, where the voyager in imaginary space traveled backwards in time: through William Kent[3] he participated in the classics; through Brown, the hunting savannah; and through Uvedale Price[4] and others the primitive forests. In the 19th century the voyager went round and round in a whirlpool[5], but re-appeared in France through painters groping[6], like Monet, towards the revolutionary art of movement and experience described by Dewey[7] in his philosophy. <u>In landscape, Le Corbusier proclaimed, like a clap of thunder, a new vision of human habitation: a city of termite towers[8] poised above a romantic landscape flowing below like a Dewey river[9]</u> (Figure 2.1). This heroic concept now fills us with alarm and even horror, but it did decisively reaffirm the landscape of future as being independent of architecture. Despite his protestations[10], Le Corbusier was preoccupied more with the beauty of the machine than with humanity, and with the universal and international rather than the particular.

Le Corbusier pulled us up with a jerk. We have seen in recent years that human cannot live on geometry alone—and mostly bad geometry at that; nor, of course, can he live solely on his rich heritage of biology, for this would be to discard the divine gift of intellect. This is why I believe that our new profession is essential to the future of civilization in any part of the world, for it is the only one that can unite these two aspects. It can explore the meaning and purpose of life and express

平地一声雷，在风景园林领域，勒·柯布西耶设想了一个居住环境的新景象：一座座高楼井然有序地矗立在城市中，环绕着高楼的是千变万化的美好景致。

---

[1] piazza：广场，走廊，露天市场
[2] Villa Gamberaia：冈贝里亚庄园
[3] William Kent：肯特（1685～1748年），英国造园家
[4] Uvedale Price：尤夫德尔（1794～1801年），英国造园家
[5] whirlpool：旋涡，涡流
[6] grope：探索，摸索
[7] Dewey：杜威 John Dewey（1859～1952年），美国哲学家
[8] termite tower：杜威认为白蚁的巢穴是井然有序，构造合理的
[9] Dewey river：杜威认为世界是不断变化的，哪怕是同一条河流，每时每刻也是不同的
[10] protestation：声明，主张

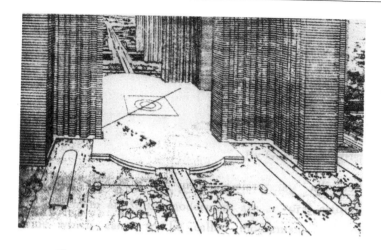

Figure 2.1 Le Corbusier's concept for city planning

its findings on a scale never before conceived; and in doing so, it can lift the spirit from the body, like the dream of the Chinese philosopher of old.

Let me end with a quotation from *The Eternal Present*, written by Professor Giedion 17 years ago:

The human organism requires equipoise[1] between its organic environment and its artificial surroundings. Separated from earth and growth, it will never obtain the equilibrium[2] necessary for life.

It is time we became human again and let the human scale rule over all our ventures. The man in equipoise we need is new only when seen contrasted with our distorted period. <u>He revives age-old demands which must be fulfilled in the terms of our own times if our civilization is not to collapse.</u>

他复兴的是人类久已拥有的需求，如果我们不想让文明衰败的话，这些需求必须随着时代的发展而得以满足。

Notes

1. Extracts from

Text

Jellicoe, Geoffrey. Contemporary Meanings in the Landscape [J]. *Landscape Architecture*, 1980 (1): 51-54.

2. Background Information

（1）Geoffrey Jellicoe 杰弗里·杰里科（1900～1996 年）。

英国景观设计（风景园林）领域的一代先驱，首届国际风景园林师联合会（IFLA）主席及 IFLA 的终身名誉主席。

（2）1979 年，在英国剑桥大学耶稣学院举行 IFLA 成立 31 周年大会。本文选自杰里科在此次大

[1] equipoise：平衡，均衡
[2] equilibrium：平衡，均衡

会上的演讲，有删节。

3. **Sources of Additional Information**

Siegfried Giedion

　　http：//eng. archinform. net/arch/11821. htm？ scrwdt＝1024

Villa Lante

　　http：//en. wikipedia. org/wiki/Villa _ Lante

　　http：//www. gardenvisit. com/ge/lante. htm

Villa Gamberaia

　　http：//www. gardens-of-tuscany. net/villa _ gamberaia-eng. htm

Further Reading A

Post Modernism Looks beyond Itself

By George Hargreaves

Modernism and post modernism are cultural attitudes rather than design styles. Within that attitude which came to be called modernism, Modern Art had many styles ranging from impressionistic[1] to cubist[2] to abstract expressionism[3] (and there are styles within these styles). The same can be said for modern architecture—deStijl[4], futuristic, international, brutalistic[5] and high-tech[6]—all are modernist styles. While landscape architecture probably had fewer styles within modernism, the two predominant styles—asymmetrically geometric[7] and fluidly amorphous[8]—revolve around an internal organization system, sharing with architecture and sculpture an idealized and invented space.

Art critic Kim Levin describes the aspirations of modernism: "For the modernist period believed in scientific objectivity, scientific invention; its art had the logic of structure, the logic of dreams, the logic of gesture or material. It longed for perfection and demanded purity, clarity, order. And it denied everything else, especially the past: idealistic, ideological, and optimistic, modernism was predicated on the glorious future, the new and improved. Like technology, it was based all along on the invention of man-made forms, or as Meyer Schapiro has said, 'a thing made rather than a scene represented.'"

Post modernism seems to be a direct reversal of modernistic philosophy and its attendant idealization. "It knows about shortages. It knows about inflation and devaluation. It is aware of the increased costs of objects. And so it quotes, scavenges[9], ransacks[10], recycles the past. Its method is synthesis rather than analysis. It is style-free and free-style. Playful and full of doubt, it denies nothing. Tolerant of ambiguity[11], contradiction, complexity, incoherence, it is eccentrically inclusive. It mimics[12] life, accepts awkwardness and crudity[13], takes an amateur stance. Structured

---

[1] impressionistic: 印象主义的
[2] cubist: 立体主义
[3] abstract expressionism: 抽象表现主义
[4] deStijl: 荷兰风格派（建筑）
[5] brutalistic: 野兽派
[6] high-tech: 高技派
[7] asymmetrically geometric: 不对称几何形
[8] fluidly amorphous: 自由曲线形
[9] scavenge: 提取
[10] ransack: 搜索
[11] ambiguity: 含糊，不明确
[12] mimic: 模仿
[13] crudity: 生硬

by time rather than form, concerned with context instead of style, it uses memory, research, confession, fiction—with irony, whimsy[1], and disbelief." Levin's descriptions of modernism and post modernism are clearly about cultural attitudes rather than specific styles.

The failures of modern landscape architecture are not so enormous nor so visible as steel boxes on every street corner. The groundbreaking work of Thomas Church established modern landscape architecture on much the same rules as modern art, but in doing so he at least broke away from the age-old formal/informal style of the country place era. In back gardens of the San Francisco Bay Area, Church developed the asymmetric geometric style that became the model for future works at an urban scale. The bond that these works of the last 40 years share is that they are modeled after internal, psychological space: in other words, the design "hangs together."

That the failures of modernist landscape architecture are less visible lies in the profession's history of working with site. The traditional process (*a*) understands the site (emphasis on natural phenomena); (*b*) fits the functional program of land use; and (*c*) makes it beautiful. At best, it has produced site-adjusted works which could be said to straddle[2] site-dominated[3] and site-specific[4] sculpture.

Church's Donnell Garden[5] draws its inspiration from fluctuating[6] tidal flats directly off-site[7]. The shape of the pool as perceived at eye level can became one and the same with off-site water. If the Donnell pool complex were to be site-specific, that experience of on-site/off-site water would determine the entire concept of the garden. Instead, Ade Kent's sculpture refocuses the experience on its own compositional core.

Where modern landscape architecture went awry[8] parallels[9] modern architecture's problem, in the hands of modernist imitators. Fluidly shaped designs occur on rooftops, courtyards, backyards, and parks, where they could originate only from the idealized, psychological space of the landscape architect. In the hands of true masters, the asymmetrical geometric style was acceptable and sometimes magnificent, as at Portland Forecourt[10], but again the imitators have blown this model up and down in scale to fit plaza, courtyard, backyard, and bus stop.

To make matters worse, reliance on historical style, and its combination with fluid or geometric styles, created a group of cartoon solutions. For urban plaza, there is the pale blonde/dark brown color scheme with an asymmetric ground plane pattern incorporating level changes, and always centered on a feature (water, sculpture, etc.). The park or grounds have to be English picturesque[11]: mounds (often the wrong

---

[1] whimsy：奇想

[2] straddle：跨骑

[3] site-dominated：受控场地，含义与特定场地相对

[4] site-specific：特定场地。site-specific sculpture 是一种存在于特定场地的雕塑，它把周围场地看成雕塑作品的有机组成部分

[5] Donnell Garden：唐纳花园，托马斯·丘奇的代表作品之一

[6] fluctuating：上下波动

[7] off-site：外部场地

[8] awry：歪曲的，错误的

[9] parallels：轨道

[10] Portland Forecourt：波特兰市前庭花园，又称"演讲堂前广场"，劳伦斯·哈普林代表作之一

[11] English picturesque：英国风景园

scale); spatially clumped planting; a colorful understory; and of course, a curvilinear water body. What developed out of modernism and landscape history in the hands of misguided imitators was a version of site-dominant landscape architecture. No matter how much site analysis was done, built works of landscape architecture fell into a few easily defined categories. These set images, in concert with idealized internal space, became the litany[1] of professional landscape architecture.

In 1969, Ian McHarg issued a call for responsible regional and environmental planning. In response, landscape architecture began fervently[2] mapping and analyzing natural phenomena on regional areas and development sites. At the regional scale, the need for data base and public awareness has become a part of the political process. At the scale of individual sites, two things occurred: a new version of formal/informal and a resurgence[3] of imitative naturalism. The new formal/informal left as much of the site natural as possible and placed an internalized design in the core area of intensive human use. And if areas were disturbed, by grading for instance, the landscape architect sought to reinstitute[4] what was there before.

Hidden amongst the maps, charts, and diagrams of McHarg's *Design with Nature* was the philosophy of a utopian society where man is "in nature rather than against nature." Ian McHarg seems to be calling for the cultural attitudes of post modernism. One is reminded of Robert Irwin when McHarg says, "The Naturalists will not change pre-existing conditions unless they can demonstrate that such changes are creative," or Frank Stella and Don Judd when McHarg says, "the Naturalists have turned to the world at large in order to find laws and forms of government that might work satisfactorily."

Gas Works Park[5] by Richard Haag exhibits many post-modern qualities. As a design, the park is specific to the site and its industrial relics[6]. With a small construction budget, Haag features the Gas Works as a poetry of steel forms hard by the industrial edge of Lake Washington. A "prehistoric" mound, a couple of hundred feet in width and 30-40 feet in height, is covered with rough grasses and processional walks. Clipped grass, small ponds, curvilinear walkways, and colorful picturesque plantings were forsaken[7] as a typology. Seattle's gray weather elements—fog, rain, drizzle—emphasize the hardness of steel, the industrial use of a lake edge, and the primitive landscape. One seems to be in the presence of a recent culture that has gone archaic[8]. Only in the barn, where bright colors are supposed to create a children's atmosphere, does this project return to the tried and worn.

Recently, two projects of differing participatory natures—Strawberry Creek Park in Berkeley, California, and the Necco/Tire Garden at MIT in Cambridge, Massachusetts (Figure 2.2)—begin to define the range of styles in post-modern landscapes. Strawberry Creek Park by Doug Wolfe relies

[1] litany：枯燥重复的连续不断的叙述
[2] fervently：热情的
[3] resurgence：复兴
[4] reinstitute：重新建立
[5] Gas Works Park：西雅图煤气厂公园，理查德·黑格代表作之一
[6] relics：遗迹
[7] forsaken：被抛弃的
[8] archaic：陈旧的

solely on the participation of neighborhood residents for its meaning. The park celebrates the unearthing of the creek to let nature back into society. Concrete debris[1] from the culvert was used to reinforce the creek's bank. Yet this project falls short of lasting meaning. Once the immediate participants leave the neighborhood, how many people will recognize the reversal of nature being driven out of the city? This glorious idea should be developed in such a way that it communicates to the human spirit years from now. Experts can only guess at the origins of prehistoric Avebury[2] in England or Machu Picchu[3] in Peru, yet these sites communicate man's relation to nature and the universe for all time.

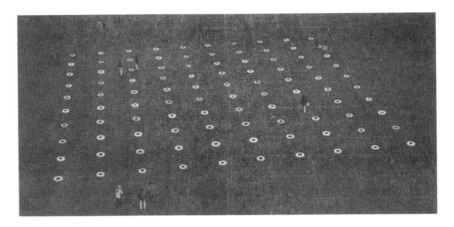

Figure 2.2 Necco/Tire Garden

The Necco[4]/Tire Garden was installed by a mixed group of students, artists, and professionals in one morning to honor May Day. Pastel candies produced by the neighboring New England Confectionary Company establish a contextual dialogue. The placement of two grids, neccos and tires, respects on one hand the formal, neo-classic Killian's Court and, by twisting out of this formality, the other grid enacts a directional pull across the Charles River to downtown Boston. The perceived effect is one of strong on-site/off-site relationships with materials specific to the area in a whimsical[5] way. Even its brief duration exhibits yet another post-modern value—temporality.

Gas Works Park, Strawberry Creek Park, and Necco/Tire Garden have many elements of post modernism in common and had little-to-no construction budget. But this doesn't mean that all post-modernist work comes about because of lack of funds.

Lack of actual area was probably the greatest contributing factor to Harlequin Plaza (Figure 2.3). Extruding mechanical systems are enclosed in mirrors and angled to bring the off-site in. "Vibrating" ground pattern and strong colors combine with multi-reflections to create a rich environment in contrast to the context of suburban office park. Harlequin Plaza uses as inspiration the distant Rocky Mountains and, through human abstraction, attempts to insert liveliness into a lifeless nine-to-five land use. That this plaza draws art groups for public performance and uses, sightseers,

---

[1] debris：碎片
[2] Avebury：埃夫伯里（世界文化遗产，位于英格兰西南威尔特郡，世界上最著名的巨石古迹）
[3] Machu Picchu：马丘比丘（世界文化遗产，秘鲁印加遗址）
[4] Necco："New England Confectionary Company" 的缩写，美国最老的糖果生产公司
[5] whimsical：古怪的

and weekend picnickers is contrary to the dogmatic notion that users of landscapes must sit down in the shade and eat lunch.

Less dramatic and startling, is the canal and weir[1] system which connects two lakes at Lakewood Hills, a residential development in Sonoma County, California (Phase I-SWA Group; Phase II-Hargreaves, Allen, Sinkosky & Loomis). The canal is 700 feet long and 5 feet wide. Three water weirs, set at regular distances perpendicular[2] to the canal, decrease in length as they march toward the foothill and vineyard vista.

Figure 2.3 Harlequin Plaza

A walk curves around the rim of an earth basin located at the canal's midpoint; a crescent[3] of cypress trees hugs the walk's edge. Water flows down the canal, creating two-foot waterfalls at each weir only during heavy rainfall. Simple evidence of a shift from internal space to external concerns is found in the non-ending or non-closed composition, which creates a foundation to the hills; forced perspective makes the hills seem to stand up and come on-site.

The foregoing is a small group of known projects that demonstrate post-modernist attitudes. With society's shift from belief in the new to its recognition of limits and interest in what exists, the stylistic explorations in post-modernist landscape architecture have just begun. Pluralism is appropriate. The expression of symbolism[4], mysticism[5], and humanism will become a preoccupation[6]. Time, nature, and culture will serve as physical media and subject. Light, shadow, sky, rain, plants, dirt, debris, people of all types, and manmade elements that intensify and abstract what is already here, will become the focus of simpler, more receptive compositions and non-compositions. Where this will lead and what landscapes may come to look like is, of course, open ended.

Notes

1. Extracts from

Text

Hargreaves, George. Post Modernism Looks Beyond Itself [J]. *Landscape Architecture*, 1983 (7): 60~65.

Pictures

Figure 2.1：大卫·路德林，尼古拉斯·福克著．王健，单燕华译．营造21世纪的家园 [M]. 北京：中国建筑工业出版社，2005：39.

Figure 2.2：Tom Richardson. *The Vanguard Landscapes and Gardens of Martha Schwartz*

---

[1] weir：堰，拦河坝
[2] perpendicular：垂直的，正交的
[3] crescent：新月状的
[4] symbolism：象征主义
[5] mysticism：神秘主义
[6] preoccupation：当务之急

[M]. London: Thames & Hudson, 2004: 47.

Figure 2.3: Hargreaves Associates. Harlequin Plaza [EB/OL]. http://www.hargreaves.com/projects/harlequin/index.html, 2007-04-28.

2. Background Information

乔治·哈格里夫斯（George Hargreaves, 1952年～）。

著名风景园林设计大师，曾担任哈佛设计学院风景园林系主任，宾夕法尼亚大学、哈佛大学、维吉尼亚大学、伊利诺斯大学等多所著名大学客座教授，一般被认为是风景园林设计后现代主义代表人物，代表作品有澳大利亚2000年悉尼奥运会公共区域环境设计、圣·何塞市广场公园等。

本文选自哈格里夫斯发表于 *Landscape Architecture* 杂志上的文章 *post modernism looks beyond itself* 的 landscape architecture 部分。

3. Sources of Additional Information

Strawberry Creek Park

  http://www.shopinberkeley.com/parks/strawberrycreek/index.html

  http://www.ci.berkeley.ca.us/parks/parkspages/StrawberryCreek.html

Harlequin Plaza

  http://www.hargreaves.com/projects/harlequin/index.html

Further Reading B

Post-Postmodernism

By Tom Turner

Giving names to periods is difficult. As cultural terms, "Classicism", "Neoclassicism[1]", "Romanticism", "Impressionism[2]" and "Post-Impressionism" are imprecise[3]. Yet we all find the terms useful, and arguing over their meaning keeps many scholars in gainful employment. With regard to time periods, "Ancient World" is fairly secure; "Middle Ages" is becoming progressively inaccurate; "Modern Age" keeps moving forwards. "Modernism" is partly a cultural term and partly a time word. Should its time reference make "Modern" unusable as a cultural term, what might take its place? In art history, a case can be made for Age of Abstraction. In a wider context, "Age of Analysis", "Age of Reduction" or "Age of Science" might serve. Each highlights a key characteristic of twentieth century thought: the endeavour[4] to analyses everything into essential constituents[5]. Abstract artists reduced art to shapes and forms; music was reduced to tones; novels became streams of consciousness; chemists hunted for the smallest components of matter; physicists looked for a single explanation of the universe. It is too soon for us to know what period label will take the place of "Modern", but something will.

"Postmodern" may survive longer than "Modern" because of its very eccentricity[6]. It could however be replaced by Post-Abstract if Age of Abstraction came into use. In the sixth edition of *The Language of Post-Modern Architecture*, Jencks takes heart from his critics' proclamation of the death of postmodernism and classifies them, deftly[7], as Neo-Moderns[8]. This places them after modernism yet before postmodernism. Jencks sees their criticism as proof of architectural postmodernism's continued vitality, "for who is going to waste time flogging[9] a dead style?" Actually, it is a very popular activity. Postmodern architecture can be seen as inherently trivial, glitzy and stuntish, appealing to the wallet, not to the mind and not to the soul. It belongs in shop windows and in cinemas, in Madison Avenue[10] and in Tinseltown. No one who uses retail shops or watches movies should despise the great products of these great industries. But an "anything goes"

---

[1] Neoclassicism：新古典主义
[2] Impressionism：印象派
[3] imprecise：不严密的
[4] endeavour：尽力，竭力
[5] constituent：要素
[6] eccentricity：古怪
[7] deftly：巧妙地
[8] Neo-Moderns：新现代主义
[9] flogging：鞭打
[10] Madison Avenue：麦迪逊大街（美国纽约市的一条街，美国广告中心）；美国广告业

pluralist[1] approach to urban design gives us the equivalent of a junk shop with, perhaps, an empty chocolate box, a kettle[2] and an old TV set (Figure 2.4). Alone, each might be elegant, stylish or beautiful; together they are jumble. As a direct consequence of pluralism, the postmodern city street resembles an out-of-step chorus line. If anything goes, then nothing goes.

Figure 2.4 "Anything goes" postmodernism gives us the pluralism of the junk

But there are signs of post-postmodern life, in urban design, architecture and elsewhere. They are strongest in those who place their hands on their hearts and are willing to assert "I believe". Faith always was the strongest competitor for reason: faith in a God; faith in a tradition; faith in an institution; faith in a person; faith in a nation. The built environment professions are witnessing the gradual dawn of a post-postmodernism that seeks to temper reason with faith. Designers and planners are taking to the rostrum[3] and the pulpit. Christopher Day has written a book on *Place of The Soul* (Day, 1990). Christopher Alexander's work is discussed at greater length later in this book, in essays on design methods and the Pattern Language.

As a youngster, Alexander was a mainstream technocratic[4] modernist. When disillusion[5] set in, he set forth on the road to San Francisco. Once here, he gathered a community of designers, read Taoist philosophy, and published books on the *Timeless Way of Building* (Alexander, 1979). Jencks classifies him as a postmodern ad hoc[6] urbanist. Alexander, rightly in my view, rejects the label "postmodern" (Alexander and Eisenman, 1984). The pattern language rests on deep faith as much as it does upon reason. It is post-postmodern, or pre-Modern. Alexander starts and finishes the first chapter of the *Timeless Way* with a traditionalist creed[7]:

There is one timeless way of building.

It is thousands of years old, and the same today as it has always been.

The great traditional buildings of the past, the villages and tents and temples in which man feels at home, have always been made by people who were very close to the centre of this way...

To purge ourselves of these illusions, to become free of all the artificial images of order which distort the nature that is in us, we must first learn a discipline which teaches us the true relationship between ourselves and our surroundings.

Then, once this discipline has done its work, and pricked the bubbles of illusion which we cling

---

[1] pluralist：兼职者
[2] kettle：水壶
[3] rostrum：讲坛
[4] technocratic：技术统治论者
[5] disillusion：醒悟
[6] ad hoc：来源于拉丁语，意为"特别的"(for this purpose)
[7] creed：信条

to now, we will be ready to give up the discipline, and act as nature does.

This is the timeless way of building: learning the discipline and shedding[1] it.

The "artificial images of order" that Alexander criticizes were rational[2], modernist and utopian[3]. Postmodern "planning" was anti-planning. When the hoped-for urban paradise turned into a hated "concrete jungle", with streets in the air, criminal gangs, tall blocks and vacant open spaces, planners lost heart. Post-postmodern planning is a sign of returning self-confidence. Traditions are being rediscovered. In place of the old singular zones, for housing, industry, commerce and recreation, a plural[4] zoning, resembling a pile of rubber bands, is being founded on belief and sentiment[5]. Plural zoning has a greater similarity to natural habitats than singular zoning (Figure 2.5).

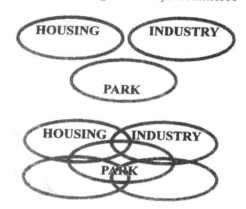

Figure 2.5 Singular and plural zoning

The waterfronts of the world are becoming Zones of Waterfront Character, with special regulations. Old high streets are now themed shopping areas, dominated by antique[6] bistros[7]. Landon and San Francisco have Chinatowns.

In New York City, these generative rules are legion: a special district controls the recycling of Union Square as a luxury enclave[8]; new contextual[9] zoning is abetting[10] the development of the Upper West Side in a regenerated 1930s Art Deco[11] format; while great parts of Manhattan stand cordoned[12] off behind the boundaries of historic districts as large as Greenwich Village and the Upper East Side[13] (Boyer, 1990).

The zones are cultural, not functional. They overlap and there are other possibilities. Central Paris is a zone of low buildings. Bavaria[14] has zones for timber buildings. Many cities now have ecological habitat policies. When London's Isle of Dogs[15] was designated as an Enterprise Zone, it could also have become a "Willow World", using *Salix*[16] as the major tree species. With dock basins and high walls of mirror glass, the willows would have been beautiful and the symbolism would not

---

[1] shedding: 蜕落
[2] rational: 推理的
[3] utopian: 乌托邦的，理想化的
[4] plural: 复数的，多于一个的
[5] sentiment: 情感，情绪
[6] antique: 古老的
[7] bistros: 小酒馆
[8] enclave: 被包围的领土
[9] contextual: 文脉上的，前后关系的
[10] abet: 教唆，煽动，帮助，支持
[11] Art Deco: 艺术装饰派
[12] cordon: 用警戒线围住
[13] Upper East Side: 阳光上东，美国纽约曼哈顿区的一个街区，在中央公园和东江之间
[14] Bavaria: 巴伐利亚
[15] London's Isle of Dogs: 英国伦敦东区的一个半岛
[16] *Salix*: (植物) 柳属

have been inappropriate.

New zones can be visual, historic, ecological, cultural, or they can give a spatial dimension to belief. Los Angeles[1] has Koreatown, Little India, Little Saigon and Gaytown, which could become self-managed communities (Figure 2.6). There could also be a Green Town, based on conservationist principles, and an Esperanto[2] Town, which uses the international language. As post-postmodernism is a preposterous[3] term, we must hope for something better. The Age of Synthesis[4] is a possibility. Coherent, beautiful and functional environments are wonderful things, which can be produced in different ways. The modernist age, of "one way, one truth, one city", is dead and gone. The postmodernist age of "anything goes" is on the way out. Reason can take us a long way, but it has limits. Let us embrace post-postmodernism—and pray for a better name.

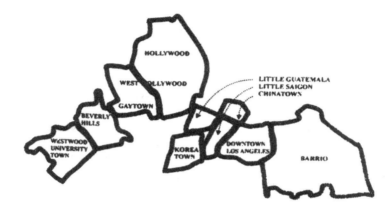

Figure 2.6 Cultural zones in central Los Angeles, based on Jencks' *Heteropplis*

Notes

1. Extracts from

Text

Turner, Tom. *City as Landscape: A Post-postmodern View of Design and Planning* [M]. London: E &F. N. Spon Ltd., 1996: 8-10.

Pictures

Figure 2.4: Tom Turner. *City as Landscape: A Post-postmodern View of Design and Planning* [M]. London: E &F. N. Spon Ltd., 1996: 9.

Figure 2.5: Tom Turner. *City as Landscape: A Post-postmodern View of Design and Planning* [M]. London: E &F. N. Spon Ltd., 1996: 9.

Figure 2.6: Tom Turner. *City as Landscape: A Post-postmodern View of Design and Planning* [M]. London: E &F. N. Spon Ltd., 1996: 10.

2. Background Information

汤姆·特纳（Tom Turner），英国伦敦格林尼治大学教授，著名的风景园林教师和评论家，在城市规划设计、风景园林设计、花园设计、园林史等方面著作颇丰。代表作品有 *Garden History*：

---

[1] Los Angeles：美国加利福尼亚州洛杉矶市
[2] Esperanto：世界语
[3] preposterous：荒谬的
[4] synthesis：综合

Philosophy and Design 2000 *BC-*2000 *AD*（2005）等。

本文选自他的重要著作 *City as Landscape*：*A Post-postmodern View of Design and Planning*，为麻省理工学院建筑学院必读书之一。

3. **Sources of Additional Information**

http：//www.gardenvisit.com/garden_history/garden_history_tom_turner.htm

Further Reading C

Landscape for Living

By Garrett Eckbo

Here in the middle of the 20th century we are left with, not the sterile[1] dichotomy[2] of the official academic theory, but a rich and many-sided octagon of landscape tradition. Here are its parts:

(1) The formal tradition of Renaissance and Baroque[3] Europe and the Moslem world, with its subcurrent of Greek and Gothic[4] irregularity.

(2) The informal[5] romantic tradition of China, Japan, and eighteenth-century England.

(3) The overriding[6] fascination with plants for their own sake, based on the horticultural and botanical advances of the nineteenth and twentieth centuries.

(4) The conservation movement, with its emphasis on the value and wonder of the indigenous[7] primeval[8] landscape, expressed in our field in the American park movement.

(5) The urban and regional planning movement, with its compulsion toward reexamination of the relations between buildings and open space, town and country.

(6) The modern movement in the arts, in architecture, and in landscape design since the mid-1930s.

(7) The rural tradition, and the folk or little garden tradition, two notes of 20th century social realism.

If we examine these streams for their relevance to our work in the balance of the century, we will find that they boil back down to another dichotomy, broader, richer, and more fertile than the academic formal: informal dichotomy.

The great problem and the great opportunity of our times is to rebuild, on an infinitely higher plane, the unity and solidarity between man and nature which existed and still exists in primitive communal societies, and which was broken and shattered by the great sweep of history through slavery and feudalism[9] to capitalism. This we can work toward every day on every job and every project, no matter how small or inconsequential[10] it may seem.

---

[1] sterile：缺乏想象力的，缺乏新意的
[2] dichotomy：二分法
[3] Baroque：巴洛克式
[4] Gothic：哥特式的
[5] informal：不规则的
[6] overriding：最重要的，高于一切的
[7] indigenous：本土的
[8] primeval：原始的
[9] feudalism：灭亡
[10] inconsequential：不合逻辑的，不合理的

On the mention of theory, two questions are apt to arise: one on the need for theory, the other on the nature of the theory needed. We must be able to answer these questions, the other on the nature of the theory needed. We must be able to answer these questions, especially here in practical America, where so much of our environment is built on the sole theory that no theory for its planning is needed—we just go out and build it.

Theory is a generalization of social experience in any particular field, or in all fields. It is at one and the same time a generalization of the past, a vitalizer[1] of the present, and a projection of the future. If it is any one without the others it tend toward sterility[2], decadence[3] or frivolity[4]. Only positive exploratory theory can take us beyond the precedents of yesterday. Theory is the vehicle which guarantees the continuous growth and expansion of tradition. Theory and tradition develop together and grow continuously, however unevenly or erratically[5], through any number of struggles with contradictions. To try to freeze them at any given time in a system of academic rules and proportions is like trying to dam a strong stream with no spillway[6] for overflow control. Sooner or later the stream will find its way over or around the dam, whatever the joints are weakest, and come forth with a burst of vigor equal to the length of time it has been impounded[7].

Theory, by analyzing the past while working in the present, can project the length and character of the next step into the future. This is the process which has been responsible for all human progress of every sort. Every step forward, technically, culturally, or socially, had to be an idea—a theory—in one or many heads before it could be taken. The whole long chain of development of human tools, from the first flint[8] ax to the most delicate and powerful machinery of today's industry, has come about through this process of analyzing the past in the present toward the future. Theory is theory, whether it is an idea in a clever mechanic' head, or five hundred pages of windy discourse. New shelter and new clothing, steam engine and electricity, Magna Charta and Declaration of Independence and Bill of Rights; all began as theories, as ideas, some of which were called radical[9]. The scientific process of building theory and constantly developing it by analysis, hypothesis, and experiment is basic to our twentieth-century civilization.

Theory in the arts is, of course, the stumbling[10] block for those practical souls who have gone along with us so far. Yet art is only a process of trying to extract the maximum potential human experience out of necessary practical activities. Painting, sculpture, music, architecture, landscape design have all grown from sound, practical, functional roots in the living activities necessary to people. They have grown to cultural heights by the exact process of imaginative building on the past

---

[1] vitalizer: 激发活力者
[2] sterility: 思想贫乏
[3] decadence: 颓废
[4] frivolity: 无聊的举动
[5] erratically: 不规律地
[6] spillway: 溢洪道,泄洪道
[7] impound: 关在栏中
[8] flint: 燧石
[9] radical: 激进的
[10] stumble: 绊倒,使困惑

that we have been describing. The architect today can plan a better house than the carpenter, the brickmason[1], or the general contractor[2], because, if he is abreast of the possibilities of his profession, he is more aware of the maximum potential for an interior harmony of space, size, and form; for an exterior harmony of open and solid wall; for the most satisfying combinations of materials. Before he can produce this he must, obviously, solve all the practical functional problems in a way which also makes possible a maximum contribution from carpenter, brickmason, and general contractor. Theory—as idea—is not developed for its own sake, even though it precedes practice. It must come from practical necessity, and be based on constant observation and experience.

A good theory of landscape design, then, must be a theory of form as well as of function. It must be artistic as well as practical, in order to produce the maximum for those who will experience work influenced by it. Every work of landscape design, conscious or unconscious, whether it be the utility garden of the southern sharecropper or the Central Composition of Washington, D.C., produces an arrangements of forms, colors, and textures in space which results in some sort of cumulative[3] effect, good or bad, on those who pass through it. We cannot avoid the problem of producing form in the landscape. From the formal western school which went after it with axis and vistas[4], through the informal eastern school which avoided it with poetry, rationalizations[5], and subjective grotesqueries[6], to the horticulturists[7] and the naturalists, who bury it in collections or hide it behind nature—all have produced arrangements of cumulative effect, good or bad, on us who experience them, whether or not we know their literary rationalizations. It should be noted that the good and bad is not necessarily between schools but within schools; all produce good work (pleasing to us) and bad work (unpleasant to us). The goodness or badness, for us, is not necessarily based on the theory of the particular school, but rather on certain questions of the arrangements of spaces, the development of sites, the use of materials, unity and variety[8], scale and proportion, rhythm and repetition, which we know or can determine are basic to our experiences in our environment.

Our theory, then, must point the way toward good form in the landscape, but it cannot define it rigidly, on an exclusive, selective basis, with dogma[9] and formulae, rules and regulations, precedents and measured drawings. We must base ourselves upon a flexible understanding and assimilation[10] of those basic questions of scale, proportion, unity, variety, rhythm, repetition, which have been the primary guides for good men in all fields in all times and places. We must remember that most landscape problems are so plastic, so little under the control of functional requirements, that any number of solutions is possible. For most, the final best solution is probably

---

[1] brickmason：砌砖工人，泥水匠
[2] contractor：承包人
[3] cumulative：累积的
[4] vista：深景，尤指人透过如两排建筑或树木之间空隙看到的远景或视觉感受
[5] rationalization：合理化，合于经济原则
[6] grotesquery：怪诞，古怪
[7] horticulturist：园艺家
[8] unity and variety：统一和变化（形式美原则之一）
[9] dogma：教条
[10] assimilation：同化

as unreachable as the final best solution for a square of canvas on an easel[1], or a block of stone in the sculptor's yard. Design, like life, has no limits to its development.

It will be said, then, on what shall we base our forms? Where shall we find them? And the answer is, in the world which is around you in space, and behind you in time. If you understand it and love it and enjoy it there is your inspiration. The more you are a part of your world the more inspired you will be, if you find those parts which are streaming steadily forward, rather than the many stagnant[2] backwaters which exist to trap the unwary[3].

It must be remembered that the great pre-industrial styles of the past were produced by societies of a certain stability, a certain established structure and discipline within which artist and designer found enough security, orientation and direction to produce their best work. The nineteenth and twentieth centuries have been a period of tremendous historical acceleration, of great flux and movement throughout the world, of huge contradictory[4] struggles, of the rise of the common man and the democratic idea. The old relation of the artist to a clientele of the social elite[5] has gradually receded; the new relation of artist to a democratic mass clientele[6] is barely visible over the horizon; in between is the no man's land of commercialism[7], eclecticism[8], egocentrism[9], and escapism[10] in which the artist has been wandering for lo[11], these many moons. Our theory must be oriented within the social, as well as the technical and aesthetic, potential of the times, if it is to be relevant to the artist as producer and the people as consumers.

Our theory of landscape design for the balance of the twentieth century must be concerned with the realities of the now engrossing problems of the overall outdoor environment of the American people, rather than with abstractions about systems of axes, or poetic subjectivities about nature. We have tremendous problems, of unprecedented[12] social and esthetic potential, ahead of us. As we prepare for them and work on them we can absorb and assimilate[13] the old ideas, build on the strong base of our rich octagon[14] of landscape tradition, and go on to a unified expression of integrated social and natural landscape such as has never been seen before.

Notes

1. Extracts from
Text
Eckbo, Garrett. *Landscape for Living* [M]. New York: Architectural Record with Duell,

[1] easel：画架
[2] stagnant：停滞的，迟钝的
[3] unwary：不注意的，粗心的，不警惕的
[4] contradictory：矛盾
[5] elite：精英，杰出人物
[6] clientele：客户
[7] commercialism：商业主义，重商主义
[8] eclecticism：折衷主义
[9] egocentrism：中间路线
[10] escapism：逃避现实主义
[11] lo：看，瞧
[12] unprecedented：空前的
[13] assimilate：吸收
[14] octagon：八边形，八角形

Sloan, & Pearce, 1950: 57-60.

2. Background Information

盖瑞特·埃克博（Garrett Eckbo, 1910~2000 年）。

20 世纪最重要的风景园林师之一。1910 年生于纽约，1932 年入加州大学伯克利分校学习风景园林，1936 年秋进入哈佛大学设计研究生院学习。在那里，他与丹·克雷、詹姆士·罗斯一起共同发起了反对学院派理论的"哈佛革命"。毕业后他规划设计了大量的花园、公园、广场、社区和城镇。1950 年，出版了第一部著作《为生活而建的风景园林》（*Landscape for Living*）。1964 年，他和其他三位设计师一起成立了 EDAW 公司（易道环境规划设计有限公司）。1963 年，埃克博回到加州大学伯克利分校任教，并担任了四年的风景园林系主任。1998 年他出版著作《风景中的人》（*People in a Landscape*），2000 年 5 月在美国旧金山去世。

《为生活而建的风景园林》（*Landscape for Living*）。

1950 年出版发行，是现代风景园林史上的重要著作，曾在 1999 年 ASLA 成立 100 周年回顾中被评为"100 年来最具影响的十本书"之一。在书中，他主要强调了风景园林设计中功能和艺术的和谐。

3. Sources of Additional Information

Garreftt Eckbo

  http://en.wikipedia.org/wiki/Garrett_Eckbo

"The elements of Western landscaping: a visit with Garrett Eckbo - interview"

  http://findarticles.com/p/articles/mi_m1216/is_n5_v186/ai_10663252

Exercises

1. Questions（阅读思考）

认真阅读本单元文章，并讨论以下问题。
（1）风景园林中的现代主义有哪些主要观点？
（2）风景园林设计中的后现代主义和现代主义有哪些不同？
（3）汤姆·特纳关于当代园林设计发展的主要观点是什么？

2. Skill Training（技能训练）

（1）查阅现代主义园林设计英文文献，记录其主要观点。
（2）查阅英文文献，介绍"哈佛革命"的主要内容。

Tips

风景园林主要英文网站

近年来，随着技术的发展，互联网上出现了越来越多的风景园林相关信息和资源，网络已经成为不容忽视的信息媒介。对学习者而言，英文网页正越来越成为重要的窗口——帮助我们在第一时间了解世界风景园林动态，阅读西方风景园林新思潮、新作品；获取专业经典英文文献，游览世界各大名园——从而使我们得以更系统深入地理解西方风景园林。

风景园林主要英文网站可分以下四大类：专业协会和组织、教育机构、企事业及其他。

1. 专业协会和组织

最重要的专业协会网站自然是 IFLA [International Federation Of Landscape Architecture（国际风景园林师联合会），http://www.iflaonline.org]。如今的 IFLA 网站开始向门户网站方向发展，除阐明风景园林学科和职业历史、宗旨、目标，发布 IFLA 会议和竞赛通知等之外，更报道世界各国业界新闻、学术动态、重要演说等内容，主要分 news，members，resources，conference，education 五大版块。

其中 members 版块列出了各会员协会和组织的联系人、联系方式、网页地址等，世界各主要风景园林协会基本都有网站链接，是一个很不错的"网络黄页"。和此版块相类似的还有 resources/links 部分，列出了相关组织的网站地址，如世界建筑师联盟、联合国教科文组织等。

在各国协会官方网站中，比较有特色的是 ASLA [American society of landscape architects（美国风景园林师协会），http://www.asla.org]。该协会一年一度的优秀设计奖评选是世界风景园林界的一大盛事，获奖作品主要内容、设计特色，包括一些设计图在该网站上均有介绍。

其他主要专业协会和组织英文网站举例如下：

（1）Association of Professional Landscape Designers：http://www.apld.com/.
（2）International Society of City and Regional Planners：http://www.isocarp.org/.
（3）Canadian Society of Landscape Architects：http://www.csla.ca.
（4）The Cultural Landscapes foundation：http://www.tclf.org/.

2. 教育机构

各国高校等教育机构网站也值得一看，尤其对风景园林学习者来说，往往可以在此类网站中了解到重要的学习信息，获取学习资料。

麻省理工学院开放课堂（MIT's OpenCourseWare：http://www.core.org.cn/OcwWeb），是一个正在建设中的项目。它以服务社会为宗旨，免费向公众开放麻省理工学院网络课堂。浏览者可以了解 MIT 的课程设置，课程主要内容，学生作品等。

2006 年 DesignIntelligence 网站，排名全美风景园林研究生教育第一的哈佛大学设计研究生学院风景园林系（the Department of Landscape Architecture, Harvard University Graduate School of Design：http://www.gsd.harvard.edu/academic/la），排名全美风景园林本科教育第一的哥伦比亚大学环境设计学院（The School of Environmental Design, University of Georgia：http://www.uga.edu/sed）网站均值得一看。此外，宾夕法尼亚大学设计学院（University of Pennsylvania School of Design：http://www.design.upenn.edu），康奈尔大学风景园林系（Department of

Landscape Architecture at Cornell University：http://www.landscape.cornell.edu）等著名教育机构的网页上也有课程介绍、学生作业展览和学术资源链接等内容。

3. 企事业

相关企事业网站也是学习风景园林的重要窗口，主要有著名风景园林企业网站，著名公园绿地风景区网站和政府部门网站。前两者可以了解设计企业、公园绿地风景区的基本情况，往往有丰富的图片资源。后者可了解宏观的规划管理、人文历史等内容，如美国国家公园署（http://www.nps.gov）等。这些网站一般均可以在搜索引擎中检索到。

有的政府网站还公开部分档案材料，如西雅图政府网页链接有"舍伍德公园休闲地文件"（http://www.cityofseattle.net/parks/history/sherwood.htm），可以免费浏览和下载一些宝贵的西雅图市公园绿地建设文件档案，其中有著名的哈普林设计的西雅图高速公路绿地原始设计资料。

4. 其他

有一些风景园林网站致力于向门户网站方向发展，如有的还建有网络资源页，如加利福尼亚伯克利大学"景观在线"（http://www.lib.berkeley.edu/ENVI/lawww.html），"风景资源中心"（http://www.lih.gre.ac.uk/），"花园游览"（http://www.gardenvisit.com），"建筑报导"（http://archrecord.construction.com），"设计智慧"（http://www.ingentaconnect.com）等。这类网站内容丰富，更新速度快，是跟踪业界动态的重要窗口，也是检索资料、发现并登陆其他相关网站的有益途径。缺点是不稳定，往往出现转换域名，甚至关闭等情况。

互联网日新月异，新的网站在不断地涌现，世界景观设计（风景园林）信息交流日渐便捷，更多精彩的内容留待读者自己去探索。以英文为工具，通过网络了解世界景观设计（风景园林）历史和现状，将大大开拓读者视野，进入风景园林学习的新领域。

Unit 3

Charters & Policy

- 风景园林设计相关宪章
- 相关宣言
- 相关法规

The general theory of architecture is an integration of architecture, landscape and urban planning with the core of city design.

The Beijing Charter 1999

Text

The Venice Charter 1964

By IInd International Congress of Architects and Technicians of Historic

> 《威尼斯宪章 1964》，全称《保护文物建筑及历史地段的国际宪章》(International Charter for the Conservation and Restoration of Monuments and Sites)，是关于历史文化建筑保护的重要文献。宪章肯定了历史文物建筑的重要价值和作用，将其视为人类的共同遗产和历史的见证。全文共分定义、保护、修复、历史地段、发掘和出版6部分，共16条。

Imbued with a message from the past, the historic monuments of generations of people remain to the present day as living witnesses of their age-old traditions. People are becoming more and more conscious of the unity of human values and regard ancient monuments as a common heritage. The common responsibility to safeguard them for future generations is recognized. It is our duty to hand them on in the full richness of their authenticity.

It is essential that the principles guiding the preservation and restoration of ancient buildings should be agreed and be laid down on an international basis, with each country being responsible for applying the plan within the framework of its own culture and traditions.

By defining these basic principles for the first time, the Athens Charter of 1931 contributed towards the development of an extensive international movement which has assumed concrete form in national documents, in the work of ICOM and UNESCO and in the establishment by the latter of the International Centre for the Study of the Preservation and the Restoration of Cultural Property. Increasing awareness and critical study have been brought to bear on problems which have continually become more complex and varied; now the time has come to examine the Charter afresh in order to make a thorough study of the principles involved and to enlarge its scope in a new document.

Accordingly, the IInd International Congress of Architects and Technicians of Historic Monuments, which met in Venice from May 25th to 31st 1964, approved the following text：

世世代代人民的历史古迹，饱含着过去岁月的信息留存至今，成为人们古老的活的见证。人们越来越意识到人类价值的统一性，并把古代遗迹看做共同的遗产，认识到为后代保护这些古迹的共同责任。将它们真实、完整地传下去是我们的职责。

古代建筑的保护与修复指导原则应在国际上得到公认并作出规定，这一点至关重要。各国在各自的文化和传统范畴内负责实施这一规划。

1931年的雅典宪章第一次规定了这些基本原则，为一个国际运动的广泛发展作出了贡献，这一运动所采取的具体形式体现在各国的文件之中，体现在国际博物馆协会和联合国教育、科学及文化组织的工作之中，以及在由后者建立的国际文化财产保护与修复研究中心之中。一些已经并不断变得更为复杂和多样化的问题已越来越受到重视，并为其展开了紧急研究。现在，重新审阅宪章的时候已经来临，以便对其所含原则进行彻底研究，并在一份新文件中扩大其范围。

为此，1964 年 5 月 25～31 日在威尼斯召开了第二届历史古迹建筑师及技师国际会议，通过了以下文本：

Definitions

Article 1.

The concept of a historic monument embraces not only the single architectural work but also the urban or rural setting in which is found the evidence of a particular civilization, a significant development or a historic event. This applies not only to great works of art but also to more modest works of the past which have acquired cultural significance with the passing of time.

Article 2.

The conservation and restoration of monuments must have recourse to all the sciences and techniques which can contribute to the study and safeguarding of the architectural heritage.

Article 3.

The intention in conserving and restoring monuments is to safeguard them no less as works of art than as historical evidence.

定义

第一条　历史古迹的要领不仅包括单个建筑物，而且包括能从中找出一种独特的文明、一种有意义的发展或一个历史事件见证的城市或乡村环境。这不仅适用于伟大的艺术作品，而且亦适用于随时光逝去而获得文化意义的一些较为朴实的艺术品。

第二条　古迹的保护与修复必须求助于对研究和保护考古遗产有利的一切科学技术。

第三条　保护与修复古迹的目的旨在把它们既作为历史见证，又作为艺术品予以保护。

Conservation

Article 4.

It is essential to the conservation of monuments that they be maintained on a permanent basis.

Article 5.

The conservation of monuments is always facilitated by making use of them for some socially useful purpose. Such use is therefore desirable but it must not change the lay-out or decoration of the building. It is within these limits only that modifications demanded by a change of function should be envisaged and may be permitted.

Article 6.

The conservation of a monument implies preserving a setting which is not out of scale. Wherever the traditional setting exists, it must be kept. No new construction, demolition or modification which would alter the relations of mass and colour must be allowed.

Article 7.

A monument is inseparable from the history to which it bears witness and from the setting in which it occurs. The moving of all or part of a monument cannot be allowed except where the safeguarding of that monument demands it or where it is justified by national or international interest of paramount importance.

Article 8.

Items of sculpture, painting or decoration which form an integral part of a monument may only be removed from it if this is the sole means of ensuring their preservation.

保护

第四条　古迹的保护至关重要的一点在于长期的维护。

第五条　为社会公用之目的使用古迹永远有利于古迹的保护。因此，这种使用合乎需要，但决不能改变该建筑的布局或装饰。只有在此限度内才可考虑或允许因功能改变而需做的改动。

第六条　古迹的保护包含着对一定规模环境的保护。凡传统环境存在的地方必须予以保存，决不允许任何导致改变主体和颜色关系的新建、拆除或改动。

第七条　古迹不能与其所见证的历史和其产生的环境分离。除非出于保护古迹之需要，或因国家或国际之极为重要利益而证明有其必要，否则不得全部或局部搬迁古迹。

第八条　作为构成古迹整体一部分的雕塑、绘画或装饰品，只有在非移动而不能确保其保存的唯一办法时方可进行移动。

Restoration

Article 9.

The process of restoration is a highly specialized operation. Its aim is to preserve and reveal the aesthetic and historic value of the monument and is based on respect for original material and authentic documents. It must stop at the point where conjecture begins, and in this case moreover any extra work which is indispensable must be distinct from the architectural composition and must bear a contemporary stamp. The restoration in any case must be preceded and followed by an archaeological and historical study of the monument.

Article 10.

Where traditional techniques prove inadequate, the consolidation of a monument can be achieved by the use of any modern technique for conservation and construction, the efficacy of which has been shown by scientific data and proved by experience.

Article 11.

The valid contributions of all periods to the building of a monument must be respected, since unity of style is not the aim of a restoration. When a building includes the superimposed work of different periods, the revealing of the underlying state can only be justified in exceptional circumstances and when what is removed is of little interest and the material which is brought to light is of great historical, archaeological or aesthetic value, and its state of preservation good enough to justify the action. Evaluation of the importance of the elements involved and the decision as to what may be destroyed cannot rest solely on the individual in charge of the work.

Article 12.

Replacements of missing parts must integrate harmoniously with the whole, but at the same time must be distinguishable from the original so that restoration does not falsify the artistic or historic evidence.

Article 13.

Additions cannot be allowed except in so far as they do not detract from the interesting parts of the building, its traditional setting, the balance of its composition and its relation with its surroundings.

修复

第九条　修复过程是一个高度专业性的工作，其目的旨在保存和展示古迹的美学与历史价值，并以尊重原始材料和确凿文献为依据。一旦出现臆测，必须立即予以停止。此外，即使如此，任何不可避免的添加都必须与该建筑的构成有所区别，并且必须要有现代标记。无论在任何情况下，修复之前及之后必须对古迹进行考古及历史研究。

第十条　当传统技术被证明为不适用时，可采用任何经科学数据和经验证明为有效的现代建筑及保护技术来加固古迹。

第十一条　各个时代为一古迹之建筑物所做的正当贡献必须予以尊重，因为修复的目的不是追求风格的统一。当一座建筑物含有不同时期的重叠作品时，揭示底层只有在特殊情况下，在被去掉的东西价值甚微，而被显示的东西具有很高的历史、考古或美学价值，并且保存完好足以说明这么做的理由时才能证明其具有正当理由。评估由此涉及的各部分的重要性及决定毁掉什么内容不能仅仅依赖于负责此项工作的个人。

第十二条　缺失部分的修补必须与整体保持和谐，但同时须区别于原作，以使修复不歪曲其艺术或历史见证。

第十三条　任何添加均不允许，除非它们不致于贬低该建筑物的有趣部分、传统环境、布局平衡及其与周围环境的关系。

Historic Sites
历史地段

Article 14.

The sites of monuments must be the object of special care in order to safeguard their integrity and ensure that they are cleared and presented in a seemly manner. The work of conservation and restoration carried out in such places should be inspired by the principles set forth in the foregoing articles.

第十四条　古迹遗址必须成为专门照管对象，以保护其完整性，并确保用恰当的方式进行清理和开放。在这类地点开展的保护与修复工作应得到上述条款所规定之原则的鼓励。

Excavations

Article 15.

Excavations should be carried out in accordance with scientific standards and the recommendation defining international principles to be applied in the case of archaeological excavation adopted by UNESCO in 1956.

Ruins must be maintained and measures necessary for the permanent conservation and protection of architectural features and of objects discovered must be taken. Furthermore, every means must be taken to facilitate the understanding of the monument and to reveal it without ever distorting its meaning.

All reconstruction work should however be ruled out "a priori". Only anastylosis, that is to say, the reassembling of existing but dismembered parts can be permitted. The material used for integration should always be recognizable and its use should be the least that will ensure the conservation of a monument and the reinstatement of its form.

发掘

第十五条　发掘应按照科学标准和联合国教育、科学及文化组织1956年通过的适用于考古发掘国际原则的建议予以进行。遗址必须予以保存，并且必须采取必要措施，永久地保存和保护建筑风貌及其所发现的物品。此外，必须采取一切方法促进对古迹的了解，使它得以再现而不曲解其意。然而对任何重建都应事先予以制止，只允许重修，也就是说，把现存但已解体的部分重新组合。所用粘结材料应永远可以辨别，并应尽量少用，只须确保古迹的保护和其形状的恢复之用便可。

Publication

Article 16.

In all works of preservation, restoration or excavation, there should always be precise documentation in the form of analytical and critical reports, illustrated with drawings and photographs. Every stage of the work of clearing, consolidation, rearrangement and integration, as well as technical and formal features identified during the course of the work, should be included. This record should be placed in the archives of a public institution and made available to research workers. It is recommended that the report should be published.

出版

第十六条　一切保护、修复或发掘工作永远应有用配以插图和照片的分析及评论报告这一形式所做的准确的记录。清理、加固、重新整理与组合的每一阶段，以及工作过程中所确认的技术及形态特征均应包括在内。这一记录应存放于一公共机构的档案馆内，使研究人员都能查到。建议该记录应出版。

Notes

1. Extracts from

Texts

（1）IInd International Congress of Architects and Technicians of Historic Monument. The Venice Charter 1964 ［EB/OL］. http：//www. iflaonline. org/resources/policy/pdf/charter/venice _ charter. pdf，2007 - 09 - 22.

（2）张京祥. 西方城市规划思想史纲［M］. 南京：东南大学出版社，2005：272 - 275.

2. Background Information

《威尼斯宪章1964》，全称《保护文物建筑及历史地段的国际宪章》。1964 年 5 月 31 日，从事历史文物建筑工作的建筑师和技术人员国际会议第二次会议在威尼斯通过的决议，故简称"威尼斯宪章"。宪章肯定了历史文物建筑的重要价值和作用，将其视为人类的共同遗产和历史的见证。明确了历史文物建筑的概念，同时要求，必须利用一切科学技术保护与修复文物建筑。强调修复是一种高度专门化的技术，必须尊重原始资料和确凿的文献，决不能有丝毫臆测。其目的是完全保护和再现历史文物建筑的审美和价值，还强调对历史文物建筑的一切保护、修复和发掘工作都要有准确的记录、插图和照片。

威尼斯宪章规定了关于历史文化建筑保护和修复的基本原则，多年来一直被世界各国建筑、城市规划、风景园林等学科引为重要专业文献。

3. Sources of Additional Information

http：//www. international. icomos. org/charters/venice _ e. htm

http：//www. iflaonline. org/policy. php

Further Reading A

The Convention on Biological Diversity

Preamble

The Contracting Parties

Conscious of the intrinsic value of biological diversity and of the ecological, genetic, social, economic, scientific, educational, cultural, recreational and aesthetic values of biological diversity and its components.

Conscious also of the importance of biological diversity for evolution and for maintaining life sustaining systems of the biosphere.

Affirming that the conservation of biological diversity is a common concern of humankind.

Reaffirming that States have sovereign rights over their own biological resources.

Reaffirming also that States are responsible for conserving their biological diversity and for using their biological resources in a sustainable manner.

Concerned that biological diversity is being significantly reduced by certain human activities.

序言

缔约国

意识到生物多样性的内在价值和生物多样性及其组成部分的生态、遗传、社会、经济、科学、教育、文化、娱乐和美学价值。

还意识到生物多样性对进化和保持生物圈的生命维持系统的重要性。

确认生物多样性的保护是全人类共同关切的事项。

重申各国对自己的生物资源拥有主权权利。

也重申各国有责任保护自己的生物多样性并以可持久的方式使用自己的生物资源。

关切一些人类活动正在导致生物多样性的严重减少。

Aware of the general lack of information and knowledge regarding biological diversity and of the urgent need to develop scientific, technical and institutional capacities to provide the basic understanding upon which to plan and implement appropriate measures.

Noting that it is vital to anticipate, prevent and attack the causes of significant reduction or loss of biological diversity at source.

Noting also that where there is a threat of significant reduction or loss of biological diversity, lack of full scientific certainty should not be used as a reason for postponing measures to avoid or minimize such a threat.

Noting further that the fundamental requirement for the conservation of biological diversity is the in-situ conservation of ecosystems and natural habitats and the maintenance and recovery of viable populations of species in their natural surroundings.

Noting further that ex-situ measures, preferably in the country of origin, also have an important role to play.

意识到普遍缺乏关于生物多样性的资料和知识，亟需开发科学、技术和机构能力，从而提供基本理解，据以策划与执行适当措施。

注意到预测、预防和从根源上消除导致生物多样性严重减少或丧失的原因至关重要。

并注意到生物多样性遭受严重减少或损失的威胁时，不应以缺乏充分的科学定论为理由，而推迟采取旨在避免或尽量减轻此种威胁的措施。

注意到保护生物多样性的基本要求，是就地保护生态系统和自然生境，维持恢复物种在其自然环境中有生存力的群体。

并注意到移地措施，最好在原产国内实行，也可发挥重要作用。

Recognizing the close and traditional dependence of many indigenous and local communities embodying traditional lifestyles on biological resources, and the desirability of sharing equitably benefits arising from the use of traditional knowledge, innovations and practices relevant to the conservation of biological diversity and the sustainable use of its components.

Recognizing also the vital role that women play in the conservation and sustainable use of biological diversity and affirming the need for the full participation of women at all levels of policy-making and implementation for biological diversity conservation.

Stressing the importance of, and the need to promote, international, regional and global cooperation among States and intergovernmental organizations and the non-governmental sector for the conservation of biological diversity and the sustainable use of its components.

Acknowledging that the provision of new and additional financial resources and appropriate access to relevant technologies can be expected to make a substantial difference in the world's ability to address the loss of biological diversity.

Acknowledging further that special provision is required to meet the needs of developing countries, including the provision of new and additional financial resources and appropriate access to relevant technologies.

Noting in this regard the special conditions of the least developed countries and small island States.

Acknowledging that substantial investments are required to conserve biological diversity and that there is the expectation of a broad range of environmental, economic and social benefits from those investments.

Recognizing that economic and social development and poverty eradication are the first and overriding priorities of developing countries.

认识到许多体现传统生活方式的土著和地方社区同生物资源有着密切和传统的依存关系，应公平分享从利用与保护生物资源及持久使用其组成部分有关的传统知识、创新和做法而产生的惠益。

并认识到妇女在保护和持久使用生物多样性中发挥的极其重要作用，并确认妇女必须充分参与保护生物多样性的各级政策的制定和执行。

强调为了生物多样性的保护及其组成部分的持久使用，促进国家、政府间组织和非政府部门之间的国际、区域和全球性合作的重要性和必要性。

承认提供新的和额外的资金和适当取得有关的技术，可对全世界处理生物多样性丧失问题的能力产生重大影响。

进一步承认有必要订立特别规定，以满足发展中国家的需要，包括提供新的和额外的资金和适当取得有关的技术。

注意到最不发达国家和小岛屿国家这方面的特殊情况。

承认有必要大量投资以保护生物多样性，而且这些投资可望产生广泛的环境、经济和社会惠益。

认识到经济和社会发展及根除贫困是发展中国家第一和压倒一切的优先事务。

Aware that conservation and sustainable use of biological diversity is of critical importance for meeting the food, health and other needs of the growing world population, for which purpose access to and sharing of both genetic resources and technologies are essential.

Noting that, ultimately, the conservation and sustainable use of biological diversity will strengthen friendly relations among States and contribute to peace for humankind.

Desiring to enhance and complement existing international arrangements for the conservation of biological diversity and sustainable use of its components, and determined to conserve and sustainably use biological diversity for the benefit of present and future generations.

Have agreed as follows:

意识到保护和持久使用生物多样性对满足世界日益增加的人口的粮食、健康和其他需求至为重要，而为此目的取得和分享遗传资源和遗传技术是必不可少的。

注意到保护和持久使用生物多样性终必增强国家间的友好关系，并有助于实现人类和平。

期望加强和补充现有保护生物多样性和持久使用其组成部分的各项国际安排；并决心为今世后代的利益，保护和持久使用生物多样性。

兹协议如下：

Article 1. Objectives

The objectives of this Convention, to be pursued in accordance with its relevant provisions, are the conservation of biological diversity, the sustainable use of its components and the fair and equitable sharing of the benefits arising out of the utilization of genetic resources, including by appropriate access to genetic resources and by appropriate transfer of relevant technologies, taking into account all rights over those resources and to technologies, and by appropriate funding.

第一条　目标

本公约的目标是按照本公约有关条款从事保护生物多样性、持久使用其组成部分以及公平合理分享由利用遗传资源而产生的惠益；实现手段包括遗传资源的适当取得及有关技术的适当转让，但需顾及对这些资源和技术的一切权利，以及提供适当资金。

Article 2. Use of Terms

For the purposes of this Convention:

"**Biological diversity**" means the variability among living organisms from all sources including, inter alia, terrestrial, marine and other aquatic ecosystems and the ecological complexes of which they are part; this includes diversity within species, between species and of ecosystems.

"**Biological resources**" includes genetic resources, organisms or parts thereof, populations, or any other biotic component of ecosystems with actual or potential use or value for humanity.

"**Biotechnology**" means any technological application that uses biological systems, living organisms, or derivatives thereof, to make or modify products or processes for specific use.

第二条　用语

为本公约的目的：

"**生物多样性**"是指所有来源的形形色色生物体，这些来源除其他外包括陆地、海洋和其他水生生态系统及其所构成的生态综合体；这包括物种内部、物种之间和生态系统的多样性。

"**生物资源**"是指对人类具有实际或潜在用途或价值的遗传资源、生物体或其部分、生物群体、或生态系统中任何其他生物组成部分。

"**生物技术**"是指使用生物系统、生物体或其衍生物的任何技术应用，以制作或改变产品或过程以供特定用途。

"**Country of origin of genetic resources**" means the country which possesses those genetic resources in in-situ conditions.

"**Country providing genetic resources**" means the country supplying genetic resources collected from in-situ sources, including populations of both wild and domesticated species, or taken from ex-situ sources, which may or may not have originated in that country.

"**Domesticated or cultivated species**" means species in which the evolutionary process has been influenced by humans to meet their needs.

"**Ecosystem**" means a dynamic complex of plant, animal and micro-organism communities and their non-living environment interacting as a functional unit.

"**Ex-situ conservation**" means the conservation of components of biological diversity outside their natural habitats.

"**Genetic material**" means any material of plant, animal, microbial or other origin containing functional units of heredity.

"**Genetic resources**" means genetic material of actual or potential value.

"**遗传资源的原产国**"是指拥有处于原产境地的遗传资源的国家。

"**提供遗传资源的国家**"是指供应遗传资源的国家，此种遗传资源可能是取自原地来源，包括野生物种和驯化物种的群体，或取自移地保护来源，不论是否原产于该国。

"**驯化或培殖物种**"是指人类为满足自身需要而影响了其演化进程的物种。

"**生态系统**"是指植物、动物和微生物群落和它们的无生命环境作为一个生态单位交互作用形成的一个动态复合体。

"**移地保护**"是指将生物多样性的组成部分移到它们的自然环境之外进行保护。

"**遗传资源**"是指具有实际或潜在价值的遗传材料。

"**Habitat**" means the place or type of site where an organism or population naturally occurs.

"**In-situ conditions**" means conditions where genetic resources exist within ecosystems and natural habitats, and, in the case of domesticated or cultivated species, in the surroundings where they have developed their distinctive properties.

"**In-situ conservation**" means the conservation of ecosystems and natural habitats and the maintenance and recovery of viable populations of species in their natural surroundings and, in the case of domesticated or cultivated species, in the surroundings where they have developed their distinctive properties.

"**Protected area**" means a geographically defined area which is designated or regulated and managed to achieve specific conservation objectives.

"**生境**"是指生物体或生物群体自然分布的地方或地点。

"**原地条件**"是指遗传资源生存于生态系统和自然生境之内的条件；对于驯化或培殖的物种而言，其环境是指它们在其中发展出其明显特性的环境。

"**就地保护**"是指保护生态系统和自然生境以及维护和恢复物种在其自然环境中有生存力的群体；对于驯化和培殖物种而言，其环境是指它们在其中发展出其明显特性的环境。

"**保护区**"是指一个划定地理界限、为达到特定保护目标而指定或实行管制和管理的地区。

"**Regional economic integration organization**" means an organization constituted by sovereign States of a given region, to which its member States have transferred competence in respect of matters governed by this Convention and which has been duly authorized, in accordance with its internal procedures, to sign, ratify, accept, approve or accede to it.

"**Sustainable use**" means the use of components of biological diversity in a way and at a rate that does not lead to the long-term decline of biological diversity, thereby maintaining its potential to meet the needs and aspirations of present and future generations.

"Technology" includes biotechnology.

"区域经济一体化组织" 是指由某一区域的一些主权国家组成的组织,其成员国已将处理本公约范围内的事务的权力付托它并已按照其内部程序获得正式授权,可以签署、批准、接受、核准或加入本公约。

"持久使用" 是指使用生物多样性组成部分的方式和速度不会导致生物多样性的长期衰落,从而保持其满足今世后代的需要和期望的潜力。

"技术" 包括生物技术。

Notes

1. Extracts from

Text

United Nations Conference on Environment and Development. *The Covention on Biological Diversity* [EB/OL]. http://www.cbd.int/convention/convention.shtml,2008-03-20.

2. Background Information

《生物多样性公约》(*The Convention on Biological Diversity*):是一项保护地球生物资源的国际性公约,于1992年6月5日由签约国在巴西里约热内卢举行的联合国环境与发展大会上签署。公约于1993年12月29日正式生效。常设秘书处设在加拿大的蒙特利尔。该公约是一项有法律约束力的公约,旨在保护濒临灭绝的植物和动物,最大限度地保护地球上的多种多样的生物资源,以造福于当代和子孙后代。全文分序言和目标、用语、原则、管辖范围、合作、保护和持久使用方面的一般措施、查明与监测、就地保护、移地保护等20项条目。本文选自公约的开篇部分。

3. Sources of Additional Information

http://www.cbd.int

http://www.biodiv.org

Further Reading B

Beijing Charter: Towards An Integral Architecture

During the past 50 years, the architects of the world have met to debate over a large number of issues. These debates have much furthered our understanding in all branches of architecture. It is therefore appropriate to review the progress so far and redefine the limits, the contents, and the organisation of our discipline and profession.

1. The Theoretical Premises

Over the centuries the role of an architect is constantly modified to suit the needs and requirements of its time. Where traditional methods are shown to be inadequate, new approaches are developed to take their place. Yet without exception, each redefinition pushes the boundary of architecture outwards for a wider coverage, as well as inwards for higher degrees of specialisation in the component parts. The 20th century is perhaps the most exemplary in this regard.

A wider coverage of its contents and finer degrees of specialisation have empowered the 20th century architect with unprecedented professional opportunities and potential, yet at a personal level, an expanding profession with growing specialisation can seem elephantine. In a sense, the architects' Tower of Babel appears to have fallen: it is increasingly difficult for one architect to grasp the expertise of a fellow colleague; although the body of knowledge has grown collectively, the outlook of any single designer tends to become paradoxically narrow and fragmented. The specialist expertise is brought together through financial ties and managerial skills, rather than a coherent intellectual framework. As a result, the role of an architect continues to be marginalised in the decision making over the human habitat today.

From the point of view of an architect, his or her ability to propose creative design solutions depends critically on the intellectual and professional spheres he or she commands. Narrow and fragmented individual outlooks cannot be made to work, however wonderfully the individual designers are managed externally. Nevertheless, any given person cannot and should not attempt to master the whole body of knowledge of our profession. Quo vadis?

Classical Chinese philosophers went to great pains to pinpoint the differences between methodology (alternatively translated as Dao or Tao) which concerns an intellectual framework, and methods (Fa) which deal with specific techniques. It is useful to draw on their wisdom in this matter. Whatever professional talents, expertise, or preferences an architect may have, these techniques can only realise their true value when guided by a larger, intellectual perspective. An architect may work in a specialised area by choice or chance, yet he or she must not lose sight of the profession as a whole and of the vast sphere of knowledge which is potentially at his or her disposal.

Past and contemporary masters have shown how their understanding of the Dao of architecture has helped them to achieve magnificent heights in design and planning. However although such understanding could be regarded as a luxury enjoyed by the masters in the past, it will increasingly become goods of necessity for all architects in the age of information explosion. In the rapidly expanding professional universe, an intellectual orientation that organises the body of knowledge and expertise and relates architecture to the wider processes that give shape to the built environment, is paramount.

So what does this methodology contain?

近百年来，世界建筑师聚首讨论了许多课题，深化了对建筑学的理解。如今，重新回顾这些讨论，并对建筑学的范围、内涵及其学科和专业体系重新定义，当大有裨益。

1. 基本前提

建筑学的内容和建筑师的业务一直随着时代而横向拓展，纵向深化。旧方法一旦不合时宜，新方法就会取而代之。每一次革新都使建筑学更广大，也更精彩，20世纪建筑的发展就充分证明了这一点。

建筑学广阔而纵深的拓展赋予20世纪的建筑师前所未有的用武之地。然而，学科的扩大与专门化也难免让从事活动的个人觉得建筑学如盲人摸象，一时不能把握全局。学科知识的总体在扩张，设计师个人的视野却趋向狭窄和破碎，专门的设计知识和技术仅仅依靠投资和开发组织来维系，学科自身缺乏完整的知识框架，其结果，建筑师参与人居环境建设决策的作用却日见削弱。

建筑师的设计创造仰仗其对学科知识的把握，只有在统领全局的学科观、专业观的指导下，才能真正发挥个人的才干、技能和天分。纵览古今大师们的成就，更感到他们对建筑之高瞻远瞩弥足珍贵。在过去，这样的全面建筑观堪称大师们私藏的瑰宝，然而在信息爆炸的今天，全面的、广义的建筑观应当成为所有建筑专业人员之必备。一旦领悟了设计的基本哲理，具体的技术、形式问题就不难努力以赴，正如中国古人所云："一法得道，变法万千。"

2. A Fusion of Architecture, Landscape Architecture and City Planning

The professional identity of an architect in the wider world is focused on the built forms that are ultimately created.

Basically, the general theory of architecture is an integration of architecture, landscape and urban planning with the core of city design. However, the increasing scale and scope of modern development provide architects with great opportunities to deal with architecture, landscape and urban planning as a whole. This tripartite composition enables the designer to search for solutions within a wider sphere.

2. 融合建筑、地景与城市规划

建筑学与大千世界的辩证关系，归根到底，集中在建筑的空间组合与形式的创造。

现代工程规模日益扩大，建设周期缩短，建筑师可以在较为广阔的范域内，从场地选择到规划设计，直至室内外空间的协调，寻求设计的答案。

广义建筑学，就其学科内涵来说，是通过城市设计的核心作用，从观念上和理论基础上把建筑、地景和城市规划学科的精髓整合为一体，将我们关注的焦点从建筑单体、结构最终转换到建筑环境上来。如果说，过去主要局限于一些先驱者，那么现在则已涉及整个建筑领域。

3. Architecture as a Process for Human Habitat

Metabolism is one of the fundamental rules in the development of human settlement. Architecture is the discipline that deals with human settlement, so it should regard the physical objectives of construction as a system of circulation. The life cycle of buildings should be regarded as a fundamental factor of design.

The life cycle of buildings not only includes the construction and running phases, but also includes processes aiming at lower resource costs, less pollution and grey energy consumption, recycling as much as possible, and reformation of environments.

On the aspect of urban settlement, factors such as planning, architectural design, historical preservation, adapted re-use of old buildings, urban rehabilitation, city renewal and reconstruction, utilisation of underground facilities, etc., should be integrated into a dynamic circulation system. This is a system for better architecture in the modern space-times of architecture. It is also an exemplification of the sustainable approach in urban planning and architecture design.

3. 建筑学的循环体系

新陈代谢是人居环境发展的客观规律，建筑单体极其环境经历一个规划、设计、建设、维修、保护、政治、更新的过程。建设环境的寿命周期恒长持久，因而更依赖建筑师的远见卓识。将建筑循环过程的各个阶段统筹规划，将新区规划设计、旧城整治、更新与重建等纳入一个动态的、生生不息的循环体系之中，在失控因素作用下，不断提高环境质量，这也是实现可持续发展战略的关键。

4. Multiple Technology Rooted in Indigenous Cultures

To utilise technological innovation to its full extent is one of our basic tasks in the coming century.

Firstly, in the 21st century, various presentations of technology will co-exist, based on the fact that there are regional contrasts and imbalance in the development of technology.

Theoretically, it is necessary to adopt new technology from foreign sources and integrate it with local conditions to improve the local technological standards. If architects themselves can realise the ecological challenges mankind is facing, and adopt advanced technology creatively, then the buildings they design are bound to be sustainable and healthy.

Because of technological complexity, low-tech, light-tech and high-tech are different in scale and level. For each project, the choice of technological approach should be made according to the specific conditions. In other words, for the progress of every building project, different forms of technology should be integrated, utilised and improved.

As for the utilisation of technology, considerations on humanist, ecological, economic and regional aspects should be integrated. Different levels of innovation should be carried out in order to improve the level of architectural creativity. Many theoretical and practical examples are available today, and it is obvious that much more progress will be made in the next century.

Secondly, today's progress includes both science and technology. The development of technology must be related to human factors. As Alvar Aalto said, "the preservation of difference should also be strengthened. The development of architecture should be rooted in the regional background, and take the local conditions as its starting point in the search for better solutions. Based upon this, foreign ideas can be integrated into our own. This would finally lead to a human society showing both integrity and variety".

4. 植根于地方文化的多层次技术建构

充分发挥技术对人类社会文明进步的促进作用是新世纪的重要使命。地域差异预示着21世纪仍将是多种技术并存的时代。高新技术革新能迅猛地推动生产力的发展，但是成功的关键仍然有赖于技术与地方文化、地方经济的创造性结合。不同国度和地区之间的经验交流，不是解决方案的简单移植，而是激发地方想象力的一种手段。

技术功能的内涵要从科学和工程方面加以扩展，直至覆盖心理范畴。

5. Architecture of Harmony Instead of Monotony

Architecture is by definition a regional product: buildings serve, and derive their significance from local contacts. Regional architecture is yet by no means a mere product of a region's past. Rather, it is derived from the concerns for its future. The significance of our profession lies in the creative designs that bridge the past and the future. We use our professional knowledge to guide an informed choice amongst the options that are increasingly opened to local communities. "The sharing of experiences among various countries and geographical regions must never be seen as a simple transfer of ready-made solutions, but as a means of stimulating local imaginations".

The localisation of modern architecture and the modernisation of local architecture is a common approach to be shared by all in the progress toward architectural proliferation.

5. 建筑文化的和而不同

建筑学是地区的产物，建筑形式的意义与地方文脉相连，并成为地方文脉的诠释。但是，地区建筑学并非只是地区历史的产物，它更关系到地区的未来。建筑物相对永久的存在成为人们日常生活中的感情寄托，然而地方社区的演进过程中最终限定了建筑师工作的背景，我们职业的深远意义就在于以创造性的设计联系过去和未来。地方社区对未来的选择方案日渐增多，我们要运用专业知识找到真正符合当时当地情况的建筑发展方向。

我们在为地方传统所鼓舞的同时，不能忘记我们的任务是创造一个和而不同的未来建筑环境。现代建筑的地区化，乡土建筑的现代化，殊途同归，推动世界和地区的进步与丰富多彩。

6. Art for the Sake of the Built Environment

After the industrial revolution, urbanisation of increasing speed resulted in dramatic changes in urban structure and architecture forms (Figure 3.1). The physical environment has been led to anarchy. We should try to find order in the anarchy, to find beauty and harmony in the chaos.

Figure 3.1 Art for the sake of the built environment

To consider the relationship between architecture and its environments with traditional design methods is far from adequate. We have to look at architecture from a massive and urban view. Architectural thoughts should shift from single buildings to building complexes, to urban and rural regional planning. The holistic relationship with nature is another important factor that should be considered.

In the histories of all cultures, architecture became the ultimate manifestation of inseparable parts in fine arts, such as sculpture, painting, craftsmanship, etc. This should be one of our goals.

6. 建筑作为艺术形式的最终表现

当今，城市建设规模浩大，速度空前，城市以往的表面完整性遭到破坏，建筑环境的整体艺术成为新的追求，宜用城市的观念看建筑，重视建筑群的整体和城市全局的协调，以及建筑与自然的关系，在动态的建设发展中追求相对的整体的协调美和"秩序的真谛"。

综观各种文化发展史，建筑最终成为美术与手工艺的表现。如今，工业发展为艺术创造提供了前所未有的技术可能性，我们应为建筑、工艺和美术在更高层次上的结合而努力。

Notes

1. Extracts from

Text

吴良镛. 国际建协《北京宪章》——建筑学的未来（中英文版）[M]. 北京：清华大学出版社，2002：3-14, 177-184.

Picture

Figure 3.1：吴良镛. 国际建协《北京宪章》——建筑学的未来（中英文版）[M]. 北京：清华大学出版社，2002：81.

2. Background Information

1999年，国际建协第20届世界建筑师大会在北京召开，大会一致通过了由吴良镛教授起草的《北京宪章》。《北京宪章》总结了百年来建筑发展的历程，并在剖析和整合20世纪的历史与现实、理论与实践、成就与问题以及各种新思路和新观点的基础上，展望了21世纪建筑学的前进方向。

本文选自《北京宪章》第三部分"从传统建筑学走向广义建筑学"，主要阐述了一个中心观点：用广义建筑学的观点融合建筑、地景、城市规划学科，来解决目前城乡建设中存在的复杂的问题。

3. Sources of Additional Information

http://chinaasc.org/html/80/23580-21082.html

Further Reading C

Federal & State Regulations

By Nicholas T. Dines, Kyle D. Brown and Jeffrey D. Blankenship

Environmental Impact Analysis

The National Environmental Policy Act (NEPA) requires the preparation of an Environmental Impact Statement (EIS) to assess the impact of a major federal action on the quality of the environment. The EIS's discussion of environmental impacts forms the scientific and analytic basis for the comparisons of alternatives by decision makers, which are the heart of the EIS process. NEPA applies only to projects and programs entirely or partly financed, assisted, conducted, regulated or approved by federal agencies.

Determination of Significance

Actions triggering[1] an EIS are defined by statute or determined to be significant by agency officials. While the determination of significance is discretionary[2], the Council on Environmental Quality (CEQ) provides guidelines that stress analysis of context and intensity. The regulatory definition of context provides that the significance of an action must be analyzed in several contexts such as society as a whole (human, national), the affected region, the affected interests, and the locality. Significance varies with the setting of the proposed action. Intensity is defined as the severity[3] of impact from the proposal. Table 3.1 outlines considerations in evaluating intensity, as defined by CEQ.

Table 3.1 Considerations for Evaluating Intensity of Environmental Impacts

No.	Impacts
1	Impacts that may be both beneficial and adverse. A significant effect may exist even if the Federal agency believes that on balance the effect will be beneficial
2	The degree to which the proposed action affects public health or safety
3	Unique characteristics of the geographic area such as proximity[4] to historic or cultural resources, park lands, prime farmlands, wetlands, wild and scenic rivers, or ecologically critical areas
4	The degree to which the effects on the quality of the human environment are likely to be highly controversial[5]
5	The degree to which the possible effects on the human environment are highly uncertain or involve unique or unknown risks
6	The degree to which the action may establish a precedent for future actions with significant effects or represents a decision in principle about a future consideration

---

[1] triggering：触发，控制

[2] discretionary：任意的，自由决定的

[3] severity：严肃，严格，严重，激烈

[4] proximity：接近

[5] controversial：争议的

No.	Impacts
7	Whether the action is related to other actions with individually insignificant but cumulatively significant impacts. Significance exists if it is reasonable to anticipate[1] a cumulatively significant impact on the environment. Significance cannot be avoided by terming an action temporary or by breaking it down into small component parts
8	The degree to which the action may adversely[2] affect districts, sites, highways, structures, or objects listed in or eligible for listing in the National Register of Historic Places or may cause loss or destruction of significant scientific, cultural, or historical resources
9	The degree to which the action may adversely affect an endangered or threatened species or its habitat that has been determined to be critical under the Endangered Species Act of 1973
10	Whether the action threatens a violation of Federal, State, or local law or requirements imposed for the protection of the environment

Source: Adapted from *Council on Environmental Quality*, Regulations for Implementing NEPA.

Analysis of Impacts

Although NEPA primarily focuses on environmental planning, the statute[3] includes several provisions[4] that encourage agencies to consider the social, economic, cultural and historic impacts of their actions, as well as the environmental and human health impacts. The discussion must include both direct and indirect effects of a proposed project. The requisite[5] level of detail for an EIS depends on the nature and scope of the proposed action. The discussion of environmental effects of alternatives need not be exhaustive[6]. What is required is information sufficient to permit a reasoned choice of alternatives as far as environmental aspects are concerned.

Structure of Process

The development of an EIS consists of three major stages: ① scoping; ② preparation of draft EIS; ③ preparation of final EIS. The scoping process determines issues to be addressed and identifies the significant issues related to a proposed action. The agency must publish a notice of intent to prepare an EIS and solicit[7] input from the public regarding the scope of the issues and alternatives to be considered.

After an agency completes the scoping process for an EIS, the agency prepares a draft EIS in accordance with the scope that was developed through the scoping process. The intent of the draft is to solicit feedback from appropriate federal, state and local agencies, as well as other interested parties. CEQ guidelines require an interdisciplinary[8] approach to authoring an EIS, which insure the integrated use of the natural and social sciences, as well as the environmental design arts. Reports are

---

[1] anticipate: 预期，期望
[2] adversely: 逆地，反对地
[3] statute: 法令，条例
[4] provisions: 规定
[5] requisite: 需要的，必不可少的，必备的
[6] exhaustive: 彻底的，详尽的
[7] solicit: 恳求
[8] interdisciplinary: 各学科间的

encouraged to be less than 150 pages in most circumstances and for proposals of unusual scope or complexity shall normally be less than 300 pages.

After the close of the comment period on a draft EIS, the authoring agency will evaluate comments and prepare a final EIS. The agency shall discuss at appropriate points in the final statement any responsible opposing view that was not adequately discussed in the draft statement and shall indicate the agency's response to the issues raised.

SEPAs

As a result of NEPA, a number of states adopted State Environmental Policy Acts (SEPAs) to evaluate impacts of development for projects requiring state and/or local governmental action. While statutes vary from state to state, many are modeled after NEPA regulations, and require the preparation of an Environmental Impact Report (EIR) for projects that exceed established thresholds related to size, scope, or significance of impact. SEPA regulations typically affect many large projects that are exempt from NEPA regulations because they do not require federal involvement.

Water Resource Protection

The Clean Water Act and other federal policies require permits for altering or impacting water bodies, wetlands, and water quality resulting from stormwater runoff. These include:

SECTION 404 PERMIT—requires a project involving the discharge of dredged[1] or fill material into waters of the United States (including federally defined wetlands) to obtain a permit from the Army Corps of Engineers. The permit may be a programmatic general permit, an individual permit or an official letter of permission. Many states have adopted similar regulations requiring state-issued permits for activities affecting wetlands and water bodies.

SECTION 10 PERMIT—requires a federal permit from the Army Corps of Engineers for dredging, filling or obstruction of navigable waters. The application for the Section 10 Permit is often submitted in conjunction with an application for a Section 404 Permit.

EPA STORMWATER NOTICES OF INTENT AND/OR NPDES PERMITS—required for stormwater discharges associated with certain industrial activities. Industrial activity was recently redefined to include construction activity including clearing, grading, and excavation[2] activities except operations that result in the disturbance of less than five acres of total land area, which are not part of a larger common plan of development or sale. Many states or local communities have approved NPDES General Permits for stormwater systems that provide coverage for construction projects. The project owner and operator is required to file a Notice of Intent (NOI) and pollution abatement plan with the Environmental Protection Agency to be included under the NPDES general permit.

Flood Protection

National Flood Insurance Act and Disaster Protection Act Certification prohibits banks from issuing loans secured by improved real estate located in an area having flood hazards, and in which

---

[1] dredge: 挖掘，疏浚
[2] excavation: 挖掘，发掘

flood insurance[1] is available, unless the building securing the loan is covered by flood insurance. For insurance purposes, special flood hazards are defined as 100-year flood plains. This usually requires review and certification of building plans by the lender. Flood Insurance Rate Maps (FIRMs) are published by FEMA and delineate[2] flood hazards for local communities.

Historic Preservation

National Historic Preservation Act Section 106 requires that certain federally assisted, permitted and licensed activities that might have an adverse effect on properties listed with or eligible[3] for listing with, the national Register of Historic Places be reviewed concerning that effect and its consequences. Many states and local communities have adopted similar legislation[4].

Notes

1. Extracts from

Text

Dines, Nicholas T. et al. *Landscape Architect's Portable Handbook* [M]. New York: McGraw-Hill Publishing Company, 2003: 402 – 407.

2. Background Information

《风景园林师便携手册》(*Landscape Architect's Portable Handbook*)是一本有广泛影响的风景园林工具书,该书列出了风景园林规划设计分析步骤、标准、常用尺寸、材料和园林工程技术等,是风景园林师设计资料集,也是重要的风景园林设计教学参考书。本书多次修订再版,本文选自2003年第三版,第四部分 Administration 中第 21 章 Permitting Process。本书中文版译名为《景观设计师便携手册》。

Nicholas T. Dines,1968 年哈佛大学风景园林硕士研究生毕业,马萨诸塞州大学风景园林教授,长期从事风景园林规划设计和园林工程教学,曾出版多部设计手册,*Landscape Architect's Portable Handbook* 是其最新著作。他目前也是美国 *Landscape Architecture* 杂志编辑。

3. Sources of Additional Information

(1) National Environmental Policy Act (NEPA)

http://www.epa.gov/compliance/nepa/index.html

(2) The Council on Environmental Quality

http://www.whitehouse.gov/ceq/

(3) Clean Water Act

http://www.epa.gov/r5water/cwa.htm

(4) National Historic Preservation Act

http://www.achp.gov/nhpp.html

[1] insurance:保险
[2] delineate:描绘
[3] eligible:符合条件的,合格的
[4] legislation:立法,法律的制定(或通过)

Exercises

1. Questions（阅读思考）

认真阅读本单元文章，并讨论以下问题。
(1)《威尼斯宪章（1964）》在文物保护方面有哪些特点？
(2)"生物多样性保护"的目的、内容和一般方法？
(3) 对照《北京宣言》，谈谈风景园林师如何在改善人居环境方面发挥作用？
(4) 对照英美风景园林法律法规，谈谈如何加强我国风景园林师的法律法规意识？

2. Skill Training（技能训练）

(1) 查找并收集风景园林学科有关国际宪章、宣言等，分门别类并加注释，编成电子书。
(2) 查阅文献，介绍关于文化保护和生态保护的主要宪章和宣言。

Tips

检索和获取风景园林英文资料方法

现代网络技术的发展，使得人们可以更为便捷地获取信息资源。然而面对海量的网络资源，如何准确快捷地找到需要的资料，这里面需要一定的技巧。尤其是风景园林学科，从业人员、社会影响不如医疗、法律等学科广泛，在检索和获得风景园林英文资料有较大难度。但是如果掌握一些检索技巧，就会发现网络资源还是比较充足的，甚至常有意想不到的收获。在此谈一点笔者的心得和体会，供读者参考。

风景园林英文专业资料网络检索和获取一般可通过以下五种方式：搜索引擎、文献数据库、电子期刊、共享资源和求助。

1. 搜索引擎

搜索引擎无疑是检索资料的主要利器。目前最常用的英文搜索引擎是 google，它的功能非常强大，尤其是搜索英文风景园林学术资源。

在 google 检索方式中，使用专题检索更有效率，检索图片、图书、学术论文乃至专利，均可以进入相关专题搜索页面。其中 google 图书（http://books.google.com/）是一个很好的资源，能检索到非常丰富的风景园林英文专著，很多图书可以全文浏览甚至全文下载，如美国早期著名园林师唐宁（A. J. Downing）的《论造园学的理论和实践》（*A Treatise on the Theory and Practice of Landscape Gardening*）。

我们注意到，从网上找到的国外论文大部分是 pdf 格式，因此可以尝试用"key words"+"pdf"的模式搜索英文文献，效果很好！比如输入"landscape design pdf"，出现许多全文文献。同样原理，也可以加"doc"、"ppt"等检索。如果为了选题或希望了解本研究领域的进展，可以用例如"pdf+a survey of landscape design"作为关键词的模式进行搜索，寻找相关文献综述。如果想寻找某个特定文献，可以复制一段原文放到搜索引擎里试试。

除 google 搜索外，Ixquick（http://www.ixquick.com）、中图公司 cnpLINKer 在线数据库检索系统（http://cnplinker.cnpeak.com/）等搜索平台也各具特色。总之，搜索引擎是我们在互联网上检索资料的重要工具，仔细阅读服务说明，不断总结经验，巧妙选择检索关键词和检索方式，将可以从互联网上获取更多更好的外文专业资料。

2. 文献数据库

国外期刊会议论文、学位论文、研究报告、working paper 等可以通过文献数据库检索与获取。我们可以把数据库分为付费综合、免费综合和专题数据库三类。

（1）付费综合数据库。

在国内高等院校普遍购买的 SpringerLink、ProQuest、EBSCO、ScienceDirect 等数据库中，可以检索到不少风景园林英文文献，尤其是期刊学术论文、会议论文和学位论文等。这类数据库收录的文献比较偏向学术性，收录如 *Landscape and Urban Planning*（《风景和城市规划》），*Landscape Ecology*（《景观生态》）等期刊论文（两者均是 SCI 收录期刊）和各校硕博士学位论文，但在收录美国 *Landscape Architecture*、欧洲 *Topos* 等本学科常见期刊方面有所欠缺。当然，如果以学术研究为目的，使用付费综合数据库肯定是一个不错的选择。

(2) 免费综合数据库。

如果暂无条件获得付费数据库，可使用免费数据库。免费数据库收录内容繁多，特色明显，常常在某类文献方面颇具特色，使用方便。缺点是资料收录不够全面，站点容易变化。收藏一批好的免费综合数据库站点，对查找风景园林文献是有益的补充。以下列举一些目前较好的免费综合数据库。

1）英国 White horse press 全文期刊：http：//www.ingentaconnect.com/content/whp。
2）开放获取期刊指南：http：//www.doaj.org/。
3）免费全文期刊：http：//www.freefulltext.com/A.htm。
4）Look Smart Find articles：http：//findarticles.com/。
5）宾西法尼亚大学在线 e 书网：http：//onlinebooks.library.upenn.edu/。
6）电子书检索网（Digital Book Index）：http：//www.digitalbookindex.org/。
7）新闻（Information Today Search）：http：//search.infotoday.com/。

(3) 专题数据库。

综合数据库信息量大，内容广泛，但其缺点也是显而易见的：内容过于庞杂，在收录风景园林学科文献方面没有针对性。国外风景园林科研教育机构、协会、专业网站等往往建有专题数据库，收集专业文献。可惜的是，风景园林专题数据库十分少见，需要多方获取信息。

有些风景园林网站会披露或收藏本学科专题数据库，如http：//www.lih.gre.ac.uk/，http：//www.gardenvisit.com/，http：//www.thegardeningwebsite.co.uk/等。一些研究机构和高校链接有该机构收藏的文献，如美国哈佛大学 Dumbarton Oaks 研究所（The Dumbarton Oaks Research Library and Collection）：http：//www.doaks.org/，收藏很多风景园林研究论文。康奈尔大学图书馆有"城市规划 1794～1918 文献库"（http：//www.library.cornell.edu/Reps/DOCS），可全文免费浏览。

3. 风景园林电子期刊

目前大多数风景园林英文杂志都有收费电子版本，可以在网上订购，并在客户端或邮箱中接收。当然，对大多数学生来说，免费风景园林期刊最值得关注。

一些风景园林期刊，如 Topos（《世界风景园林设计都市规划评论》The International Review of Landscape Architecture and Urban Design）：http：//topos.de/index.php，每期有内容介绍，部分文章可供全文浏览。而澳大利亚风景园林师联合会电子期刊http：//www.aila.org.au/laonline/，国际风景园林师联合会电子期刊http：//www.iflajournal.org/，哈佛大学阿诺德植物园电子期刊http：//arboretum.harvard.edu/publications/publications.html，哥伦比亚大学《哥伦比亚风景》电子期刊http：//www.uga.edu/sed/publications/galandscp.htm 等是最好的资源，提供全文免费下载！此外，加拿大林业学会全文期刊 http：//pubs.nrc-cnrc.gc.ca/tfc/，美国农业生物工程师学会全文期刊 http：//asae.frymulti.com/toc.asp 等相关领域免费电子期刊也值得关注。

就目前情况而言，免费风景园林电子期刊一般由各国风景园林协会、相关教育科研机构网站发布，并且有越来越多的趋势，读者可在各网站 Publications 专题下查找收集。

4. 共享资源

有些外国年轻人喜欢把制作的电子书或专业资料存放于一些文件存储共享网站，如 Rapidshare、Oxyshare、badongo 等，检索这些网站也能获得一些资料。具体方法还是使用搜索引擎，使用"书名＋rapidshare/oxyshare/banongo"、"书名"rapidshare. de/files""、"书名 inurl：rapidshare"等模式检索。

另外，也可以使用电骡（eMule）和 BT（BitTorrent）检索并下载，上面的风景园林英文文献也不少，尤其是关于花园设计和施工的内容。

5. 求助

如果急需某个文献，而使用以上方法都无法找到，不妨试试去国内"文献检索"相关论坛向高手求助，利用群众的力量来找到你需要的。还可以尝试搜寻文献作者的主页，或者通过作者的邮件地址直接写信索取，一般西方国家专家是比较热情的，愿意提供相关帮助。

Unit 4

Urban Landscapes Planning & Design

- 城市规划设计
- 城市绿地设计

The city, the suburbs, and the countryside must be viewed as a single, evolving system within nature, as must every individual park and building within that larger whole.

By Anne Whiston Spirn

Text

The Granite Garden

By Anne Whiston Spirn

> 从太空看,地球本是一个花园的世界,是一颗生命的行星。但随着人类社会的介入,整个世界已经被钢筋混凝土所包裹,环境遭到了破坏,绿色在消失。
>
> 本文阐述了应该将自然与城市协调发展,充分体现和利用自然的社会价值,在保护自然的基础上,进行城市的建设。如果要有效地利用存在于城市中的自然环境,要跳过短期成本和效益,要洞察城市日常生活中无数似乎毫不相干的行为所产生的结果,要协调城市中日新月异的变化,就必须树立一个崭新的观点来对待城市及其形成。

Seen from space, the earth is a garden world, a planet of life, a sphere of blues and greens sheathed in a moist atmosphere. At night, lights of the cities twinkle far below, forming constellations[1] as distinct and varied as those of the heavens beyond. The dark spaces that their arcs embrace, however, are not the voids[2] of space, but are replete with forests and farms, prairies and deserts. As the new day breaks, the city lights fade, overpowered by the light of the sun; blue seas and green forests and grasslands emerge, surrounding and penetrating the vast urban constellations. Even from this great distance above the earth, the cities are a gray mosaic permeated by tendrils and specks of green, the large rivers and great parks within them.

Homing in on a single constellation from hundreds of miles up, one cannot yet discern the buildings. But the fingers and patches of green-stream valley, steep hillsides, parks, and fields-swell and ponds catch the sunlight and shimmer[3]. Swinging in, now only a few miles up, the view is filled by a single city. Tall buildings spring up toward the sky,

从太空看,地球是一个花园的世界,一颗生命行星,一颗被潮湿的大气层所包裹的蓝绿色的星球。

新的一天来临了,城市的灯光在阳光的照射下显得暗淡,蓝色的海洋和绿色的森林、草原显现出来,蔓延在繁星般的城市灯火周围。即便从太空中俯视,地球上的城市看上去也就像灰色的马赛克,其中镶嵌着斑斑点点的绿色,这些就是树林、河流和公园。

从数百英里的上空俯瞰城市,没人能辨明这些建筑物。

一幢幢摩天大楼高耸入云,一

---

[1] constellation:星座,星群
[2] void:空隙
[3] shimmer:微光

outcrops[1] of rock and steel, and smaller homes poke up out of the suburban forest. Greens differentiate themselves into many hues. Silver ribbons of roadway flash across the landscape, and stream meanders interrupt and soften the edges of the city's angular grid.

Flying low, one skims over a city teeming with life. The amount of green in the densest part of city is astonishing[2]; trees and gardens grow atop buildings and in tiny plots of soil. On the ground, a tree-of-heaven sapling is thriving[3] in the crack between pavement[4] and building, and a hardy weed thrusts itself up between curb and sidewalk. Its roots fan out beneath the soil in search of nutrients and water. Beneath the pavement, underground rivers roar through the sewers.

The city is a granite[5] garden, composed of many smaller gardens, set in a garden world. Parts of the granite garden are cultivated[6] intensively, but the greater part is unrecognized and neglected.

To the idle eye, trees and parks are the sole remnants of nature in the city. But nature in the city is far more than trees and gardens, and weeds in sidewalk cracks and vacant lots. It is the air we breathe, the earth we stand on, the water we drink and excrete, and the organisms with which we share our habitat. Nature in the city is the powerful force that can shake the earth and cause it to slide, heave, or crumple. It is a broad flash of exposed rock strata[7] on a hillside, the overgrown outcrops in an abandoned quarry, the millions of organisms[8] cemented in fossiliferous[9] limestone of downtown building. It is rain and the rushing sound of underground rivers buried in storm sewers. It is water from a faucet, delivered by pipes from some outlying river or reservoir, then used and washed away into the sewer, returned to the waters of river and sea. Nature in the city is an evening breeze, a corkscrew eddy swirling down the face of a building, the sun and the sky. Nature in the city is

[1] outcrop：（矿脉）露出地面的部分，露头
[2] astonishing：令人惊讶的，惊人的
[3] thrive：繁茂，蔓延
[4] pavement：人行道，铺过的道路
[5] granite：花岗岩，坚硬的
[6] cultivated：耕种的，耕植的
[7] strata：地层
[8] organisms：有机体，生物体
[9] fossiliferous：含有化石的

dogs and cats, rats in the basement, pigeons on the sidewalks, raccoons in culverts, and falcons crouched on skyscrapers. It is the consequence[1] of a complex interaction between the multiple purpose and activities of human being and other living creatures and of the natural processes that govern the transfer of energy, the movement of air, the erosion of the earth, and the hydrologic[2] cycle. The city is part of nature.

自然是人和其他生物的各种目的和活动相互作用的结果，是支配能量转化、空气流动、大地侵蚀以及水循环的自然过程的结果。

Nature is a continuum[3], with wilderness at one pole and the city at the other. The same natural processes operate in the wilderness and in the city. Air, however contaminated, is always a mixture of gasses and suspended particles. Paving and building stone are composed of rock, and they affect heat gain and water runoff just as exposed rock surfaces do anywhere. Plants, whether exotic or native, invariably seek a combination of light, water, and air to survive. The city is neither wholly natural nor wholly contrived. It is not "unnatural" but, rather, a transformation of "wild" nature by humankind to serve its own needs, just as agricultural fields are managed for food production and forests for timber[4]. Scarcely a spot on the earth, however remote[5], is free from the impact of human activity. The human needs and the environmental issues[6] that arise from them are thousands of years old, as old as the oldest city, repeated in every generation, in cities on every continent.

自然是个连续的统一体，原野和城市分别位于这个统一体的两端。

植物，无论外来的还是本土的，总是寻求光、水和空气来维持生存。

这不是"非自然"，应该说是人类为满足其自身需要对"未开发"的自然进行的一个改造，就像人类开垦农田以生产粮食，管理森林以生产木材等。

人们的需要和环境之间的问题在几千年前就出现了，从最早的城市产生开始就已经出现，而且在每一片大陆的每一座城市里一直延续下来。

The realization that nature is ubiquitous[7], a whole that embraces the city, has powerful implications for how the city is built and maintained and for the health, safety, and welfare of every resident. Unfortunately, tradition has set the city against nature, and nature against the city. The belief that the city is an entity[8] apart from nature and even antithetical to it has dominated the way in which the city is perceived and continues to affect how it is built. This attitude has aggravated and even created many of the city's environmental problems: poisoned air and water; depleted or irretrievable resources; more

自然是一个无处不在，并且环拥着城市的整体。这一理念对城市的建造和维护以及每个居民的健康、安全和福利都有相当大的含义。

[1] consequence：结果，影响
[2] hydrologic：水文的
[3] continuum：连续
[4] timber：木材
[5] remote：很久以前的
[6] issue：问题
[7] ubiquitous：无所不在的
[8] entity：实体，本质

frequent and more destructive floods; increased energy demands and higher construction and maintenance costs than existed prior to urbanization[1]; and, in many cities, a pervasive ugliness. Modern urban problems are no different, in essence, from those that plagued[2] ancient cities, except in degree, in the toxicity and persistence of new contaminants, and in the extent of the earth that is now urbanized. As cities grow, these issues have become more pressing. Yet they continue to be treated as isolated phenomena, rather than as related phenomena arising from common human activities, exacerbated by a disregard for the processes of nature.

除了程度，新污染物的毒性和持久性以及地球的城市化范围等方面，那些困扰着古代城市的问题与现代都市中的问题从本质上来说是没有不同的。

Nature has been seen as a superficial embellishment, as a luxury, rather than as an essential force that permeates[3] the city. Even those who have sought to introduce nature to the city in the form of parks and gardens have frequently viewed the city as something foreign to nature, have seen themselves as bringing a piece of nature to the city.

有些人用公园和庭园的形式力图将自然引入城市，但即便是这些人也常常认为城市和自然是格格不入的，而他们自己则是将自然的少许部分带给城市的人。

To seize the opportunities[4] inherent in the city's natural environment, to see beyond short-term costs and benefits, to perceive the consequences of the myriad[5], seemingly unrelated actions that make up daily city life, and to coordinate thousands of incremental improvement, a fresh attitude to the city and the molding of it is necessary. The city must be recognized as part of nature and designed accordingly. The city, the suburbs, and the countryside must be viewed as a single, evolving system within nature, as must every individual park and building within that larger whole. The social value of nature must be recognized and its power harnessed, rather than resisted. Nature in the city must be cultivated, like a garden, rather than ignored or subdued.

要有效地利用存在于城市中的自然环境，要跳过短期成本和效益，要洞察城市日常生活中无数似乎毫不相干的行为所产生的结果，要协调城市中日新月异的变化，有必要树立一个崭新的观点来对待城市及其形成。我们必须把城市看作自然的一部分并进行因地制宜的设计。

自然的社会价值必须得到认可，对自然的力量必须加以控制而不是抵制。对城市中的自然必须像庭园一样加以养护，而不是忽略或抑制。

Notes

1. Extracts from
Text
Whiston Spirn, Anne. *The Granite Garden: Urban Nature and Human Design* [M]. New York: Basic Books, Inc. Publishers, 1984: 3-5.

---

[1] urbanization：城市化
[2] plague：困扰
[3] permeate：渗透，渗入
[4] opportunity：机会；有利环境
[5] myriad：无数

2. Background Information

(1) Anne Whiston Spirn 安妮·惠斯顿·史必恩。

女，现为美国 MIT（麻省理工学院）景观设计（风景园林）教授，在国际上因致力于景观设计和环境规划的交叉研究享有很高的声誉。她将在生态设计方面的理论与方法的研究成果应用到城市领域作出了重要贡献。自 1987 年起，她主持西费城景观规划项目，这个项目在 1999 年被白宫最高规格的四十位"公众领域的专家学者"被评价为"最现实的模型"。她曾分别在哈佛大学、宾夕法尼亚大学任教。在著作、设计实践、教育三方面均取得令人羡慕的成就。

(2) *The Granite Garden: Urban Nature and Human Design*《花岗岩花园：城市自然与人文设计》。

1984 年出版，是安妮·惠斯顿·史必恩的第一本书，赢得 1984 年美国景观设计（风景园林）师协会（ASLA）主席大奖，被翻译成两种文字，到现在仍然是一本杰出的大学教材。

本文选自该书《序言：花岗岩花园》，有删节。

3. Sources of Additional Information

http://unjobs.org/authors/anne-whiston-spirn

Further Reading A

The Growth of Suburbs

By Yi-Fu Tuan

The suburban image is dominated by the comfortable residential setting of the upper middle and middle classes. Suburbs, however, come in a variety of types which reflect the socioeconomic status of the residents, the presence or absence of industry, and their age—since the suburb is commonly a step in the transformation form rural to urban life styles. To understand suburban values and attitudes, a brief excursus into the growth of suburbs is necessary. The historical perspective helps us to appreciate the range of meaning possible in the term "suburb".

Archaeological[1] evidence suggests that by the second millennium[2] B. C., population had already spilled beyond the walls of Ur. This is one of the earliest known examples of extramural development. If we think of the suburb as simply the growing edge of a city, then it is a phenomenon that has happened repeatedly: whenever, in fact, cities were rapidly expanding into the countryside. But, for lack of documentary evidence we can rarely say whether the distinction between "city" and "suburb" was applicable to urban areas that had long since returned to dust. Walls, what remain of them, are a means for drawing the distinction: the wall was the clearest expression of what the city builders took to be the limits of their domain. In ancient China most cities had walls. The concentric[3] ramparts[4] defined the successive stages of incorporation of suburban communities into the urban sphere. Tradesmen and artisans[5] crowded outside the city gates, and in time their numbers reached a size that justified the protection of a wall. Suburban growth could be rapid. For example, within decades of the completion of Peking's rampart, in the 1420s, a large suburb of more than one hundred thousand families sprang up beyond the southern wall. Traders from all parts of the empire and from foreign countries established shops and homes there. By the middle of the 16th century (ca. A. D. 1552), a new wall fenced in the suburban sprawl to from the Outer or Southern City. Concentric enclosures were also built around old European cities like Paris which, since the late Medieval[6] period, consistently outgrew the limits imposed by its successive walls.

Suburbs housed the poorer elements of the population including traders, craftsmen, innkeepers, and foreigners, but wealthy people found quarters there too. Some Italian cities, for example, had acquired extramural[7] outskirts of cottages and villas, with ample gardens, as early as the 13th

---

[1] Archaeological: 考古学
[2] millennium: 千年
[3] concentric: 同心
[4] ramparts: 垒
[5] artisans: 工匠
[6] Medieval: 中世纪
[7] extramural: 围墙外

century. Land for a belt of three miles around Florence was occupied by rich estates with costly mansions; Venetian[1] families had their villas on the Brenta. The suburb, Mumford observes, "might almost be described as the collective urban form of the country house—the house in the park." The suburban way of life is "a derivative of the relaxed, playful, goods-consuming aristocratic life that developed out of the rough, bellicose, strenuous existence of the feudal stronghold."

At the beginning of the 18th century, the regular commuter appeared on the English scene. A place like Epsom was not only a rural market town and spa but a suburb of London some 15 miles away. Businessmen set up their families in Epson; they themselves commuted to the city daily. This adds a new meaning to the suburb. Whereas in the past the aristocrats maintained suburban villas which they occupied for extended seasons, by the 18th century middle-class merchants could live permanently in the suburb and do their business in town. Improvements in transportation had made commuting possible. Before 1700 the suburb embraced two extreme life styles: that of the poor who lived and worked there and that of the leisured rich who repaired to their country estates in summer. Time given over to traveling was minimal. As roads and carriers improved, villas and summer homes could be established at scenic spots with little regard to distance from the city, while suburbs, housing daily commuters, sprang up at the urban fringe within reach to the central places of business. Residential suburbs soon acquired not only respectability but a reputation for self-importance and a penchant for rural fantasies that William Cowper, in 1782, could already mock in verse[2].

Suburban villas, highway-side retreats,
That dread the encroachment of our growing streets,
Tight boxes, neatly sash'd, and in a blaze,
With all a July sun's collected rays,
Delight the citizen, who, gasping there,
Breathes clouds of dust, and calls it country air.

By the middle of the 18th century, London was growing so fast that one of Tobias Smollett's characters was made to say (1771), "London is gone out of town." Prosperous suburbs also sprang up to serve England's wealthier commercial and manufacturing cities. Birkenhead, for example, came into existence after the Napoleonic Wars as a place of residence for the rich merchants of Liverpool; families were lured there "by pleasant country scenery, fine river views and the wonderful ease with which they were able to pass from the bustle of the town to the quiet of the country." Southport became another residential satellite of Liverpool, and bloomed especially after the opening of the railway in 1848. Although urban growth was rapid in the 18th and in the early part of the 19th century, it could not match the "explosive" expansion of the great metropolises into the countryside in the late Victorian era and since. Two major innovations in transport made this expansion possible: first the railroad, then the motor car.

Railroad and the development of mass transit broadened the economic base of people who could afford the move to the country. The middle classes followed their social betters in the exodus. Railroads radiating from the city influenced the direction of suburban growth. New housing at first

---

[1] Venetian: 威尼斯式
[2] verse: 诗歌

clustered neatly around the train stations that were placed some 3~5 miles apart. This early type of suburb was small in size, seldom housing as many as 10,000 people because, other than the wealthy who could retain horse and carriage, accessibility meant being within easy walking distance to the train station. Upper-middle and middle-class residential suburbs were strung out like beads on the railway line. Each was surrounded by country greenery. The houses themselves were spacious, set in their own grounds and, from 1850 onward, both the houses and the street plans showed an increasing tendency to abandon rectilinear urban forms in favor of romantic eccentric styles for the residences, and curved naturalistic lines for the streets. Although these affluent suburbs have been criticized as havens of social irresponsibility, environmentally they had great appeal. The railroad and the tramway also enabled the working men and their families to move out of the central cities. Unfortunately the house built for them at the urban fringe showed little more imagination than the crowded accommodations they had abandoned. The air over the suburb was cleaner but otherwise the working class lived in mass-produced, dreary housing estates, often large enough to make nature seem as remote as when the workers were packed deep in the urban core. Even sanitation[1] was not much better. The idea behind suburban living was its healthier environment: in reality the advantages of country air were offset by faulty building, bad drainage[2], and inadequate water supply beyond the town boundaries.

The trend in metropolitan expansion initiated in the railroad era continued and accelerated with the mass production of the automobile. The motorized car, at first a toy of the rich, within three decades became a major means of transportation available to the people. This was primarily an American success story. The number of automobiles in the United States increased from 9 million in 1920 to 26.5 million in 1930 and to about 40 million in 1950. The figures imply a vast gain in people's mobility as well as an overall rise in economic welfare despite major setbacks such as the depression in the thirties. In late Victorian England, working-class people were able to make the suburban move not only because of the railways but, as important, because they had steady jobs, shorter hours, and better wages—in other words, the economic means. Similarly in the twenties and in the post-World War II period, occurred during the phases of economic boom.

The most singular feature in modern metropolitan expansion is its speed and scale. Suburbs appeared "overnight." They have the character of a "rush." Consider Toronto. In 1941 the combined population of the three outlying townships of Etobicoke, Scarborough, and North York was 66,244. In 1956 it was 413,475. 5 years later it had become 643,280. Areas which had only three or four rural families might be covered, a year later, by as many as 500 to 1,000 suburban homes. In 1961 greater Toronto had about two million people, more than one-half of which lived beyond the city borders. For every person domiciled in an old built-up area in 1961, one other was living in a residential area not more than 15 years old. Statistics of dramatic suburban growth are commonplace, and indeed there is no need to appeal to figures for a phenomenon so aggressively visible as the ubiquitous[3] suburban sprawl. The word "sprawl" is descriptive. In contrast to the neatly planned communities of the upper-middle class, the suburbs of the lower-middle and the newly-affluent

---

[1] sanitation：卫生
[2] drainage：排水设施
[3] ubiquitous：普遍存在

working classes are a sea of undifferentiated dwellings, street blocks, and subdivisions[1], beginning and ending nowhere that could be seen clearly. The suburban estates of the rich and of the high-income professional people are utopian enclaves in the vast brickish skirt that surrounds the central city.

Notes

1. **Extracts from**

   Text

   Tuan, Yi-Fu. *Topophilia: a study of Environment Perception, Attitudes, and Values* [M]. New York: Columbia University Press, 1990: 230-234.

2. **Background Information**

   Yi-Fu Tuan（段义孚），美国科学院院士、英国科学院院士、美国 Wisconsin-Madison 大学教授，是世界著名的地理大师。他的早期著作包括《新墨西哥州的气候》、《亚利桑纳州的碛原》等。自 20 世纪 70 年代，他开始转向对社会文化地理现象的分析，并以发表在《美国地理联合会年报》上的《人本主义地理学》一文为标志，进入了新的研究阶段。人本主义地理学（humanistic geography），是从人的感觉、心理、社会文化、伦理和道德的角度来认识人与地理环境的关系。他的人本主义地理学思想在西方地理学界，以及与西方关系密切的其他地方的地理学界，产生了重大影响。他将人的种种主观情性与客观地理环境的丰富关系进行了极具智慧的阐发，吸引了众多学者的目光。从此，Yi-Fu Tuan（他的英文名字）一名，蜚声于世界人文地理论坛。

   段义孚 1930 年生于天津，后随家到澳大利亚、菲律宾。在牛津大学读大学。1951 年入美国伯克利大学为研究生，1957 年获博士学位。之后，分别在印地安那大学、芝加哥大学、新墨西哥大学、多伦多大学、明尼苏达大学、威斯康星大学教书。自任教于明尼苏达大学始，段义孚在地理学上贡献陡增，声名鹊起。1973 年，获得美国地理学家协会授予的地理学贡献奖，1987 年美国地理学会授予他 Cullum 地理学勋章。

   段义孚本来学的是地貌，但他后来却完全关注人的问题。他注重人性、人情，称自己研究的是"系统的人本主义地理学"（systematic humanistic geography），以人为本，就是他的"地学"的特征。大概因为有在不同文化中生活的经验，段义孚深知文化的影响力。他指出，爱与怕是人类情感的基本内容，而被文化转化为种种形式。就"爱好"与"惧怕"这两个重要主题在人文地理中的表现，他各写了一本书，一本是《恋地情结》（Topophilia: A Study of Environmental Perception, Attitudes and Values），另一本是《恐惧景观》（Landscapes of Fear）。其中《恋地情结》是他的成名作，此书至今仍是美国各大学景观专业的必读教材。段义孚教授著作颇丰，仅个人著作就有约 15 部之多。

3. **Sources of Additional Information**

   http://www.geography.wisc.edu/~yifutuan/disclaimers.htm

---

[1] subdivisions：细分

Further Reading B

Landscape Design in the Urban Environment

By Garrett Eckbo, Daniel U. Kiley and James C. Rose

Cities Redesigned for Living...

"the remodeling of the earth and its cities," Lewis Mumford[1] has said, "is still only at a germinal[2] stage: only in isolated works of technics[3], like a power dam or great highway, does one begin to feel the thrust and sweep of the new creative imagination: but plainly, the day of passive acquiescence[4] to the given environment, the day of sleepy oblivion[5] to this source of life and culture, is drawing to an end. Here lies a new field..."

Certain it is that the city today stands between man and the source of recreation, consuming his free time in traveling to and from those areas which provide a means of restoring vitality dissipated in work.

The sharp division of working and recreative worlds in today's urban areas has produced a fallacious[6] weighting of their relative values. Work has become the dominant fact of human life—leisure and recreation a more or less luxurious[7] afterthought which is fitted in around it. The dreary[8] preponderance[9] of building, the almost total absence of gardening, in our metropolis[10], is a direct reflection of this. Yet leisure and recreation, in their broadest sense, are fundamentally necessary factors of human life, especially in an industrial age. Recreation, work, and home life are fundamentally closely interdependent units, rather than entities to be segregated by wastefully attenuated transportation facilities, as they are today.

Since most production in the city takes place under roof, indoors, it is obvious that urban recreation must emphasize the out-of-doors, plant life, air, and light. In our poorly mechanized, over-centralized, and congested cities the crying need is for organized space: flexible, adaptable outdoor space in which to stretch, breath, expand, and grow.

---

[1] Lewis Mumford: 刘易斯·芒福德（1895～1995 年），城市规划理论的思想家与先驱
[2] germinal: 初期的
[3] technics: 工艺
[4] acquiescence: 默许
[5] oblivion: 遗忘，湮没
[6] fallacious: 不合理的
[7] luxurious: 奢侈的，豪华的
[8] dreary: 沉闷的
[9] preponderance: 优势，占优势
[10] metropolis: 大城市

Unit 4　Urban Landscapes Planning & Design

Trend Toward Recreational Systems

The urban dweller requires a complete, evenly distributed, and flexible[1] *system* providing all types of recreation for persons of every age, interest, and sex. The skeletal outlines of such systems are emerging in many American cities—New York, Cleveland, Washington, New Orleans, Chicago—although usually in a fragmentary and uncoordinated[2] form. Of these, the combined park systems of New York City, Westchester County and Long Island undoubtedly constitute the most advanced examples.

But, aside from their sheer inadequacy[3]—no American city boasts even minimum standards of one-acre open to each 100 population—these systems have many qualitative[4] shortcomings. Public park systems are usually quite isolated: on the one hand, from the privately owned amusement and entertainment centers—theaters, dance halls, stadia[5], and arenas; and, on the other from the school, library, and museum systems. This naturally makes a one-sided recreational environment. Even the largest elements—Lincoln Park in Chicago, Central Park in New York—are too remote from the densest population areas to service them adequately. And nearly all of these systems, or parts of systems, still labor under antiquated[6] concepts of design, seldom coming up to the contemporary plane of formal expression. Nevertheless, the trend is more and more toward considering a well-balanced system essential, such a system including the following types.

(1) *Play lot*—a small area within each block or group of dwellings for preschool children. One unit for every 30 to 60 families; 1,500 to 2,500 sq. ft. minimum. A few pieces of simple, safe but attractive apparatus[7]—chair swings, low regular swings, low slide, sand box, simple play materials, jungle gym, playhouse. Open space for running. Enclosure by low fence or hedge, some shade. Pergola[8], benches for mothers, parking for baby carriages.

(2) *Chlidren's playground*—for children 6 to 15 years. At or near center of neighborhood, with safe and easy access. 1-acre playground for each 1,000 total population; 3 to 5 acres minimum area in one playground. Chief features: apparatus area; open space for informal play; fields and courts for games of older boys and girls; area for quiet games, crafts, dramatics[9], storytelling; wading pool.

(3) *Distrtct playfield*—for young people and adults. 1/2-to 1-mile radius; 10 acres minimum size, 20 desirable[10]. One playfield for every 20,000 population, one acre for each 800 people.

(4) *Urban park*—large area which may include any or all of above activities plus "beauty of landscape." Organized for intensive use by crowds—zoos, museums, amusement, and entertainment

---

[1]　flexible：灵活的
[2]　uncoordinated：不协调的
[3]　inadequacy：不充分
[4]　qualitative：质量上的
[5]　stadia：(复数) 露天大型运动场
[6]　antiquated：陈旧的
[7]　apparatus：器械，设备，仪器
[8]　Pergola：(藤本植物的) 棚架，藤架，绿廊，凉棚
[9]　dramatics：业余演出
[10]　20 desirable：20 acres desirable size 的简写

zones.

(5) *Country park and green belts*—for "a day in the country"—larger area, less intensive use, merely nature trimmed up a bit. Foot and bridle paths, drives, picnic grills[1], comfort stations.

(6) *Special areas*—golf course, bathing beach, municipal camp, swimming pool, athletic field, stadium.

(7) *Parkways and freeways*—increasingly used (1) to connect the units listed above into an integrated system and (2) to provide quick, easy, and pleasant access to rural and primeval areas.

But Quantity Is Not Enough...

But the types listed above constitute only the barest outlines of a recreational *system*; provision of all of them does not in any way guarantee[2] a *successful recreational environment*. In other words, the problem is qualitative[3] as well as quantitative—not only how *much* recreational facilities, but *what kind*. Here the element of design is vital, and success is dependent upon accurate analyses of the needs of the people to be environed. These needs are both individual and collective.

Every individual has a certain optimum[4] space relation—that is, he requires a certain volume of space around him for the greatest contentment and development of body and soul. This space has to be organized three-dimensionally[5] to become comprehensible and important to man. This need falls into the intangible[6] group of invisible elements in human life which have been largely disregarded in the past. Privacy out-of-doors means relaxation, emotional release from contact, reunion with nature and the soil.

Collectively, urban populations show marked characteristics. Not only do their recreational needs vary widely with age, sex, and previous habits and customs (national groups are still an important item in planning); but they are also constantly shifting—influenced by immigration, work, and living conditions. Recent studies by Professor Frederick J. Adams of M.I.T[7]. indicate constantly changing types of activity within definite age groups, and a gradual broadening of the ages during which persons participate[8] most actively in sports. Organized recreation is spreading steadily downward (to include the very young in kindergarten and nursery) and upward (to provide the elderly with passive recreation and quiet sports).

In addition, the urban population uses a constantly increasing variety of recreation forms—active and passive sports, amusements, games, and hobbies. Old forms are being revived (folk dancing, marionettes[9]); new forms are being introduced (radio, television, motoring, etc.); foreign forms

---

[1] picnic grills: 野餐烧烤屋
[2] guarantee: 保证
[3] qualitative: 性质上的，定性的
[4] optimum: 最适宜的
[5] three-dimensionally: 三维的
[6] intangible: 无形的
[7] M.I.T: 麻省理工学院
[8] participate: 参加
[9] marionettes: 牵线木偶

imported (skiing, fencing, archery[1]).

Consideration of the above factors imply certain design qualities for the recreational environment which are generally absent from all but the very best of current work. Design in the recreational environment of tomorrow must ①integrate landscape and building; ②be flexible; ③be multiutile; ④exploit mechanization; ⑤be social, not individual, in its approach.

(1) *Integration*. The most urgent need is for the establishment of a biologic relationship between outdoor and indoor volumes which will automatically control density. This implies the integration of indoors and outdoors, of living space, working space, play space, of whole social units whose size is determined by the accessibility of its parts. Thus landscape cannot exist as an isolated phenomenon, but must become an integral part of a complex environmental control. It is quite possible, with contemporary knowledge and technics, to produce environments of sufficient plasticity[2] as to make them constantly renewable, reflecting the organic social development. It is possible to integrate landscape again with building— *on a newer and higher plane*—and thus achieve that sense of being environed[3] in great and pleasantly organized space which characterized the great landscapes of the past.

(2) *Multiple-use*. Most types of recreation are seasonal and, within the season, can be participated in only during certain hours of the day or evening. In addition, different age and occupational[4] groups have free time at different hours, and a great variety of recreational interests exists within the same groups. The trend toward multiple-use planning reflects needs which permeate all forms of contemporary, and saving of time in unnecessary travel.

(3) *Greater flexibility* in building design—to provide for wider varieties of use and greater adaptability to changing conditions—can be extended into the landscape. The construction creates a skeleton of volumes which are perforated[5] enough to permit air and sunlight for plant growth. Plants now replace the interior partitions[6], and divide space for outdoor use. When building and landscape achieve this flexibility, we discover that the only difference between indoor and outdoor design is in the materials and the technical problems involved. Indoors and outdoors become one— interchangeable and indistinguishable except in the degree of protection from the elements.

Flexibility in design expresses in a graphic way the internal growth and development of society. For this reason, the great tree-lined avenues and memorial parks terminating the axes are not satisfactory, though they may have twice the open area per person above that which might be called an optimum. Once such a scheme is built, it is a dead weight on the community because it is static and inflexible. It is neither biologic nor organic, and neither serves nor expresses the lives of the people in its environs.

(4) If scientific and technical advance has created the urban environment of today, it—and it alone—has also made possible the urban environment of tomorrow. This implies a

---

[1] archery: 射箭
[2] plasticity: 可塑性
[3] environed: 环绕
[4] occupational: 职业的
[5] perforated: 有孔的或多孔的
[6] partition: 隔离物

frank[1] recognition, on the part of landscape designers particularly, of the decisive importance of "the machine"; it must be met and mastered, not fled from. Indeed, the only way in which landscape design can be made flexible, multi-utile, and integral with building is by the widest use of modern materials, equipments, and methods.

As a matter of fact, this is already pretty generally recognized, though, again, in a fragmentary fashion. The great parkway systems of America are the best example of new landscape forms evolved to meet a purely contemporary demand. The sheer[2] pressure of a mobile population forced their creation; and archaic[3] design standards fell by the wayside almost unnoticed. The landscapings of the New York and San Francisco fairs are other examples, though perhaps more advanced in the construction methods employed than in the finished form. The use of modern lighting and sound systems, mobile theatrical units (WPA caravan theaters, Randall Island rubber-tired stages, St. Louis outdoor opera theater) is already widespread. Throughout America, advances in agriculture, silviculture[4], horticulture, and engineering are constantly being employed by the landscape designer.

But these are, as yet, few examples which exploit the full potentialities or achieve the finished form which truly expresses them. Nor does the use of the fluorescent light[5], microphone, or automobile alone guarantee a successful design. The real issue will be the use to which such developments are put. The parkway, for example, can either serve as a means of integrating living, working, and recreation into an organic whole; or it can be used in an effort to sustain their continued segregation. The theoretical 150-mile radius at the disposal of all urban dwellers for recreation is dream of the drafting board, which, for the majority of people, is blocked at every turn by the inconvenience and cost of transportation. Only if the parkway reduces the time, money, and effort involved in getting from home to work to play, will it justify its original outlay.

In building the recreational environment of tomorrow even our most advanced forms must be extended and perfected. Man reorganizes materials consciously; their form effect is produced consciously; any effort to avoid the problem of form will produce an equally consciously developed form. Nothing in the world "just happens." A natural scene is the result of a very complicated and delicately balanced reaction of very numerous natural ecological forces. Man, himself a natural force, has power to control these environmental factors to a degree, and his reorganizations of them are directed by a conscious purpose toward a conscious objective. To endeavor to make the result of such a process "unconscious"[6] or "natural" is to deny man's natural place in the biological scheme.

(5) While the individual garden remains the ancestor of most landscape design, and while it will continue to be an important source of individual recreation, the fact remains that most urbanites do not nor cannot have access to one. And even when (or if) each dwelling unit has its private garden, the most important aspects of an urban recreational environment will lie outside its boundaries. The

---

[1] frank：率直的
[2] sheer：全然的，纯粹的
[3] archaic：古老的，陈旧的
[4] silviculture：造林术，森林学
[5] fluorescent light：明亮的荧光色，荧光灯
[6] unconscious：不省人事的，无意识的

recreation of the city, like its work and its life, remains essentially a social problem.

Landscape—Like Building—Moves Forward

Landscape design is going through the same reconstruction in ideology and method that has changed every other form of planning since the industrial revolution. The grand manner of axes, vistas, and facades has been found out for what it is—a decorative covering for, but no solution to, the real problem. Contemporary landscape design is finding its standards in relation to the new needs of urban society. The approach has shifted, as in building, from the grand manner of axes and facades to specific needs and specific forms to express those needs.

Plants have inherent quality, as do brick, wood, concrete, and other building materials, but their quality is infinitely more complex. To use plants intelligently, one must know, for every plant, its form, height at maturity[1] rate of growth, hardiness, soil requirements, deciduousness[2], color texture, and time of bloom. To express this complex of inherent quality, it is necessary to separate the individual from the mass, and arrange different types in organic relation to use, circulation, topography[3], and existing elements in the landscape. The technics are more complicated than in the Beaux Arts[4] patterns, but we thereby achieve volumes of organized space in which people live and play, rather than stand and look.

Notes

1. Extracts from

Text

Eckbo, Garrett. et al. Landscape Design in the Urban Environment [C]. Marc Treib. *A Critical Review. London*: The MIT Press: 1993: 78-82.

2. Background Information

盖瑞特·埃克博（Garreftt Eckbo，1910～2000年）、丹·凯利（Daniel U. Kiley，1921～2004年）和詹姆士·罗斯（James C. Rose，1910～1991年）：美国现代主义景观设计（风景园林）运动的领军人物，1938～1941年，他们三人就读于哈佛大学景观设计（风景园林）系，在此期间发表了一系列文章，动摇并最终导致了的"巴黎美术学院派"教条的解体和现代设计思想的建立，并推动美国的景观设计行业向适合时代精神的方向发展。这就是今天被人们津津乐道的"哈佛革命"。

《城市环境中的景观设计》（*Landscape Design in the Urban Environment*）：1938年，《建筑评论》（*Architectural Record*）杂志的编辑找到三人，请他们写一些文章谈谈现代环境中的景观设计问题。本文即为这一系列文章中比较重要的一篇。

3. Sources of Additional Information

盖瑞特·埃克博

http://en.wikipedia.org/wiki/Garrett_Eckbo

"The elements of Western landscaping: a visit with Garrett Eckbo - interview"

http://findarticles.com/p/articles/mi_m1216/is_n5_v186/ai_10663252

丹·凯利

---

[1] maturity：成熟
[2] deciduousness：落叶的
[3] topography：地形学
[4] Beaux Arts：巴黎美术学院派艺术，注重测绘、临摹和背诵古典艺术的经典著作

http：//en. wikipedia. org/wiki/Dan _ Kiley

An Appreciation of Landscape Architect Dan Kiley

http：//www. asla. org/land/kiley32204. html

The World by Kiley

http：//query. nytimes. com/gst/fullpage. html? res ＝ 9F05E3DF1E39F932A35750C0A960958260&sec＝&spon＝&pagewanted＝print

詹姆士·罗斯

http：//www. jamesrosecenter. org/jamesrose/bibliography/index. html

James Rose and the Modern American Garden

http：//www-unix. oit. umass. edu/～cardasis/content/publications/maverick. pdf

Further Reading C

The 100 Mile City

By Deyan Sudjic

Imagine the force field around a high-tension power line, crackling[1] with energy and ready to flash over and discharge 20000 volts at any point along its length, and you have some idea of the nature of the modern city as it enters the last decade of the century.

The city's force field is not a linear[2] one, however. Rather, it stretches for a hundred miles in each direction, over towns and villages and across vast tracts of what appears to be open country, far from any existing settlement that could conventionally[3] be called a city. Without any warning, a flash of energy short-circuits the field, and precipitates[4] a shopping centre so big that it needs three or five million people within reach to make it pay. Just as the dust has settled, there is another discharge of energy, and an office park erupts[5] out of nothing, its thirty-and forty-storey towers rising sheer[6] out of what had previously been farmland. The two have no visible connection, yet they are part of the same city, linked only by the energy field, just like the housing compounds that crop up here and there, and the airport, and the cloverleaf on the freeway, and the corporate headquarters with its own lake in the middle of park.

Somewhere, in a remote corner, there is no doubt a little enclave of pedestrian[7] streets, a fringe[8] of terraced[9] houses that circles the crop of office towers that marks downtown. There will be a sandblasted old market hall recycled for recreational shopping. And somewhere else, there will be the social derelicts[10], the casualties, trapped in welfare housing or worse.

The energy that powers the force field is of course mobility. The wider ownership of the car that has come since the 1960s has finally transformed the nature of the city. The old certainties of urban geography have vanished, and in their place is this edgy and apparently amorphous new kind of settlement. The chances are that the force field couldn't have come into being without a downtown, or historic crust, because massive amounts of resources are needed to achieve the critical mass required by this kind of city as a trigger. But in its present incarnation, the old centre is just another piece on

---

[1] crackling: 爆炸
[2] linear: 线形
[3] conventionally: 常规地
[4] precipitates: 沉淀物
[5] erupts: 喷发
[6] sheer: 偏离
[7] pedestrian: 步行者
[8] fringe: 边缘
[9] terraced: 露台
[10] derelicts: 流浪汉

the board, a counter that has perhaps the same weight as the airport, or the medical centre, or the museum complex. They all swim in a soup of shopping malls, hypermarkets and warehouses, driven-in restaurants and anonymous industrial sheds, beltways and motorway boxes.

The words to describe it vary in different cultures, but the pattern is no longer confined to North America. You can see the same phenomenon around Atlanta, or in the stretch of France that connects Barcelona with Milan, or along the M_4 motorway running westward out of London. As Europe has become more mobile, so it too has taken on the same characteristics, despite its older city centers and their traditional patterns based on the pedestrian.

At first sight, all sense of coherence has gone. In this new city, it's no longer necessary to go to the centre to work or to shop. So the geography becomes disorientating, especially to newcomers who need far longer to pick up the signals that we all need to make sense of an unfamiliar city. Some traditional functions have been taken out of the equation altogether. In many cities there is no main street that anybody would care to visit given the choice. So there is Disney instead, already installed on both sides of America and in Japan, and due to arrive shortly in Paris, which offers the trappings of the convivial[1] old city, instantly available for pilgrims and free of the sense of threat that is a constant subtext to the big city as we know it.

In the force field city, nothing is unself-conscious, any urban gesture is calculated.

For the affluent[2], the home is the centre of life—though given the astonishing increase in household mobility, it's not likely to be the same home for very long. From it, the city radiates outwards as a star-shaped pattern of overlapping routes to and from the workplace, the shopping centre, and the school. They are all self-contained abstractions that function as free-floating elements. Each destination caters to a certain range of the needs of urban life, but they have no physical or spatial[3] connection with each other in the way that we have been conditioned to except of the city.

Architecture and urban space has been overwhelmed by the sheer scale of new buildings. A new race of giant boxes has descended, warehouses, discount superstores, shopping malls and leisure centres that it is hardly possible to differentiate one from another. Despite their anonymity, they have swallowed up most civic functions.

Mobility means overlapping force fields. Cities compete with each other in a grimly determined struggle to maintain the energy that keeps them working. But the lives of their citizens take in other cities to an extent that is unparallelled in history. They move from one to another constantly, to live, to visit, to do business.

In most economic areas, one or two individual cities have come to monopolise the world market. So New York, Tokyo and London are the world's financial capitals, Los Angeles is its entertainment centre. Perhaps the hundred-mile city has already become the thousand-mile city.

The other side to the equation is the widening gap between those cities that are successful, and

---

[1] convivial: 欢乐
[2] affluent: 富有
[3] spatial: 空间

those that are not. Some cities are clearly failing. Their treasuries teeter on the edge of bankruptcy, they fail to attract new employers, they have little to recommend them culturally. But beyond that there is the widening difference between metropolitan cities which draw in the ambitious and the gifted, and provincial cities which lose them.

The hundred-mile city is a model of urban life which many people find threatening even as they embrace it. By and large, those who have a choice do not care to live in cramped[1] city centre homes, nor do they see much economic future in the old downtowns. The historic pattern of the city is undoubtedly a high point in civilisation, one that is worth preserving. But even if masonry[2] endures, its meaning has been irrevocably altered.

There are many issues posed by this new kind of city. The most alarming of course is the prospect of what happens if the supply of energy is shut off—if the private car on which so many cities depend is made obsolete[3] by future energy crises. Will exurbia have to be abandoned just like the empty Indian city of Fatephur Sikhri? It's an issue that is already preoccupying the most notorious car-orientated city, Los Angeles. It's not just the prospect of shortages that is worrying the city, but the more immediate prospect of gridlock[4] on its freeways. Clearly there are many settlements in which it is already too late to do anything about, but any responsible planning system must address the issue, which is at its most acute in the location of housing. But what the planner cannot do is to cut across the direction of events. The only plausible strategy is to attempt to harness the dynamics of development to move things in the direction that you want. For the planner or the architect to ignore the currents that are shaping the city is clearly futile[5]. Enormous amounts of energy have been expended on means of reconstructing the traditional European city, as if this were possible by the simple exertion of will.

To accept this image of the city is to accept uncomfortable things about ourselves, and our illusions about the way we want to live. The city is as much about selfishness and fear as it is about community and civic[6] life. And yet to accept that the city has a dark side, of menace[7] and greed, does not diminish its vitality and strength. In the last analysis, it reflects man and all his potential.

Notes

1. Extracts from
Text
Sudjic, Deyan. The 100 mile City [M]. New York: Harvest/HBJ Book, 1992: 305 – 309
2. Background Information
迪耶·萨迪奇（Deyan Sudjic）：1952年生于英国伦敦，父母是南斯拉夫人。早年在爱丁堡大学学习建筑。他的生涯是写作、做助理牧师和教学相结合的。1983年，他创立《蓝图》（*Blueprint*）杂

[1] cramped：局促
[2] masonry：石工
[3] obsolete：过时
[4] gridlock：高压封锁
[5] futile：徒劳
[6] civic：民事
[7] menace：威胁

志，作为杂志的编辑和主编。1993~1997 年，他在维也纳应用艺术学院任教。2000~2004 年他成为 *Domus* 的编辑，并继续为报纸《观察》(*Observe*) 从事建筑评论。1990 年开始为《守望者》(*Guardian*) 写作。迪耶·萨迪奇写了一些很有影响的书，如《建筑和民主》(*Architecture and Democracy*，2001 年)，《100 英里城市》(*The 100 Mile City*，1993 年) 等。他最近发表的书为《建筑聚合体——富裕和权利如何塑造世界》(*The Edifice Complex-How the Rich and Powerful Shape the World*，2005 年)。2002 年他成为威尼斯建筑学院的主任。

Exercises

1. Questions（阅读思考）

认真阅读本单元文章，并讨论以下问题。

（1）*The Granite Garden* 一文文章结构，主要观点及观点演绎方式。

（2）给 *The Growth of Suburbs* 一文划分段落，归纳各段落大意。划分并分析各段落中句群结构。

（3）结合 *Landscape Design in the Urban Environmen* 一文内容，谈谈风景园林师在城市规划设计中的角色定位？

（4）联系中国实际情况谈谈风景园林师如何打破学科界限，发挥自身优势为改善人居环境服务？

2. Skill Training（技能训练）

（1）查阅英文资料，了解和介绍19世纪40年代初"哈佛革命"的主要人物、观点及代表规划设计案例。

（2）查阅建筑、城市规划相关英文论文和案例，开展英文主题研讨会（Seminar）：西方现代城市规划与设计。

Tips

欧美风景园林教学常见课堂活动形式

尽管在具体课程和教学方向上各有侧重，但欧美国家风景园林教学基本框架和我国基本类似：理论教学、参观实习、工程实验、规划设计等内容构成中外风景园林规划设计教学主体。和国内有差异的是，欧美国家风景园林教学在课堂活动组织上更丰富多样，更强调学生的学习主动性和师生教学互动。了解常见课堂活动形式，不仅有利于我们读懂国外相关资料，对教学工作也是一种有益的借鉴。

可供借鉴的欧美国家风景园林教学常见课堂活动形式主要包括以下五类：案例研究（Case Studies）、表达（Presentation）、研讨会（Seminar）、设计教学（Design Studio）和设计实习（Design Workshop）。

1. 案例研究

案例研究是法律、医学、建筑等学科常用的教学方法，如今在欧美各高校风景园林教学中也越来越流行。美国风景园林基金会（Landscape Architecture Foundation）1997~1999年相关研究报告认为：案例研究方法在风景园林学科发展中非常重要，它是一种十分有效的技术传播方式，它总结性和批判性的评论对学科发展具有重要意义。报告推荐，一个完整的案例研究应包括以下四个主要部分：Photos，Project background，Project significance and impact，Lessons learned。

Photos（图片）：包括案例图纸和照片。Project background（项目背景）：用地面积、性质、存在问题、投资等前期情况；决策者、设计师、工程建设等设计建设情况；使用者、使用反响等使用情况。Project Significance and impact（项目意义和影响）：项目在规划设计方面的特点，专家学者的评价，从学科和职业发展角度总结和评论该项目的意义和影响。

Lessons learned（启示）：该项目的成败得失对未来工作的启示或教训。

案例分析方法在本课中也可应用，分析国外经典设计作品或著名设计师等可以增强学生学习主动性，增进对欧美国家风景园林的了解，活学活用专业英语，并被教学实践证明是非常好的一种课堂活动形式。

2. 表达

所谓"表达"的课堂活动形式，指教师拟定一定主题，学生围绕此主题进行观点陈述或材料展示。无论是风景园林设计或是英语学习，"表达"都非常重要。具体方法如请学生在课上展示并介绍案例研究成果。

3. 研讨会

研讨会指就某一主题，请几个主要发言者发表观点，然后大家讨论。例如本课在每一单元结束时可安排一个研讨会，先请某几个同学准备相关材料，就本单元主题发表自己的观点，然后进行讨论。

4. 设计教学

设计教学基本等同于国内的设计课，是欧美风景园林教学体系中的重要课程。以Studio为核心的设计教学是一种集授课、讲座、研讨、设计实践、交流汇报等为一身的综合性设计课教学模式。它主要可分讲授、设计、讨论三个环节。讲授包括任课教师讲课和请专家开设专题讲座，设计除图上作业外还包括调研、查阅资料、小组协作等，讨论包括设计小组内部和小组之间的研讨和交流。

一般而言，西方国家设计教学更重视设计思想、设计理念和创新思维的培养，重视交流和研讨，重视各种思想和智慧的碰撞，重视教学过程以及学生在此过程中对风景园林思想理论、职业技能、团队协作等的全面学习。

5. 设计实习

Workshop 原指车间、工场。设计实习指就给定的设计项目或专题，在一定的时间内，一群人进行设计实践并交流讨论，相互分享知识和技能。Workshop 和 Studio 的最大区别是前者更重视实践技能，注重技术性和职业性的讨论。

在西方文献中出现的 Design Workshop 通常有两类：一类是就某个专题，一群设计师带自己的作品开个讨论会，讨论和交流职业技能；另一类面对学习者开设，给定一个实际的设计项目，几天时间内完成，练习从现场调研到团队协作、讨论到成果汇报的全过程。在这过程中通常有经验丰富的设计师给予指导，但更重视参加者的相互学习。

附：美国风景园林本科教育和研究生教育前五强及其网址

（Design Intelligence Top 5 Landscape Architecture Schools 2006）

本科教学（Undergraduate Programs）

(1) University of Georgia：The School of Environmental Design.
  http：//www. uga. edu/sed/index. htm
(2) Purdue University：Department of Horticulture & Landscape Architecture.
  http：//www. hort. purdue. edu/hort/welcome/welcome. shtml
(3) Louisiana State University：The college of art & design.
  http：//design. lsu. edu/lsuhome. htm
(4) Pennsylvania State University：Department of landscape architecture.
  http：//www. larch. psu. edu/
(5) Kansas State University：College of Architecture Planning and Design.
  http：//capd. ksu. edu/

研究生教学（Graduate Programs）

(1) Harvard University：the Department of Landscape Architecture.
  http：//www. gsd. harvard. edu/academic/la/
(2) University of Pennsylvania：the Department of Landscape Architecture.
  http：//www. design. upenn. edu/new/larp/index. php
(3) University of Georgia：The School of Environmental Design.
  http：//www. uga. edu/sed/index. htm
(4) Louisiana State University：The college of art & design.
  http：//design. lsu. edu/lsuhome. htm
(5) University of Virginia：The Department of Architecture and Landscape Architecture.
  http：//www. arch. virginia. edu/landscape/

Unit 5

Parks & Recreation

- 公园规划设计
- 风景游览地规划设计

Park projects would civilize and refine the national character, foster the love of rural beauty and increase the knowledge of and taste for rare and beautiful trees and plants.

By Andrew Jackson Downing

Text

Urban Parks and Recreation

By Michael Laurie

> 本文包括城市公园价值和城市公园及休闲地两部分内容。
> (1) 公园价值：有利于公众健康；有利于提高公民道德修养；美学价值；经济价值；教育资源。
> (2) 城市公园及休闲地。在传统体系中，休闲单元按照其规模和分布进行分类，可分为以下四个等级：街区层面；邻里单位层面；社区层面；全市性公园。

Park values

The case for public parks in the 19th century was built largely on the same concerns as that for improved housing. There seems to be five basic arguments: the first is concerned with public health, the second with morality, the third is related to the development of the Romantic movement, the fourth is concerned with economics, and the fifth with education.

A concern for public health led to reform in housing and improved sanitary[1] sewers and drains. Included in the concept of better health was the availability of parks allowing the circulation of fresh air and providing space for exercise, rest, and refreshment in a sunny landscape setting. The concern for pubic health also resulted in the building of rural suburbs for the rich, who wished to get as far away from the urban conditions as possible. Thus began the exodus[2] from the city based on the opinion that cities were ugly, unhealthy, and dangerous, a viewpoint which has grown in popularity and resulted in the critical disintegration of contemporary American cities. Now more people live in the suburbs than in the cities, with those in the cities mostly minorities.

• 公园价值

在 19 世纪，人们对改善居住条件相当关注，而城市公园的建造在很大程度上是出于相同的关注。对此，有五个基本关注点：一是公众健康；二是伦理道德；三是浪漫主义运动的发展；四是经济；五是教育。

有一种观点认为城市是丑陋的，病态的，危险的，于是出现了人们大量迁出城市的情形。这一观点现在更加流行，导致当代美国城市濒临瓦解。

---

[1] sanitary: 清洁的，卫生的
[2] exodus: 许多人离开，很多人离去

PART 1　Theory

Second, the concern for morality is associated with the idea that nature itself is a source of moral inspiration. The advocates of this concept felt that if the workers had an opportunity to study and contemplate nature, this would improve their mental stability and provide them with beliefs that would transcend day to day drudgery[1] and substitute for the oblivion[2] which the "gin palaces" provided. Parks were thus associated with this concept of morality in nature inasmuch[3] as they would provide landscape for contemplation. Later, sports facilities and allotments[4] in which vegetables could be cultivated were also considered to serve this moral purpose.

第二，对伦理道德的关注来自于自然本身是道德激励的源泉这一思想。这一理念的支持者认为，如果劳动者有机会研究和思考自然的话，他们的心绪会平稳下来，摆脱对日复一日单调工作的厌恶感以及酒精对他们的控制。

The aesthetic[5] argument was that the visual qualities of the expanding industrial cities were generally considered ugly (although there were some artists who found beauty in blast furnaces[6] and other industrial installations). Essentially, the city was equated with ugliness and public parks as large landscapes inserted into this ugliness would thus serve as an antidote[7].

美学论点认为不断扩张的工业城市是丑陋的（尽管有些艺术家从熔炉等工业设备中也发现了美）。从本质上说，人们把城市与丑陋等同起来，而公园作为大面积的风景插入到这丑陋当中去将会减少城市的丑陋性。

The economic argument derives from the first three arguments. These propositions are the basis for a money-making system whereby public parks provide qualities of health, morality, and beauty to the workers, whose productivity is thus improved. At the same time, real estate values are increased by proximity[8] to the romantic landscape in an otherwise drab[9] environment. This in turn produces revenue for the city through higher taxes. Finally, the park was seen as a place of instruction in natural science through arboreta and zoos.

同时，由于毗邻了公园浪漫的景观而使房地产价值得以提升，这些房地产的环境原本单调而乏味。这也为城市通过增加税收的方式提高了年收入。

Although the landscape style of gardening was not specifically a response to the depraved[10] urban conditions, it might well have been. The 18th century theories of landscape design based on informality, naturalism, romanticism, and the picturesque became the logical antithesis to urban conditions of the 19th century. Park designs included all of the elements one

18世纪园林设计理论依据的是不拘形式，自然主义，浪漫主义及如画的风景理论，而这和19世纪城市环境条件正好相反。

[1] drudgery：辛苦而令人讨厌的工作
[2] oblivion：被完全遗忘的状态，淹没
[3] inasmuch：由于，因为
[4] allotment：所配得的一份，（尤指英国）租来作为菜园用的一小块公地
[5] aesthetic：审美的，有审美能力的，（对于艺术等）有高尚趣味的
[6] furnace：火炉，熔炉
[7] antidote：解毒剂，抗毒药
[8] proximity：接近
[9] drab：（喻）乏味的，单调的
[10] deprave：败坏其道德，使腐败

would expect: curvilinear drives and paths, rustic[1] gates, Gothic[2] architecture, irregular lakes, and informal landscape planting. In fact, the public park was to become the only place of sufficient size where the landscape garden style could be carried out successfully and parks were thus laid out as though they were large private estates with accommodations made for the greater number of people.

Andrew Jackson Downing[3] became a strong advocate of public parks for America in the manner of Birkinhead. He advanced the revolutionary idea that, as in England, parks should be maintained at the expense of the taxpayer. He also recognized the real estate, money-making possibilities for land adjacent to public parks. In addition, he argued the moral case that "such park projects would civilize and refine the national character, foster the love of rural beauty and increase the knowledge of and taste for rare and beautiful trees and plants." He saw the park as a piece of rural scenery in which people could walk, ride, or drive in carriages. It would be a relief from the city streets: "Pedestrians[4] would find quiet and secluded[5] walks when they wished to be solitary and broad alleys filled with happy faces when they would be gay." Downing died in 1852, but his advocacy persuaded the city of New York that it should plan a major public park.

他提出一个革命性的观点：应该学习英国，由纳税人负担公园维护的费用。

此外，他认为公园具有精神上的意义："这样的公园项目将使民族性格更加文明和高尚，培养对人们田园美景的热爱，增加人们对珍稀而美丽的花草树木的知识及审美感。"

Parks and recreation areas

Since the 19th century "parks and recreation" have become a major industry with a large civil service in cities, counties, states, and regional districts. Smaller neighborhood parks and playgrounds were introduced at the end of the nineteenth century as centers of sport and physical fitness and other social programs. In an attempt to measure the effectiveness of a city's provision of parks, organizations such as the National Recreation Association formulated standards in terms of acres per unit of population. By these standards there was never enough. Although they are generalities, which must be weighed with the variables of the specific place, it is still true that cities and other agencies tend to think of recreation

● 城市公园及休闲区

自19世纪以来"城市公园和休闲地"已成为主要的产业，在市、县、州以及地区都设有相应的行政机关。

尽管这些标准是通则，也必须用特定地点的变化情况来权衡，但可见城市和其他机构已经趋向于以制定标准的方式来考虑休闲设施问题了。

---

[1] rustic：有乡村居民特色的，朴素的
[2] Gothic：哥特式建筑的（哥特式为12～16世纪常见于西欧之建筑式，以尖拱、簇柱等为特色。）
[3] Andrew Jackson Downing：唐宁（1815～1852年），美国早期风景园林师
[4] Pedestrian：步行人，走路的人
[5] secluded：（尤指地方）安静的，幽僻的

facilities in terms of such standards.

The standards and acreages[1] cited below are based on a combination of several sources. It should be noted, however, that the recommendations in the California guide for recreation and parks vary according to the climatic zones of the state. Thus larger acreages are recommended in the warmer zones to permit more extensive tree planting.

In the conventional system, recreation units are classified in terms of their scale and distribution. The play lot, or block playground for use by preschool-age children, is the smallest unit and should, according to the standards, be within walking distance of a majority of homes, located perhaps in the interior of a city block. They should be between 1/8 and 1/4 acre[2] in size and are especially important where the density is high. The recent "mini-park" building program in run-down sections of cities indicates the value of play lots in such areas; low-density suburban housing may have less need.

在传统体系中，休闲单元以其规模和区域来分类。供学前儿童使用的游戏场地或街区运动场是最小的休闲单元，按照标准应在城市街区内的大多数家庭步行可达的距离内。

The next category is at the neighborhood level: the neighborhood park, playground, or recreation center, or a combination of all three. In this case the neighborhood is usually described as the area served by an elementary school (Busing of children between neighborhoods to achieve racial balance in the schools may conflict with this concept). The facility should provide indoor and outdoor recreation for children between 5 and 14 years of age. Pre-school children and family groups should also be provided for and a landscape area of at least 2 acres is desirable. Ideally the neighborhood playground would be situated within 1/2 mile[3] of each home. Recent standards argue that it is logical to combine the park with the elementary school and its yard. The standards suggest that if the school is 10 acres, then the related park should be 6 acres. If separate, 16 acres would be needed. Another standard suggests that there should be 1 acre of neighborhood park for every 800 of population. But the facilities and parks should reflect the population. Old people who do not drive may have realistic needs for open space equivalent to the tot[4] lot close to home.

下一个类别在邻里层面：邻里公园、运动场、休闲中心或以上三者相结合。

---

[1] acreage：以英亩计算之土地面积，英亩数
[2] acre：英亩，1英亩＝0.405公顷
[3] mile：英里，1英里＝1.609千米
[4] tot：儿童

The standards also call for community recreation areas, or playfields. A community is defined as a number of neighborhoods, or a section or district of a city. It suggested that these facilities should provide a wider range of recreation possibilities than those at the neighborhood level, including fields, courts, and swimming pool, and a center in which arts and crafts, clubs, and social activities may take place.

标准还要求有社区休闲区或游戏场地。社区就是一系列的邻里单位或者就是城市的一个区域或地区。这表明这些设施所提供的休闲机会应比在邻里单位层面提供的范围更广，包括球场、游戏场地、游泳池、能够进行工艺及艺术创作、打牌和社交活动的中心。

The recommended size is 32 acres or 20 acres if connected to a school, and they should be located between 1/2 and 1 mile of every home. Another measure is that 1 acre of community park should be provided for every 800 population.

Finally, city-wide recreation areas are described as large parks providing the city dweller a chance to get away from the noise of the city, its dirt and traffic (Figure 5.1). If the automobile was permanently prohibited from the large 19th century parks this would be possible. They should provide the variety of activities and possibilities to be found in Central Park and Golden Gate Park. The desired effect can hardly be provided in less than 100 acres. There should be sports centers and facilities for golf, boating, and so on. The 1956 standards for California suggest that a city of 100,000 should have 883 acres of city park, of which 21 would be needed for parking. Beyond these park standards lie special items. Every city should have golf courses, outdoor theater, zoos, botanic[1] gardens, or similar facilities.

最后，全市性的休闲区被描述成一个个大型公园，使居民有机会远离城市的喧嚣、灰尘和繁杂的交通（见图 5.1）。这点是可以实现的，只要永远禁止机动车进入这些 19 世纪的大型公园。

加州 1956 年标准提出，一座 10 万人口的城市应有 883 英亩的城市公园，其中 21 英亩用于停车。除了这些公园标准以外还有特殊条款。每一座城市都应有高尔夫球场、户外剧院、动物园、植物园或类似的设施。

Figure 5.1 Hade Park, London. A city wide recreation area within the city but large enough to be isolated from it

---

〔1〕 botanic：植物学的

These standards are, of course, abstractions, created by the parks and recreation industry. They may not take into account the extent to which recreation patterns have changed in recent years, but in general they make sense.	当然，这些标准是由公园及休闲产业设定的抽象概念。它们也许没有考虑到休闲模式近年来已经发生了很大的变化，但总的来说，它们是有意义的。
To what extent have recreation patterns changed since the nineteenth century? Shorter working hours and automation tend to result in boredom[1]rather than exhaustion for the blue-collar worker. At the adult level, Wayne Williams has observed that the traditional leisure triangle has been reversed. Those with the most leisure time today tend to be the unskilled, the semiskilled, and the skilled, in that order. Thus it is perhaps obvious that the need is for diversion rather than for recuperation[2]and that the majority of people need and will seek challenging and active recreation and meaningful involvement. For children, Williams believes that the whole city should be considered a playground, a park, and a classroom and that ways should be found to incorporate the public and especially children into the various workshops of the city-bakeries, auto repair shops, and so on. In this concept, recreation is regarded as an integral[3]part of living, as it should be, rather than as something to be enjoyed in a playground.	缩短工时和自动化带给蓝领工作者的往往是枯燥乏味，而不是疲惫。 因此，也许是显而易见的，我们需要的是消遣而非休养，而且大多数人需要也愿意寻求具有挑战性和积极性的休闲活动及有意义的参与。
New concepts of recreation, changes in population, and patterns of work and leisure come at a time when many of the parks designed over a hundred years ago have reached maturity and are in need of renewal[4]. Major opportunities exist, therefore, for the landscape profession in collaboration with parks and recreation departments and commissions and with community participation, to reevaluate their role in society, leading toward the development of new forms rather than historical restoration.	新的休闲观念已经形成，人口、工作和休闲方式已经发生改变，这也正是很多百余年前设计的公园已经成熟，到了需要更新的时候。

Notes

1. Extracts from
Text
Laurie, Michael. *An Introduction to Landscape Architecture* [M]. New York: Elsevier Science

---

[1] boredom：厌烦，厌倦
[2] recuperation：复元，休养，恢复
[3] integral：不可或缺的，作为组成部分的
[4] renewal：更新，再始，换新

Publishing Co., Inc. 1986: 74-85.

Picture

Figure 5.1: Laurie, Michael. *An Introduction to Landscape Architecture* [M]. New York: Elsevier Science Publishing Co., Inc. 1986: 85.

2. Background Information

(1) Michael Laurie 麦克尔·劳瑞（1932~2002年）。

1932年出生于苏格兰的丹地，1553年进入雷汀大学学习风景园林专业，于1956年获得毕业文凭。并在宾夕法尼亚大学学习研究生课程，师从麦克哈格，1962年获得风景园林学硕士学位，同年被聘任为加利福尼亚大学伯克利分校风景园林系讲师，1969年晋升为副教授，1979年晋升为教授，麦克尔·劳瑞曾在三个时期（1976~1978年、1981~1982年、1991~1998年）担任风景园林系主任。于1988年获风景园林教育专家委员会（CELA）杰出教育专家奖。1966年美国风景园林师协会（ASLA）授予他布拉福威廉姆斯奖章。

(2) An Introduction to Landscape Architecture (second edition)《风景园林学导论》（第二版）。

1986年出版，该书由作者在加利福尼亚大学伯克利分校授课内容基础上编写而成，是一本著名的被国际上广泛采用的的经典教材，一经出版即在美国和大不列颠成为风景园林学标准的入门教材。随后又被翻译成日文和西班牙文。

本文选自该书第四章"城市公园和休闲地"，内容有删节。

Further Reading A

People's Parks: Design and Designers

By Hazel Conway

In a sense the design of each park was the solution to the particular factors affecting each individual project. Parks ranged in size from fractions of a hectare[1] to several hundred hectares, and they were created out of a wide variety of sites which included commons, wasteland, infill and marginal[2] land such as disused quarries[3]. Some parks were created out of sites that were already partly laid out if they had previously been private parks and gardens, while others such as Moor Park, Preston, were not fully laid out until many years after they actually opened as parks. Despite this, it is possible to see certain broad changes in design during the course of the nineteenth century.

The most important influence on landscape design at the beginning of the nineteenth century was Humphry Repton[4]. It was his theories that influenced John Nash in his revised designs for Regent's Park and his ideas were kept alive well into the nineteenth century by John Claudius Loudon. It was Repton who first attempted to draw up the fundamental principles of landscape gardening and to analyse the sources of pleasure in it. Although the problems of public park design were very different from those involved in the design of private parks, many of his principles, particularly those concerning picturesque beauty, variety, novelty, contrast, appropriation and animation[5], were influential in the early phase of urban park design in the 1840s. Picturesque beauty encouraged curiosity[6] and this in turn meant that all the significant features of a design should not be immediately visible. Novelty[7] and contrast also stimulated curiosity and were important elements of the picturesque. Appropriation involved enhancing the apparent extent of an owner's private property and this was applied to public parks, where it was important to give the appearance of as large an extent of park as possible. Other sources of pleasure identified by Repton included the changing seasons, and animation which enlivened[8] the scene by means of the movement of water or animals, of smoke from distant dwellings, although smoke was hardly a positive factor for urban park designers and park keepers.

Apart from Repton and Loudon, the other major figure during the first decades of the park

---

[1] hectare：公顷
[2] marginal：边际土地（此种土地不够肥沃，耕种无利可获，非至农产品价格高涨时不加利用）
[3] quarry：采石场
[4] Humphry Repton：亨弗利·雷普顿（1752～1818年）18世纪末英国著名风景园林师，继承了"可为布朗"的事业
[5] animation：尤指活泼，有生气
[6] curiosity：好奇心
[7] Novelty：新颖
[8] enliven：使活泼，使有生气

movement is undoubtedly Joseph Paxton. Paxton's first public park was Prince's Park, Liverpool (1842), his first municipal[1] venture Birkenhead Park (1842 - 7), but the Crystal Palace Park, Sydenham (1855), which was a speculative development, was his most influential park design. One of the problems confronting the designers of public parks was how to accommodate large numbers of people and at the same time preserve a feeling of space and quiet contact with nature. Another problem concerned the location of an increasing range of activities, such as sports and music, which had to be reconciled with the prevailing contemporary ideals of landscape design. The fullest opportunities for designers to develop their ideas occurred in those parks which were not already laid out when they were acquired and some important examples from the 1840s show how Loudon, Paxton and others resolved these problems. In the 1850s and 1860s Paxton's design for the Crystal Palace Park became a major influence, while in the 1870s the introduction of French principles of park design at Sefton Park, Liverpool, seemed to point to a solution of the problem of accommodating sports and playgrounds.

In some instances designers were chosen by competition (Appendix), but the role of competitions in this area was not nearly so important as it was in the area of public architecture. The majority of municipal parks were not designed by competition and it was only after 1875 that competitions became more important.

The Parks of the 1840s

Loudon thought that "a prevailing error in most public gardens" was the lack of unity of expression. Walks crossed in all directions, providing a puzzle to the visitor, whereas if the principle of unity was applied then each garden would feature one main walk from which "every material object in the garden may be seen in a general way". From this walk "small episodical[2] walks" would branch off to other areas. These would never be broader than one third of the main walk, so there would be no confusion between the main and the subsidiary[3] walks, and they would join the main walks at right angles "so as not to seem to invite the stranger to walk in them". At Derby Arboretum[4] he put these principles into practice.

Joseph Strutt, the donor[5], wished to preserve the existing garden, trees and buildings and minimize the expense of maintenance. A design featuring trees and shrubs "in the manner of a common pleasure-ground" would tend to become boring after one or two visits and a botanic[6] garden would prove expensive. An arboretum featuring a collection of foreign and native trees and shrubs, one example of each, with its name displayed, would provide beauty, variety, interest and education in all seasons of the year. Derby Arborerum shows how Loudon combined axial[7] symmetry[8] with

---

[1] municipal：市的，市政的，自治城市的
[2] episodical：插曲式的
[3] subsidiary：辅助的，次要的
[4] Arboretum：植物园
[5] donor：赠与者，捐赠者
[6] botanic：植物的，植物学的
[7] axial：轴的，形成轴的
[8] symmetry：对称，匀称，调和，对称美

informal paths and planting. A broad central walk provides the spine[1] of the design and at the junction of the central gravel walk and the main cross walk was a statue on a pedestal[2], since "a straight walk without a terminating object is felt to be deficient in meaning". Pavilions[3] providing seats and shelter performed the same function at the ends of the cross walks; an example is shown in Figure 5.2. The subsidiary walks took the visitor around the periphery[4] of the site, and the undulating[5] ground and planting promoted the illusion of solitude[6] and obscured the boundary, so concealing[7] the apparent extent of the site. Loudon also wanted to create hollows and winding valleys, but the soil and difficulties of drainage prevented this; however flat spaces were left for tents as Strutt wished. Unlike later park designers, Loudon was not required to incorporate a wide range of facilities within his design.

Figure 5.2 Pavilion, designed by E. R. Lamb, Derby Arborerum

There were 350 seats in the Arboretum and Loudon gave detailed instructions on how they should be placed, so that the extent of the park was not made obvious. This meant that those seated should not be able to see both exits from where they were sitting, nor should they be able to see both boundaries. For interest, comfort and security he recommended that the seats should face a view or a feature, or be positioned in graveled recesses[8] by the side of walks where passers-by[9] would provide interest (Figure 5.3). Some should be open to the sun for winter use, others should be under the shade of trees, but the majority, which would be used in the summer, should face east, west or north. Loudon's concern with comfort and security can also be seen in his suggestion that the seats placed on the grass should be backed by trees and shrubs so that no one could easily come up close to them from behind; or alternatively double seats with a common back could be provided. All fixed

---

[1] spine: 脊椎，脊柱
[2] pedestal: 柱基，塑像或艺术品的座
[3] pavilion: （尖顶）大帐篷，（公园、庭园等的）亭子
[4] periphery: 外围，表面
[5] undulate: （表面）波动，起伏
[6] solitude: 独居，单独，孤独
[7] conceal: 隐藏，隐匿，隐瞒
[8] recess: （墙壁的）凹室，壁龛，隐密地方，难进入的地方
[9] passer-by: 过路人，行人

seats should have footboards[1] for the comfort of invalids and the aged. Such attention to the detail of positioning seats was not the rule and when Kemp reported on the number and position of seats in Birkenhead Park in 1857 he only gave the most general indication as to where and how they were placed. Out of a total of 72 seats, nearly half were positioned by the walks around the lower lake (Figure 5.4).

Figure 5.3 The seats in Derby Arborerum, c. 1900, were not recessed as Loudon recommended

Figure 5.4 Cast iron seat backed by shrubs. Of the many designs available, this was among the most lavish

Loudon also considered the problem of accommodating variety, with the conflicting need for unity of expression. In order to keep visitors' interest alive there should be variety and "one kind of scene must succeed another", but to give the coherent effect necessary for unity of expression, these scenes should be related according to some principle recognisable by the visitor.

The planting of the trees in the Arboretum gave Loudon an opportunity to put this principle into practice, though there was less scope for him to create a variety of picturesque scenery. This principle was applied in many of the larger parks of the middle decades of the century, sometimes more successfully than others.

Notes

1. Extracts from

Text

Conway, Hazel. *People's Parks—The Design and Development of Victorian Parks in Britain* [M]. New York: Cambridge University Press, 1991: 76-81.

Pictures

(1) Figure 5.2: Hazel Conway. *People's Parks—The Design and Development of Victorian Parks in Britain* [M]. New York: Cambridge University Press, 1991: 78.

(2) Figure 5.3: Hazel Conway. *People's Parks—The Design and Development of Victorian Parks in Britain* [M]. New York: Cambridge University Press, 1991: 79.

(3) Figure 5.4: Hazel Conway. *People's Parks—The Design and Development of Victorian Parks in Britain* [M]. New York: Cambridge University Press, 1991: 80.

2. Background Information

(1) Hazel Conway 黑泽尔·康威。

---

[1] footboard: 踏板

建筑和风景园林历史学家，尤其在近代公园历史的研究上卓有成就。

（2）*People's Parks—The Design and Development of Victorian Parks in Britain*《人民公园：不列颠维多利亚公园的设计和发展》。

1991年出版。本书阐明了对于19世纪英国的市政公园和其他公共园林发展的国内和国际的影响，将其与公园的设计和使用相联系，明确了这些成就的重要意义。社会、经济和政治的发展状况深刻影响着人们对休闲的态度，随之而发展的市政公园对改善城市环境具有重要意义。城市公园的创设者想通过公园即通过设计、建筑、雕像、户外音乐台和绿化来方便人们的教育和娱乐。本书详细描述了公园的发展、设计和使用，关于主要的相关立法的总结，早期市政公园和其他公共园林的年代地名索引（附公园规模、公园产生经过及设计者姓名等细节），附有详实的平面设计图、照片和版画，是一本相当著名的近代公园历史著作。

本文即节选自该书第五章"设计及设计者们"部分内容。

Further Reading B

Parkways[1] and Their Offspring

By Norman T. Newton

It is doubtful that any single type of park area has been more widely misunderstood and misinterpreted[2] than the parkway. The confusion is hardly to be wondered at when one considers with what free and easy imprecision[3] the term "parkway" has been used. Unfortunately, it has even been employed by real-estate[4] developers in recent years as a sort of status label. Truthfully, it would seem that something of this kind may have been the effect, despite the most innocent of motives, when the landscape architectural pioneers-Olmsted, Vaux, Cleveland, Eliot-used the term for roadways that were simply wider and more richly furnished than ordinary streets: "boulevard[5]," the more usual early title, was of course borrowed from the French. Then Olmsted and Vaux, in about 1870 in Brooklyn, designed a handsome boulevard as "a grand approach from the east" to the Plaza of Prospect Park. They called it the Jamaica Parkway and, though its name was modified almost at once to Eastern Parkway, the designation[6] stuck. Indeed, eastern Parkway remained a pleasantly impressive combination of carriageways[7], "parked" pedestrian strips, and overarching[8] elms[9] until years later, when an extension of New York's vast subway system was run under its length and started its downfall from splendor.

A few years later in Boston, when Olmsted began working on the Back Bay Fens, he saw the possibility of establishing a long continuity of park land from the Boston Common an Public Garden along the wide boulevard that was the existing inner end of Commonwealth Avenue, then through his newly created Fens and out the Muddy River to Jamaica Pond. This, with an extension to the Arboretum and Franklin Park added later, Olmsted called "the Promenade[10]" in his 1880 report. "The Parkway" was the Boston Park Commissioners' 1887 label for the entire stretch of Muddy River from the tidegate on the Charles out to Jamaica Pond; and they authorized the sectional names that eventually became today's Charlesgate, Fenway, Riverway, Jamaicaway (through Olmsted Park), and Arborway.

---

[1] parkway：公园路，风景大道，休闲路
[2] misinterpret：误译，误解，误以为
[3] imprecision：不严密，不精确
[4] real-estate：不动产
[5] boulevard：大马路，大道（两旁常植有树木），林荫大道
[6] designation：名称，称号
[7] carriageway：车道
[8] overarch：（在……上面）造成拱形
[9] elm：榆树，榆木
[10] promenade：（为运动或散心所作的）散步，骑马，散步或骑马的地方，（尤指）海滨胜地临水的大路

Cleveland, in his 1869 pamphlet[1] on "public grounds" for Chicago, recommended a "grand avenue or boulevard" some fourteen miles long to connect the parks. He even went so far, despite his usual vehement[2] antagonism[3] to gridiron[4] systems, as to present a lengthy argument favoring straightness rather than successive curves in the boulevard. He did not use the term "parkway", but he did use it in 1883 when he submitted to Minneapolis authorities a report on "Parks and Parkways" for that city.

In the 1890s, when Eliot was advising the Metropolitan Park Commission in Boston, he repeatedly urged the creation of "parkways or boulevards" as connections between units of the park system. Even though he wrote that "parkways or boulevards... are generally merely improved highways," he seemed to have sensed the importance of a major function that parkways would in later years be called upon to serve, that of providing psychological carryover of the restful influence of one large park area to its echo in another, with little or no interruption on the way.

But on the whole early "parkways" could more accurately be described as boulevards. It was only with completion of New York's Bronx River Parkway after World War I that the modern parkway came into being with its clear set of distinguishing characteristics. The term now denoted a strip of land dedicated to recreation and the movement of pleasure vehicles (passenger, not commercial automobiles). The parkway was not itself a road, it contained a roadway. The strip of land was not just a highway with uniform grassy borders; it was of significantly varying width, depending on immediate topographic and cultural conditions. The roadway itself differed markedly from that of an ordinary highway in that it was meant for comfortable driving in pleasant surroundings, not merely for getting from one place to another as fast as possible. The alignment was accordingly one of gentle curves, designed for speeds in keeping with the times. Perhaps most important was the distinctive provision that abutting owners had no right of light, air, or access over the parkway strip. It was lack of the limited access factor that most clearly kept the early boulevards from functioning as parkways in the modern sense. To appreciate the point, one need only experience the interruptive effect of frequent access roads and private driveways from the side on such older "parkways" as the Fenway and the Riverway in Boston, Eastern Parkway in Brooklyn, Mosholu Parkway in the Bronx, the Mystic Valley Parkway in Arlington, Massachusetts, or Memorial Drive in Cambridge.

The Bronx River Parkway, introducing the new combination of characteristics, was completed at Kensico Dam in New York's Westchester County in 1923, but it represented the culmination[5] of eighteen years of effort that had begun simply as a cleanup job. In 1905, William W. Niles, a public-spirited citizen of New York deeply disturbed over the filthy[6] condition of the little Bronx River as it came down from Westchester County to enter the city's Zoological park and Botanic Garden in the Borough of the Bronx, enlisted the support of the directors of the two institutions and sought relief in the New York State Legislature for a study of the problem. Not only was the Bronx River valley an

---

[1] pamphlet：小册子
[2] vehement：强烈的，猛烈的
[3] antagonism：对抗，敌对，对立
[4] gridiron：网格状
[5] culmination：顶点，极点
[6] filthy：不洁的，污秽的

unsightly mess[1]; the polluted water of the river was injurious to wild fowl in the Zoological Park.

The Blue Ridge Parkway[2] was intended for pleasure-driving at reasonable speeds and for moderate traffic. It was in no sense an expressway, but side access was fully controlled. The roadway was a simple ribbon of pavement without a central divider. It was fitted closely to the land and every effort was made, as in the Bronx River Extension and the Taconic, to blend it into the native scene. An underlying point of view recalls Prince Pückler-Muskau's attitude over a century before: the activities of humans are a fully natural part of any total landscape. In the Blue Ridge Parkway this gave rise to effective use of the scenic easement over adjacent farmlands; the owner, instead of having to sell, merely agreed for a modest consideration to keep the land in its normal agricultural use. Thus, along many portions of the route, visual protection of the parkway did not require outright purchase of lands; and use of the scenic easement actually intensified the feeling of being in entirely native surroundings, without affection or the intrusion of anything exotic[3] (Figure 5.5).

Figure 5.5 On the Blue Ridge Parkway in Virginia: the use of scenic easements helps the parkway merge with the existing

The atmosphere was further maintained by faithful attention to local ecology, whether in vegetation or in things manmade. Only native materials were used in planting, and these with restraint, so that one never felt enclosed in a tortuous[4] man-planted bowling-alley[5], as too often happens in highways with "roadside landscaping" applied as a vegetative afterthought. Where fencing of any sort was required along the right-of-way, purely indigenous[6] methods and means were adopted: walls of local stone, post and rail barriers, zigzag snake-fences of split rails. Whenever possible, structures that had been significant in local life were retained or, as in the case of the old Mabry Mill in Virginia, carefully repaired and restored to activity.

[1] mess: 散乱，杂乱
[2] the Blue Ridge Parkway: 蓝岭风景大道，连接 Shenandoah 和 Great Smoky Mountains 两个国家公园，被誉为"美国最美的公路"
[3] exotic: 外来的
[4] tortuous: 弯曲的，多扭曲的
[5] alley: （花园或公园中的）小径
[6] indigenous: （尤指动植物等）当地出产的，土产的

The recreational character of the Blue Ridge Parkway was continually emphasized; travel on it was meant to be fun. Camping grounds and picnic areas were built at convenient intervals; there were coffee shops, service stations, and comfort facilities along the way. Occasionally there were clearly indicated halting spots from which foot trails led to nearby vantage[1] points, often for otherwise inaccessible views across to distant mountains. Although the Skyline Drive, as its name implies, stayed mainly along the crests, the Blue Ridge Parkway as a whole. By varying its elevation[2] from hollows to heights and back again, avoided the oppressive monotony[3] that so often comes from too much of a good thing.

The Blue Ridge Parkway, bringing to thousands each year the chance for a leisurely drive in unspoiled Appalachian hill country, is a notable addition to the national park system. Because of its great length and difficult terrain it took years to build but is now near completion. Abbott continued as superintendent[4] until 1949, when he was called away[5] to head the National Park Service field staff working on the Mississippi Rivev Parkway Survey. This endeavor, undertaken by the Park Service and Bureau of Pubic Roads pursuant to an Act of congress approved August 24, 1949, contemplated the feasibility of a national parkway along the Mississippi form its headwaters[6] in Minnesota's Lake Itasca all the way to the Gulf of Mexico—to be done, with federal aid, by the ten states bordering upon the river. Preliminary studies quickly indicated that such an entirely new parkway, built to national standards with an 850-foot[7] right-of-way, was utterly impracticable[8] in a region with so much development on the river front. Accordingly, it was decided to recommend that wherever possible existing state and county roads be incorporated into the line of a modified parkway. A Parkway Commission of one hundred representatives from the ten "river states" was organized at St. Louis in November 1949. They and the technical staff pursued with enthusiasm the study "to marry the road and river" and in January 1952 submitted their closing report to Congress. The recommended standards, including the 220-foot right-of-way accepted as a workable compromise but proposing other protective measures, were clearly and helpfully set forth (Figure 5.6) in the diagram entitled "parkway land controls in rural areas." Meanwhile, two national parkways in addition to the Blue Ridge had been authorized by Congress in the 1930s: the Colonial Parkway linking Jamestown and Yorktown in Virginia and the Natchez Trace, on the old Indian trail from Nashville to Natchez.

Strangely enough, considering the manifest advantages of the divided roadway technique illustrated in the upper Taconic, that principle was followed slowly elsewhere. The Blue Ridge and Colonial Parkways, and the Mississippi proposals, were thought to have insufficient traffic volume to warrant the divided ways.

---

[1] vantage：优势，有利的地位
[2] elevation：起伏，小山，高处，高地
[3] monotony：单调，无聊
[4] superintendent：监督（者），管理（者）
[5] be called away：被召到别的地方
[6] headwaters：（河川的）源流
[7] foot：英尺，1英尺＝0.305米
[8] impracticable：不可行的

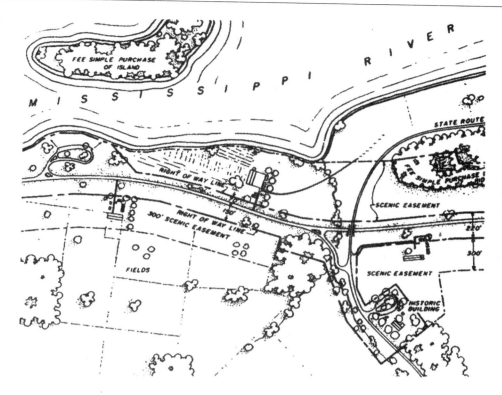

Figure 5.6 Mississippi River Parkway Survey: diagram of different tapes of land controls

Notes

1. Extracts from

Text

Newton, Norman T. *Design on the Land: The Development of Landscape Architecture* [M]. Cambridge: Harvard University Press, 1991: 596 – 615.

Pictures

Figure 5.5: Norman T. Newton. *Design on the Land: The Development of Landscape Architecture* [M]. Cambridge: Harvard University Press, 1991: 613.

Figure 5.6: Norman T. Newton. *Design on the Land: The Development of Landscape Architecture* [M]. Cambridge: Harvard University Press, 1991: 614.

2. Background Information

（1）Norman T. Newton 诺曼·托马斯·牛顿（1898~1992 年）。

景观设计师、教育家、作家、历史学家、批评家和评论家。出生于宾夕法尼亚州的科里

(Corry)。1919 年毕业于康奈尔大学，1920 年获得该校的风景园林硕士学位。在风景园林师布赖恩特·弗莱明的事务所工作了三年后，1923 年获得了罗马的美国学院颁发的罗马奖。1932 年在纽约成立了自己的事务所。1933~1939 年间担任国家公园署的助理风景园林师。1939 年牛顿进入哈佛设计学研究生院担任风景园林学副教授，除了在二战期间的服役外，他一直在该校教书，直到 1967 年退休。1957~1961 年期间任美国景观设计师协会主席。

（2）*Design on the land—The Development of Landscape Architecture.*《大地上的设计》。

1971 年出版，是关于世界风景园林史的著名著作，曾被评为"100 年来美国最具影响的十本风景园林专业书"之一。

该书回顾了风景园林规划设计历史；附有大量图纸、规划案例和照片，包括众多的参考书目，为综合性的书作。一经出版便被奉为此行业发展史的经典描述，并一直被当作教科书。在书中，牛顿把风景园林学和资源保护联系在一起，强调两者的原则都关注于"地块和及其用途之间的合理关系：简单来说，就是土地的明智使用。"

3. Sources of Additional Information

http://www.blueridgeparkway.org/
http://www.virginiablueridge.org
http://www.tfhrc.gov/pubrds/marapr00/along.htm

Further Reading C

National Parks & National Forests

By Raymond L. Freeman & Wayne D. Iverson

National Parks

The world's first national park was established in 1872 when Congress designated more than two million acres in Wyoming as Yellowstone National Park[1]. By the 1980s more than 300 diverse parks had been incorporated into America's world-renowned national park system. The concept of such reserves, pioneered in this country was a significant contribution to world civilization, and other nations eventually followed this inspiring model.

The history of our national parks and the profession of landscape architecture have long been intertwined[2]. Today, about 200 landscape architects have a vital role in providing stewardship for many of our nation's most cherished[3] natural and cultural resources through the National Park Service, a part of the U. S. Department of the Interior[4].

In 1864 President Abraham Lincoln signed legislation setting aside the magnificent Yosemite Valley and Mariposa Big Tree Groves to be held by the state of California for "public use, resort and recreation inalienable for all time." Frederick Law Olmsted was appointed a commissioner for these reservations and supervised the preparation of an influential report for their administration. In addition to his skillful plan for managing this park, he advocated a policy of establishing national parks across the nation and laid the foundation for our current national park system.

To fill the need for a separate division to oversee these parks, the secretary of the interior in 1910 recommended creating a Bureau of National Parks and Resorts specifically to employ landscape architects for their expertise in planning park development. Three years later the new position of general superintendent[5] of the national parks was created. Mark Daniels, a practicing landscape architect from California, filled this job for two years. His most significant accomplishment was bringing sensitive design into park administration and planning.

It was not until 1916, however that the National Park Service was formally established. That year the annual meeting of the American Society of Landscape Architects had passed a resolution supporting the National Park Service bill. The ASLA also addressed park issues requiring landscape

---

[1] Yellowstone National Park: 美国黄石国家公园，是世界上第一个国家公园
[2] intertwined: 使缠绕在一起
[3] cherish: 珍爱
[4] Department of the Interior: 内政部
[5] superintendent: 负责人

architectural expertise, including the delineation[1] of boundaries in consonance[2] with topography and landscape units and the development of comprehensive plans for managing natural and developed areas.

Stephen Mather, the first National Park Service director, took the ASLA recommendations to heart. He and subsequent directors, relying[3] heavily on landscape architects in guiding the development of the national park system together with an array of consultants including Frederick Law Olmsted, Jr., James Pray, Warren H. Manning, Harold Caparn and James Greenleaf (all former ASLA presidents), continued the accelerated work of establishing boundaries, campgrounds, buildings, roadways, bridges and other park facilities. Built in the early 1920s, the Ranger Club House, designed by staff landscape architect Charles P. Punchard, Jr., for the Yosemite National Park[4] in California, and Hull's Administration Building, now a museum in the Sequoia[5] National Park in California, became important symbols of this look for park buildings.

By 1922 the landscape architecture division of the National Park Service, originally located at Yosemite, was moved to San Francisco with Thomas Vint as its influential director. Under Vint the task of creating master plans for each park began in earnest, often with staff landscape architects taking the lead.

During the Depression[6], many landscape architects were employed temporarily to carry out Civilian Conservation Corps programs managed by landscape architect Conrad L. wirth, later to become Park Service director. Programs during this period shifted from a focus on natural sites to historic and parkways such as the Blue Ridge Parkway in Virginia and North Carolina and battlefields and historical parks such as Gettysburg in Pennsylvania.

Years of low funding and lack of interest left the national park system in a state of disrepair after World War II. Conrad Wirth proposed a major conservation program that would implement a nationwide effort to assess, reorganize and restore all parts of the park system. Known as Mission 66 and initiated in 1956 as a 10-year national park renewal, it led to significant federal support, which in turn raised the standards of all national parks. A planning staff, including three landscape architects headed by William Carnes, developed the program. Campgrounds[7] were restored and sometimes even relocated to more appropriate sites (a campsite in Yellowstone National Park was discovered to have been located on the path of migrating grizzly bears[8]), new roads and trails were built, visitor centers for the first time were established and management policies were instituted. By 1966 the park system had evolved fully into a nationally recognized network of parks.

In this time of explosive demands for use of the parks, the job of protecting fragile natural, historic, cultural and scenic resources has become increasingly complex for the National Park Service.

---

[1] delineation：画……的轮廓
[2] consonance：协调，和谐
[3] rely：依靠
[4] Yosemite National Park：（美国加利福尼亚州中部）约塞米蒂国家公园
[5] Sequoia：巨杉，红杉
[6] Depression：美国经济大萧条时期（1929～1933 年）
[7] Campground：野营地，露营场所
[8] grizzly bear：灰熊

There is probably no other governmental agency in which landscape architects have had more influence in this process.

National Forests

Although a federal Division of Forestry was established in 1886, it was not until 1891 that forest reserves were actually set aside. In 1905 they were transferred to the U. S. Department of Agriculture's new Forest Service under Gifford Pinchot's leadership and renamed collectively the national forests. These lands, incorporating national forests such as the San Gabriel Forest Reserve near Los Angeles and the Yellowstone Forest Reserve in Wyoming, both among the first set aside, now provide timber, water, wildlife, range and recreation for millions of people.

In 1917, after the Forest Service lost administration of Grand Canyon and other treasured lands to the newly established National Park Service (an action that has spurred rivalry[1] between the two agencies ever since), Frank Waugh was retained as a landscape architectural consultant. Waugh strongly advocated the employment of "landscape engineers" to prepare plans for recreational areas such as campgrounds, picnic areas and even resorts to increase public use of the national forests, which, previously, had been limited to timber harvesting, grazing[2] and water and power production. Arthur Carhart was hired by the Rocky Mountain regional office in 1919 as the first landscape architect for the Forest Service. He initiated forest recreation plans and pioneered wilderness area concepts, but, finding the Forest Service too slow to incorporate ideas, Carhart resigned in 1922.

With the advent of public works programs in the early 1930s, the Forest Service sought to improve its design skills, and in 1934 the agency began hiring landscape architects to design recreational sites and ranger stations for the national forests. Numerous examples of these sturdy rustic-style[3] developments still exist.

Among the 1934 recruits was R. D. Arcy Bonnet, a Harvard graduate who rose within two years from a position with the Monongehela National Forest in West Virginia to one at headquarters in Washington, D. C. Bonnet accompanied consultant Albert Davis Taylor on his second inspection tour of the national forests in 1936. With a decrease in programs, Bonnet was reassigned to the California region in 1939, where he became recognized as the dean of Forest Service landscape architects after more than 30 years' service.

In 1958 Operation Outdoors, a major five-year program to rehabilitate recreational sites, became a significant turning point for landscape architecture in the national forests. For the next two decades the number of landscape architects concentrating on developing the forests' recreational areas steadily increased to a peak of nearly 300 in 1980. In the 1960s several moved into recreation staff positions. By the 1970s some had assumed responsibilities as district rangers and land management planners, and by the 1980s two advanced to forest supervisor positions.

Influential in this expansion and integration of landscape architects was Edward H. Stone II, who

---

[1] rivalry: 竞争，敌对
[2] grazing: 放牧
[3] rustic-style: 乡村风格

become chief landscape architect in 1965. Spurred by the environmental movement and controversies over clear-cutting forests, and with support from top management, he launched an innovative visual resource management program, including training, land management and monitoring functions. The program became a model for other federal agencies. The research of R. Burton Litton, professor of landscape architecture at the University of California, Berkeley, played an important role in this new program, which maintained a set of performance standards set up by Forest Service landscape architect Warren Bacon and others to manage scenic quality and to protect natural resources. For instance, standards guiding the harvesting of trees dictated the size and shape of the area to be cut. The system became a cornerstone[1] in developing other visual resource management systems in the United States and abroad and was the basis for a popular series of Forest Service handbooks. Another major advance brought about by the Forest Service was the development of sophisticated computer applications for "seen area" mapping and computer perspectives to determine visual impacts.

Landscape architecture in the Forest Service has evolved from a few young men engaged primarily in site planning to several hundred men and women actively involved in managing all the resources of 185 million acres of prime American landscape. Today, no new forest lands are being added to the national stock, but where once national forests were broken up to make management easier, some now are being combined to make management more efficient.

Notes

1. Extracts from

Text

Freeman, Raymond L. & Wayne D. Iverson. National Parks & National Forests [C]. 南京林业大学园林学院. 园林专业英语. 南京：南京林业大学自编讲义：2005：7-13.

2. Background Information

(1) *National Parks*, Raymond L. Freeman（美国著名风景园林师和理论家，获 1984 年"ASLA 主席勋章"）。

(2) *National Forests*, Wayne D. Iverson（美国著名风景园林师）。

以上两篇短文均选自美国风景园林杂志 Landscape Architecture，分别回顾了美国国家公园和国家森林的发展历史。

3. Sources of Additional Information

http://www.nationalforest.org/

http://gorp.away.com/gorp/resource/us_national_forest/main.htm

---

[1] cornerstone：基础

Exercises

1. Questions（阅读思考）

认真阅读本单元文章，并讨论以下问题。
（1）城市公园的价值。
（2）公园规划设计的原则。
（3）谈谈对 Parkway 的理解。
（4）联系我国情况，谈谈城市公园及休闲地系统以及根据其规模和分布的分级分类情况。

2. Skill Training（技能训练）

（1）查找并浏览有关城市公园及休闲地内容的英文网页，选择感兴趣的主题仔细阅读，并做图文并茂的 PowerPoint 文件，用英文向同学介绍。
（2）以"城市公园及休闲地"为主题，组织交流和讨论。

Tips

城市公园的发展历程

　　世界造园已有 6000 多年的历史，而现代公园的出现只是近一二百年的事。无论是旧约全书中的"伊甸园"，还是可考的巴比伦空中花园（Hanging Gardens，BC. 604~562 年），均与公众的现实生活无关。但是，这并不能阻止古代城市中普通市民的游憩活动。在古希腊、古罗马城市中，公众的户外游憩活动常常利用集市、墓园、军事营地等城市空间。中世纪的欧洲城市多呈封闭型，城市利用城墙、护城河及自然地形基本上与郊野隔绝，城内布局十分紧凑密实，城市公共游憩场所除了教堂广场、市场、街道，常转向城墙以外。

　　文艺复兴时期，欧洲各国的不少皇家园林开始定期向公众开放，如伦敦的皇家花园（Royal Park）、巴黎的蒙古花园（Parc Monceau）等。1810 年，伦敦的皇家花园摄政公园（Regent Park）一部分被用于房地产开发，其余部分正式向公众开放。

　　英国工业革命开始后，资本主义迅猛发展，工业化大生产导致城市人口急剧增加，城市用地不断扩大，在社会财富迅速积聚的同时，卫生与健康环境严重地恶化。在这样的社会条件下，资产阶级对城市环境进行了某些改善，把若干私人或专用的园林绿地划作公共使用，或新辟一些公共绿地，称之为公共花园和公园。从 1833 年起，英国议会颁布了一系列法案，开始准许动用税收建造城市公园和其他城市基础设施，1843 年，利物浦动用税收建造了公众可免费使用的伯肯海德公园（Birkinhead Park），标志着第一个城市公园正式诞生。海德公园坐落在伦敦市区的西部，面积有 159.86 公顷，是伦敦皇家公园（Royal Park）中最大、最著名的一个，原是威斯敏斯特教堂的一块采邑庄园，18 世纪前是英王的狩鹿场，19 世纪后期，伦敦发展迅速，位于西郊的海德公园逐渐成为市中心区域。由于园内栽植了许多树木花草，空气新鲜，被称为"伦敦的肺脏"。

　　受英国经验的影响，法国、德国和其他国家群相效仿。在很长一段时间内，美洲大陆的欧洲后裔们的造园以殖民式为样板。直至 1858 年，在美国设计师唐宁、奥姆斯特德的竭力倡导下，美国的第一个城市公园——纽约中央公园（Central Park of New York）在曼哈顿岛诞生，传播了城市公园的思想。此后，美国的城市公园的发展取得了惊人的成就。

　　19 世纪下半叶，欧洲、北美掀起了城市公园建设的第一次高潮，称之为"公园运动"（Park Movement，1843~1887 年）。有学者对 1880 年美国统计资料的研究显示，当时美国的 210 个城市，九成以上已经记载建有城市公园，其中 20 个主要城市的城市公园尺度在 150~4000 英亩之间。在公园运动时期，各国普遍认同城市公园具有五个方面的价值，即保障公众健康、滋养道德精神、体现浪漫主义（社会思潮）、提高劳动者工作效率、促使城市地价增值。

　　1880 年，奥姆斯特德等人设计的波士顿公园体系，突破了美国城市方格网格局的限制，以河流、泥滩、荒草地所限定的自然空间为定界依据，利用 200~1500 英尺宽的带状绿化，将数个公园连成一体，在波士顿中心地区形成了景观优美、环境宜人的公园体系（Park System）。如今，该公园体系两侧分布着世界著名的学校、研究机构、学术馆和富有特色的居住区。

　　盖伦·克兰茨（Galen Granz，1982 年）认为自 19 世纪中叶以来，美国公园的发展经历了四个主要阶段：游憩园（The Pleasure Ground）、改良公园（The Reform Park）、休闲设施（The Recreation Facility）和开放空间系统（The Openspace System）。

　　游憩园流行于 1850~1900 年间，其发展至少部分起因于对新兴工业城市肮脏而拥挤的环境的反应。这类公园的典型样式就是浪漫主义时期英格兰或欧洲贵族的采邑庄园，通常坐落在城郊，是刻意为周末郊游而设计的，以大树、开阔的草地、起伏的台地、蜿蜒的步行路及自然主义风格的水景为特征，将原野及田园风光理想化。由于这些公共设施可供所有的社会群体享用，人们希望工人可以在这里通过户外游憩活动来保持健康，希望中产阶级的行为规则能够影响到贫民。

改良公园出现在 1900 年左右，是改良主义和社会工作运动的产物，为改善劳动者的生活条件而建。改良公园位于城市内部，是第一批真正意义上的邻里公园，最主要的受益者是紧邻公园的儿童和家庭，其最重要的特征是儿童游戏场。对浪漫主义风景审美的精英主义价值观的反对，导致了改良公园设计中的僵硬刻板的功能主义，其特征就是利用直线和直角构成硬质铺装、建筑和活动分区的对称布局。

随着社会工作运动的理念融入主流社会，1930 年左右休闲设施开始在美国城市和城镇中发展起来，并成为公园和社会改良目标之间的纽带。它强调体育场地、体育器械（休闲这个词变成了高档体育设施的同义词）和有组织的活动。随着郊区的扩张和小汽车数量的剧增，新型的和更大规模的公园被建设起来以提供各种各样的球场、游泳池和活动场地。公园的吸引范围和服务区域增大，小汽车成为去公园的主要交通工具。体育锻炼——特别是团体运动项目被认为对于在困难时期保持斗志是极其重要的，同时对于个体和公众健康也同样重要，这大大鼓舞了早期的改良者。

自 1965 年开始发展起来的开放空间思想，是将分散的地块，如小型公园、游戏场和城市广场等联系成一个系统——至少在理论上是如此。开放空间思想与城市复兴运动一起发展壮大，而城市复兴是城市活力的一种体现。集中分布的邻里公园只是开放空间体系中可供人们消磨休闲时光的各类地点（如娱乐园、购物中心、跳蚤市场、街道集市、州立和区域公园）中的一种。如今大部分公园都包含有上述四个历史时期公园的成分，很少有某种单纯类型的公园。

霍尔格·布劳姆（Holger Blom）认为公园在城市中的地位与作用是："公园能够打破大量冰冷的城市构筑物，作为一个系统，形成在城市结构中的网络，为市民提供必要的空气和阳光，为每一个社区提供独特的识别特征；公园为各个年龄的市民提供散步、休息、运动、游戏的消遣空间；公园是一个聚会的场所，可以举行会议、游行、跳舞，甚至宗教活动；公园是在现有自然的基础上重新创造的自然与文化的综合体"等。1938 年，布劳姆担任了瑞典斯德哥尔摩公园局的负责人，任期 34 年。他改进了前任的公园计划以增加城市公园对斯德哥尔摩市民生活的影响。布劳姆的公园计划反映了那个时代的精神：城市公园要成为完全民主的机构，公园属于所有人。在这一计划的实施过程中，斯德哥尔摩公园局成为一群优秀的年轻设计师们成长的地方，也促成了"斯德哥尔摩学派"的形成。一些人成为瑞典风景园林界的主要人物，其中最出色的一员是 E. 格莱姆（E. Glemme）。格莱姆 1936 年进入公园局，作为主要设计师一直工作到 1956 年，在形成"斯德哥尔摩学派"的大多数作品中，都留有他的手笔。

沿着 Norr Malarstrand 的湖岸步行区，可能是所有"斯德哥尔摩学派"的作品中最突出的。这是由一系列公园形成的一条长的绿带，从郁郁葱葱的乡村一直到斯德哥尔摩市中心市政厅花园结束。它的景观看起来仿佛是人们在乡间远足时经常遇到的自然环境，如弯曲的橡树底下宁静的池塘。这个公园能够用来进行斯德哥尔摩人所喜爱的休闲活动。

在许多情况下，选择作公园的基地常常是不可接近的沼泽或崎岖的山地地貌，看上去几乎没有再创造的价值。"斯德哥尔摩学派"的设计师们以加强的形式在城市的公园中再造了地区性景观的特点，如群岛的多岩石地貌、芳香的松林、开花的草地、落叶树的树林、森林中的池塘、山间的溪流等。"斯德哥尔摩学派"在瑞典风景园林历史的黄金时期出现，它是风景园林师、城市规划市、植物学家、文化地理学家和自然保护者共有的基本信念。在这个意义上，它不仅仅代表着一种风格，更是代表着一个思想的综合体。

1936～1958 年是"斯德哥尔摩学派"的顶峰时期。20 世纪 60 年代初，大量廉价、预制材料构筑的千篇一律的市郊住宅被兴建，许多土地被推平，地区的风景特征被破坏。斯德哥尔摩公园的质量后来下降了，但一些格莱姆和"斯德哥尔摩学派"其他人的作品今天仍然可以看到。如今，这些公园的植物已经长大成熟，斯德哥尔摩的市民从前一代人的伟大创举中获益无穷。

"斯德哥尔摩学派"的影响是广泛而深远的。同为斯堪的那维亚国家的丹麦和芬兰，有着与瑞典相似的社会、经济、文化状况。由于在第二次世界大战中遭到了一定的破坏，发展落后于瑞典。战后，这些国家受"瑞典模式"的影响，也成为高税收高福利国家，"斯德哥尔摩学派"很快在城市公园的发展中占据了主导地位。同时，丹麦的风景园林师在城市广场和建筑庭院等小型园林中又创造了自己的风格，他们的设计概念简单而清晰。最著名的设计师是 C. Th. 索伦森（C. Th. Sorensen），他善于在平面中使用一些简单几何体的连续图案。

"斯德哥尔摩学派"通过丹麦，又影响到德国等其他一些高福利国家。第二次世界大战后，大批德国年轻的风景园林师到斯堪的那维亚半岛学习，尤其是到丹麦，带回了斯堪的那维亚国家公园设计的思想和手法，通过每两年举办一次"联邦园林展"的方式，到1995年在联邦德国的大城市建造了20余个城市公园，著名的有慕尼黑的西园（Westpark）和波恩的莱茵公园（Rheinpark）。

城市公园在诞生后，经过了一系列的发展过程，诸如造园新风格的酝酿、城市公园体系的确立、城市公园运动等，其中城市公园运动为城市公园体系的确立和城市公园系统的规划奠定了坚实的基础。波士顿公园体系的成功，对城市绿地的发展产生了深远的影响。如1883年的双子城（Minneapolis，H. Cleveland）公园体系规划等。受此影响，1900年的华盛顿城市规划、1903年的西雅图城市规划，以城市中的河谷、台地、山脊为依托，形成了城市绿地的自然框架体系。"斯德哥尔摩学派"的影响也是广泛而深远的。

1993年，国际公园与游憩管理联合会（IFPRA）亚洲太平洋地区大会和日本公园绿地协会第35次全国大会在日本茨城县水产市联合召开，大会讨论的主题是"公园能动论"（Park Dynamism），其主要论点是：公园不仅应为游人提供游憩设施，满足居民消费需要，而且应主动地向群众宣传当前人类环境面临的问题，以及公园绿地在健全生态和提高城市化地区对社会的经济、人口增加所造成压力的支撑力方面的重要作用。通过提高群众对公园绿地系统重要性的认识，进而达到从舆论、政治、规划、法律等方面，为公园绿地的保护与发展，为城市与自然的协调共存创造有利的条件。即城市公园在城市中应具有多样的价值体系，如生态价值、环境保护价值、保健休养价值、游览价值、文化娱乐价值、美学价值、社会公益价值与经济价值等。

Unit 6

Historical & Cultural Landscape

- 风景园林历史文化保护原则
- 保护方法
- 实例

The heritage of garden art can make a considerable contribution towards general cultural education. It is of quite special importance for professional training since it contains the grammar of landscape design from the earliest beginnings up to modern times.

By Dušan Ogrin

The Conservation Policy

By Sheena Mackellar Goulty

> 花园艺术是国家以及世界历史文化遗产的重要组成部分，比其他艺术形式所包含的文化和社会历史方面的内容更多。许多园林是有历史价值的建筑的背景，还有的其本身即是历史名胜。它们既是休闲资源又是教育资源，在世界范围内越来越被视为重要的国家财富。
>
> 本文论述了对遗产文化花园的保护政策。全文可分保存、保护、复原、重造、重建及保护政策的制定六部分内容。

Conservation

The conservation[1] of gardens is the deliberate, planned and thoughtful treatment of gardens, managed to balance their protection and upkeep with the realization of their full potential for enjoyment and education. The value and attraction of any one garden depends on the approach taken towards its conservation, and the method chosen will depend on its type and size, its state of repair, the availability of historical information, its potential as a recreational or educational resource and how it is to be used. Some degree of restoration, whether full or partial[2] reinstatement[3] of an original design from an earlier period or periods, or repair or renewal[4] of some features of the garden, or just improvement to the general condition of the garden, will usually be necessary, but a "conserve as found" policy is increasingly being advocated.

● 保存

保存花园就是周密地、有计划地、细致地对待花园，想方设法平衡其保护和维护，实现花园供人欣赏、给人教益的全部潜力。

某种程度的恢复，或是全部或部分恢复原来早期的设计，或是修缮更新花园的某些特色，或只是改善花园的一般状况，通常是必要的，但"依所发现的样子保存"这一政策正得到越来越多的提倡。

---

[1] conservation：保护，保存（以免损失、浪费、损坏等）
[2] partial：部分的，不完全的
[3] reinstatement：使复原位或原状，恢复
[4] renewal：更新，再始，换新

Preservation

In garden terms a strict "conserve as found", or preservation[1], policy is rarely appropriate because of the ephemerality[2] of the principal medium, the plants. Where a garden exists only as a monument and is no longer actively form of the garden is still used, preservation is possible.

One example is the King's Knot, Stirling[3], where the visible in the contours[4] of the ground, which is now grassed over. The garden is preserved from further decay by simple maintenance of the greensward[5]. To "conserve as found" in a garden still in active use involves sympathetic care and maintenance, but if planted elements are not renewed its character will gradually be eroded. This may be acceptable where a garden's main features are architectural or where restoration measures would threaten the garden's value as a historical document. It may even be acceptable for a limited number of years until major planted elements such as avenues are no longer viable, but in general some renewal or repair is essential in a creative conservation programme.

Restoration

Restoration, the attempt to return a garden to its original form or forms and to a viable state of upkeep within a normal maintenance routine, is usually the preferred means of conservation where a garden has suffered from neglect or from changes due to inadequate management or lack of resources. The standard of historical restoration often leaves a lot to be desired. The reasons are many and various. First, there is often a lack of historical documentation and evidence. Second, information obtained from historical research and survey work is subject to mistakes and misinterpretation[6]. Third, field survey work, although an essential tool, usually gives only approximate clues to historical fact, and the correlation and interpretation of information require imaginative guesswork backed by persistent cross-checking. Fourth, our technical

---

[1] preservation：保护，储藏，维持，留存
[2] ephemerality：短暂，瞬息
[3] King's Knot, Stirling：位于苏格兰中部斯特林城堡西侧，呈八边形土台状，是中世纪皇家花园遗址
[4] contour：（海岸、山脉等的）轮廓、周线、围线，（地图、图案等）各着色区域间的区分线
[5] greensward：草皮
[6] misinterpretation：误译，误解，误以为

knowledge pertaining[1] to the re-creation of authentic[2] detail is limited, particularly in planting design. Fifth, historical interests have to be balanced with others-horticultural, artistic, recreational, educational and ecological-and this balancing act is often short-sighted in its policy and implementation. Finally, all these problems are compounded by inadequate resources of labour and finance, and in view of the difficulties involved, it is wise to proceed with caution, essential to keep clear records, and desirable, as far as possible, to make restoration measures reversible[3] where there is any doubt as to their accuracy.

最后，所有这些问题还要加上资金和劳动力资源的不足的困难。鉴于所涉及的困难，一定要审慎行事，保持清楚的纪录是绝对必要的，在对其准确性有疑问的时候，最好尽可能采取可逆的复原措施。

The most straightforward form of restoration is the repair of structural fabric[4], and renewal of planting. Where simple repairs can be made, authentic details and materials should be used and matched to the original fabric. Where details cannot be matched because of the loss of original fabric and lack of historical information, the repair should be dearly[5] recorded. Sometimes a repair can be delineated[6] without detriment[7] to the overall appearance so that there is no mistaking new fabric for the original. In the rebuilding of the boundary wall to the Charles Carroll garden, Annapolis[8], for example, the old brick surface stands proud of the new. Where planting needs attention the issues are always complicated. The cost of maintenance, for instance of 19century bedding schemes, can be prohibitive and lead to a decision to restore only the basic form of the garden without replanting herbaceous or smaller shrub material, Usually such a decision is reversible but the importance of accurate recording with planting plans and photographs both before and after restoration, to allow the decision to be reviewed in future, cannot be over-emphasized.

最直接的复原形式是修缮结构框架，以及更新植物。如果可以简单修缮，必须根据可靠的翔实资料来匹配最初的结构框架。若因缺失原有的结构及历史资料而无法获得相匹配的翔实资料，则应明确记录维修情况。

The treatment of avenues is a good example of the difficulties which must be faced in deciding to what degree the replanting of the more permanent features is appropriate. The

为了使植物具有更稳固的特征，究竟要进行多大程度的重新种植，这是一个难题，如何处理林荫大道就是一个很好的例子。

---

[1] pertain：(与 to 连用) 属于，关于，适合于
[2] authentic：真实的，可信的，可靠的
[3] reversible：可逆的
[4] fabric：结构，构造，构造物
[5] dearly：极，非常
[6] delineate：描画，描绘，描写
[7] detriment：损害，伤害
[8] Charles Carroll garden, Annapolis：位于美国马里兰州首府安纳波利斯，是 Charles Carroll 的私家花园，是仅存的 15 个美国独立宣言签署者出生地住宅

form of an avenue depends for its effect on the straight, parallel lines of regularly spaced and even-aged trunks. When trees die or otherwise need replacement, there are various alternatives to consider. The first alternative is to leave the gaps in the line until all the trees have to be felled, in order to preserve the actual historic component as long as possible. A toothless hag[1] has character but lacks the beauty of youth. The second alternative therefore is to replant the gaps, but in this case the regular appearance of trees of the same age will be lost. The third alternative is to replant on either side of the existing avenue, and fell the old trees when the new avenue is sufficiently mature, but the loss of both the historic line and the proportions of the avenue will result. The fourth alternative is simply to clear fell and replant. Except where the trees have been subject to disease, it is usually feasible to use the same species, or even to propagate from the existing stock so that in time the effect will be as near that of the original conception as possible. The decision rests on whether it is more important to try to retain the garden's artistic or historic integrity and each case will depend on its merits.

The restoration of a garden may be compromised for reasons other than the nature of the plant material or the limitations on research and available historical information. Few heritage gardens remain completely private; most have to adapt to an influx[2] of visitors as they open to the public. This may lead to a policy decision to use different materials to reduce wear and tear, or to alterations to reduce the visual impact of large numbers of visitors. In other gardens the loss of one feature, which it is not practical to replace, may alter the balance of the design so that changes to the layout are deemed more in keeping with the spirit of the original. At Giverny[3] the main paths used by visitors have been resurfaced[4] with concrete instead of gravel, but using the same gravel aggregate[5] so that the visual intrusion[6] is minimized. At Stourhead[7] the path has been moved away from the lake and

---

[1] hag: 女巫，老丑婆
[2] influx: 流入，注入
[3] Giverny: 法国巴黎西北的一个村庄，以著名印象派画家莫奈住宅及花园而著称
[4] resurface: 铺（路等）之新表面，换装新面
[5] aggregate: 聚集，聚集成团
[6] intrusion: 闯入，侵扰
[7] Stourhead: 位于英国威尔特郡，曾经是 Stourhead 家族庄园，建有典型的18世纪英国自然风景园

up into the wood, both to counter erosion of the lakeside and to hide visitors from each other, so that the peace and tranquility[1] of the eighteenth-century vision remains intact.

Re-Creation

In the examples described above, the restoration has been consciously modified. Modifications should always be clearly recorded, not only for future use by professionals involved in restoration and upkeep[2], but also as interpretative information available to the public, so that changes are not passed off as part of the original design. When well presented, the reasons behind particular restoration decisions and information on the process of restoration itself add considerably to the interest of the garden as a whole, and aid the imaginative interpretation of what the garden once was. In its most extreme form a modified restoration is sometimes termed a "restoration-in-spirit". The use of the term "restoration" is scarcely justified here because the primary objective is not accuracy, but rather an attempt to reinterpret the original intentions of the designer. This is of dubious[3] value in conservation terms, and is closer in concept to a re-creation. The garden at Edzell Castle[4] is a recreation which attempts to express something of the spirit of an early seventeenth-century garden. At Pitmedden[5] the re-creation purports[6] to be a more accurate representation of a 17th century garden, and as such is a "period garden" re-created on an existing site. The dividing line is thin, and, even when its origins are well presented, the re-creation of a garden, particularly a period garden, on an existing site can be misleading. A period garden should properly be seen as one means of explaining to the public the features of a garden of a particular historic period. They are liable to become "heritage gardens" only in terms of the perception of future generations or their intrinsic[7] artistic merit, if at all, and their creation on a genuine historic site can only be confusing.

---

[1] tranquility：安静，平静，宁静
[2] upkeep：保养
[3] dubious：（指事物、动作等）可疑的，其价值、真实性等有问题的
[4] Edzell Castle：位于苏格兰东部安格斯郡，以城堡遗址和17世纪初花园而著称
[5] Pitmedden：位于苏格兰阿伯丁郡，以17世纪法国规则式园林风格的花园而著称
[6] purports：意谓，声称，声言
[7] intrinsic：（指价值或性质）固有的，内在的

Reconstruction

Where a garden has been completely, or almost completely, destroyed but is both judged to be of great significance and sufficiently well-recorded, restoration may take the form of a reconstruction[1]. A reconstruction differs from a re-creation in its attempt to be a historically accurate re-modelling of a garden which was known to exist on the site. The recent development of new techniques of archaeological investigation has facilitated reconstructions where previously they would have proved impossible. The William Paca Garden[2] in Annapolis, Maryland, has been reconstructed in just such a manner, from archaeological evidence and little else. Before restoration the garden was buried under a hotel and three metres of fill. It is inevitable, in a complete reconstruction of this type, that the detail will not replicate the original garden exactly, unless there is a wealth of documentary evidence. How gratifying[3], therefore, when a clump of bulrushes[4] appeared in the "eye" of the reconstructed fish-shaped pond! It is often planting details which are the most difficult to reproduce. Other details can at least be copied from known examples authentic to the period.

At Het Loo[5] however it is the planting which has been painstakingly researched and copied, while the construction details are an unashamed deception, re-creating the appearance of a historic garden by using modern construction techniques and materials under the surface. This reconstruction is unarguably modern, but has been justified in terms of its setting for the palace as a museum and the numbers of visitors it has already attracted.

Policy Decisions

Wherever reconstruction, re-creation or a modified restoration are implemented the destruction of archaeological and field evidence is the inevitable result. Careful consideration must therefore be given to the importance of the site as a

● 重建

　　如果一座花园已完全或几乎完全被毁，但具有重大意义而又有充分翔实的记载，则可以采取重建的形式进行复原。重建是努力按照史实精确地重塑一座曾经存在于此地的花园，因此不同于重造。

● 制定政策

　　无论在哪里重建、重造或复原改造，都不可避免地会破坏考古和场地的证据。

---

〔1〕 reconstruction：重建，再建
〔2〕 William Paca Garden：位于美国马里兰州首府安纳波利斯，18世纪私家花园，一度完全消失，20世纪60～70年代重建
〔3〕 gratifying：欣喜
〔4〕 bulrush：芦苇
〔5〕 Het Loo：荷兰皇家园林，巴洛克风格

recreational, educational or archaeological resource, and whether the loss of source material can be justified by the gains to be had from the treatment proposed. <u>Where much of the garden still exists above ground and less drastic[1] action is needed to restore it to, or close to, its original form, hesitation can result in further loss. Where most of the evidence is below ground, the situation is reversed and, if it is not possible to make provision[2] for a very detailed investigation and recording of the site before undertaking restoration on a major scale, conservation of the site as it stands is preferable.</u>

若花园的大部分仍存留在地面上,又无须大动干戈使它还原或接近最初的样子,则应立刻进行复原,迟疑不决只能造成进一步的损失。若大部分证据都在地下,情况则正好相反;如果不能在进行大规模修复之前提供花园所处地点的详尽调查和记录,则最好对花园遗址进行保存。

Notes

1. Extracts from

Text

Goulty, Sheena Mackellar. *Heritage Gardens: Care, Conservation and Management* [M]. London and New York: Routledge Inc., 1993: 54-59.

2. Background Information

(1) Sheena Mackellar Goulty 希娜·麦克拉·古提。

希娜·麦克拉·古提(1952年~),于爱丁堡大学获得风景园林硕士学位,专门从事园林历史文化保护的研究。

(2) *Heritage Gardens: Care, Conservation and Management*《遗产文化园林:照管、保护和管理》。

1993年出版。全书包括遗产文化园林的背景知识、保护方法、维护及管理、实例研究四部分内容,形成一个完整有序的体系。本文节选自该书第二部分保护方法(The conservation process)内容,有删节。

3. Sources of Additional Information

King's Knot, Stirling

  http://members.aol.com/corvus1999/not.htm

  http://www.beloit.edu/~chem/ScotPic/pages/StirlingKnot.html

  http://www.clanstirling.org/Main/lib/lib.shtml

Charles Carroll garden, Annapolis

  http://www.charlescarrollhouse.com/

Giverny

  http://en.wikipedia.org/wiki/Giverny

Stourhead

  http://en.wikipedia.org/wiki/Stourhead

Edzell Castle

  http://www.rampantscotland.com/visit/blvisit_edzell.htm

Pitmedden Garden

---

[1] drastic:(指行动、方法、药品等)激烈的,猛烈的

[2] provision:准备,防备(尤指为未来的需要者)

http://www.gardens-guide.com/gardenpages/_0198.htm

William Paca Garden

http://www.annapolis.org/index.asp?pageid=51

http://www.bsos.umd.edu/anth/arch/PacaGarden/reconstruction.htm

Het Loo

http://www.uvm.edu/pss/ppp/gardens/gm0404.htm

Further Reading A

The Heritage of Garden Art

By Dušan Ogrin

Spiritualized Nature

The cultures of the Far East have evolved an entirely different pattern of adaptation of nature for man's residential requirements: a pattern that reflects a primordial[1] concept of the world and of the active forces that govern its course. It proceeds from the assumption that physical nature is an embodiment of supernatural powers that determine the sequence of natural events and influence man's life. Such beliefs in the spiritual powers of trees, rocks, mountains and water are, in fact, typical of the mythologies[2] of the most varied regions of the world. To the present day, trees have remained the object of numerous folk rituals[3], usually connected with certain seasons of the year. The best example is the Christmas tree, which represents the survival of a tradition dating from pre-Christian times: Ancient Egypt venerated[4] sacred lakes, sacred tree and holy groves[5]. These later reappeared in Ancient Greece and Rome, and then again in the Italian Renaissance, together with the grotto[6], another dwelling-place of mythical[7] beings.

But only in China and Japan has this mythological outlook persisted through the centuries and decisively marked the character of landscape design. This might be regarded as a quest for security, which only reverence for hidden spiritual powers would ensure, particularly efficacious[8] if the spirits were attracted to the vicinity of human habitations. The central motif[9] of Chinese garden art, *shanshui*, illustrates the structure of the world, with rocks as its chief component. According to Daoism, rocks represent the skeleton and water represents the life-blood of the earth. All matter is composed of tiny particles interlinked and kept in constant motion by hidden energy, the active fluid of life, *qi*. In the visual arts this principle is usually represented by vortices. Whirling, wavy[10], rhythmically[11] agitated strokes. But it is most clearly

---

[1] primordial: 原生的，原始的，最初的
[2] mythology: 神话学，神话的总称
[3] ritual: 仪式，典礼，宗教仪式
[4] venerate: 对……怀有敬意，崇敬
[5] grove: 丛树，小树林
[6] grotto: 洞穴（尤指人造的花园中的洞室）
[7] mythical: 神话的，仅存在于神话中的
[8] efficacious: （不用以指人）有效的
[9] motif: （艺术作品的）主题，主旨
[10] wavy: 有波状卷曲的
[11] rhythmically: 有节奏地，有韵律地

typified by the furrowed[1] limestone rocks that have predominantly defined the character of the Chinese garden since its early beginnings.

The concept of the presence of supernatural forces in the ground has contributed towards the establishment of the principle of *wu-wei* (noninterference with nature), which in turn requires a maximum adaptation to nature. Above all, this principle has contributed towards a design approach that consequently avoids the transformation of natural features into regular, geometric patterns. Japanese landscape design has been decisively influenced by the Shintoist belief that rocks, trees and mountains are inhabited by *kami*, good spirits. The transposition[2] of selected natural elements into the garden suggests a neighbourhood of supernatural beings and therefore a safer life. This concept predetermines the choice of the chief compositional elements of the garden as well as their formal character, which mainly keeps within the limits of the natural.

Even the architectural concept is attuned[3] to the spiritual orientation of residential space. In opposition to Western civilization, architecture is not a self-contained entity, removed from and dominating the environment. Wherever possible, in Sino-Japanese building tradition, the architectural and garden spaces interpenetrate and fuse into a whole. The building penetrates into the garden, at the same time opening towards it and embracing it. While moving through the garden, man finds himself at every step in the midst of a spiritualized world of which he is an integral part. Therefore this world is not designed for observation from the outside, from a certain viewpoint or along a certain axis, according to the tenets[4] of the European Renaissance and Baroque, but rather represents a microcosm[5] one is supposed to fuse with and experience in one's own self.

In this way, two great civilizations evolved entirely different, opposing patterns of man's arrangement of space. On the one hand is the concept of geometrized and ultimately ordered space born out of the spirit of classical antiquity[6] that imagined the Olympic gods in human shape. Thus man has raised himself above nature, confronted her and tried to dominate her. The result of such an approach is orderly, subdued, humanized nature. The East, on the other hand, has appropriated the concept of subordination[7] to nature, resulting in a pattern of spiritualized, deified nature that, in spite of all transformations and idealizations, still preserves the aspect of primordiality.

The Heritage of Garden Art

In spite of considerable losses throughout history, the architectural heritage is comparatively well preserved today. Even some buildings that are several thousand years old have survived, though not always in their original integrity. The fate of garden art is quite different. Its historical heritage is, compared to that of architecture proper, much scantier. There are numerous reasons for this situation. First of all, the structure of a garden depends on elements which are subject to growth and changes dictated by the ebb and flow of nature. As soon as gardens are denied the necessary care, the

---

[1] furrow: 犁沟，畦
[2] transposition: 转换，换置，换位，移调
[3] attune: 使调和，使一致，使适合
[4] tenet: 主义，信条，教理，教条
[5] microcosm: 被认作人类或宇宙之缩影的某事物（尤指人），小天地
[6] antiquity: 古代，古老，古代的习俗，古事，古迹，古物
[7] subordination: 下级的，次要的，附属的

vegetation goes its own way, new plants take over and in a generation the layout of the garden may be radically affected. If neglect lasts even longer, the decay of more solid components sets in, such as fountains, retaining walls, flights of stairs, paths, statues. In regions of humid climate the decay is far quicker than in dry zones, such as the Mediterranean or certain parts of Asia.

Even worse damage than neglect has been caused to garden art by the reconstruction and enlargement of villas or castles and quite particularly by the growth of towns. A great number of precious designs have been swamped by the changing taste of the time, caused by the prevailing social and political ideas, such as in 18th century England. In that period England irretrievably[1] lost practically all her rich Renaissance and Baroque legacy and this example was followed by numerous European countries.

In several instances great historical gardens have been transformed into public grounds, losing their historic authenticity[2]. Such transformed parks include Madrid's Famous Buen Retiro, the gardens of the Villa Borghese in Rome and the Villa Torlonia in Frascati.

According to the Commission for Historic Gardens, which functions as a joint body of the IFLA (International Federation of Landscape Architects) and the ICOMOS (International Committee on Monuments and Sites), some 2000 valuable historic gardens have been preserved around the world. Roughly two-thirds of this number belong to the European heritage. Considering the enormous number of gardens established throughout history in the world's creative periods, this is a rather modest remainder. The importance of the surviving heritage is therefore all the greater. It is not only the embodiment[3] of a unique artistic creativity, but also a complement to the general cultural and historical image of a period or nation. The garden art of certain countries shows a national identity, to at least the same degree as architecture, painting and all the other branches of visual art.

Gardens or landscapes marked by a historical style often supply the indispensable setting for valuable architectural monuments. With which they blend into complexes of unique cultural value. Therefore the heritage of garden art can make a considerable contribution towards general cultural education. It is of quite special importance for professional training since it contains the grammar of landscape design from the earliest beginnings up to modern times.

The above-mentioned points of importance demand the attention of various nature conservation and monument protection services at regional and national level. It is also the responsibility of UNESCO's Department of Cultural Heritage. At international level, a particularly active body is the Commission for Historic Gardens mentioned above, which has collected a vast amount of cataloguing material. By running a number of conferences and summer courses it helps to answer questions concerning the heritage of various parts of the world. An outstanding scientific centre for the study of garden heritage is the Centre for the History of Landscape Architecture at Harvard University, with headquarters at Dumbarton Oaks, USA. Where numerous important international conferences on the various national traditions in landscape architecture have been organized.

Thanks to society's increasing awareness of this heritage as well as numerous specialist

---

[1] irretrievably: 不可挽回地,不能补救地
[2] authenticity: 真实,真确,可靠
[3] embodiment: 能具体表现他物者,化身,被具体表现者

initiatives, historic garden are today better maintained than ever and even a number of gardens once considered irretrievably lost have now been restored. An encouraging example of this is the large-scale reconstruction of the Baroque gardens of the royal palace of Het Loo, near Apeldoorn in the Netherlands. Other examples include Painshill and the Renaissance gardens of Ham House in Great Britain, Leonberg near Stuttgart, the gardens of Kroměříž, Moravia, Monticello, Virginia, Petrodvorets and Tsarskoe Selo near St. Petersburg, Kuskovo near Moscow, several old gardens in Japan, and above all, the numerous gardens of Suzhou and other sites in China. However, there are still countless gardens awaiting restoration, especially in Iran, India, Korea, Vietnam and in several European countries.

There is an increased interest in cultural heritage everywhere. This is shown by, among other things, numerous book publications casting light on the so far unstudied regions or periods. In spite of this encouraging development there are some national traditions that are still too little known, such as those of Korea and Vietnam. We can say with pleasure that the last few years have seen the appearance of some reputable magazines concerned exclusively with the subject of historic gardens, such as *Journal of Garden History*, published in London, and *Gartenkunst* in Germany. Another journal, *Garden History*, is published by the Garden History Society in Great Britain, active since 1965, Material on historical heritage is also periodically included in magazines such as *Landscape Architecture Journal and Landscape Journal* in the USA. *Landscape Design* and *Apollo* in Great Britain, *Gartenarchitektur* and *garten und Landschaft* in Germany, and *Landskap* in Scandinavia.

Notes

1. Extracts from

Text

Ogrin, Dušan. *The world Heritage of Gardens* [M]. London: Thames and Hudson Ltd, 1993: P14 - 22.

2. Background Information

(1) Dušan Ogrin 杜恩。

欧洲现代风景园林教育先锋，斯洛文尼亚卡布亚那大学教授，曾在哈佛大学设计学院任教多年。

(2)《*The world Heritage of Gardens*》《世界园林文化遗产》。

这本书是一本关于风景园林历史文化的重要著作，在欧洲有很大影响。

本文节选自《世界园林文化遗产》前言部分内容，有删节。

Further Reading B

New Birth for Gettysburg?

Ambitious overhaul[1] moves forward at nation's premier battle field park

By Chris Fordney

By the time they collided[2] in the fields of Pennsylvania in the third summer of the Civil War, the main armies of the United States and the Confederacy[3] had evolved into low-tech but brutally[4] efficient killing machines. The aftermath[5] of the Battle of Gettysburg probably defies description-a blasted landscape of torn fields, sheared woods, flattened fences, and shell-and bullet-pocked barns and buildings. More than 150,000 men had struggled in extremis for three days, and a third had been killed or wounded.

Looking around Gettysburg National Military Park today, one sees little hint of that carnage[6]. While a forest of monuments and the graves in the Soldiers National Cemetery[7] testify that something big happened around this small town, it's hard to get a sense of what it was from the battlefield's groomed pastures, tidy lanes, and exhibits with scarcely any photographs of deadbodies. Sterile and bloodless as old television western, the park and town of Gettysburg seem stuck in a 1960s tourism time warp, an interesting but not particularly compelling stop for the nearly two million annual visitors who amble[8] among the monuments, curio[9] shops, and wax museum.

The National Park Service has long been aware that this national park fails to measure up to its profound meaning for the American people as the ground where the war came to its devastating[10] climax and Abraham Lincoln Summoned a new nation out of the destruction with the Gettysburg Address. Four years ago, the agency unveiled a new management plan for Gettysburg, probably the most ambitious make-over ever for a battlefield park. The goal, in the words of park superintendent John Latschar, is to make Gettysburg "just as much a shrine and pilgrimage[11] to all the peoples of the nation as the Liberty Bell."

The central-and controversial-aspect of the project is a $95-million plan to demolish the park's

---

[1] overhaul: 追上，赶上
[2] collide: 互撞，碰撞
[3] Confederacy: 州联盟，邦联
[4] brutally: 野蛮地，残忍地
[5] aftermath: 结果，后果，余波
[6] carnage: 大屠杀，残杀（人类）
[7] Cemetery: 墓地，公墓
[8] amble: （指人）骑马或走路缓缓而行
[9] curio: 希奇古怪而其价值即在此的艺术品，古董，古玩，珍品
[10] devastate: 毁坏，破坏，使荒凉，使成废墟
[11] pilgrimage: 朝圣者的旅程

current complex of visitor facilities, now perched on historic and sensitive Cemetery Ridge and including a former modernist visitor center designed by architect Richard Neutra, and to replace it with a new visitor center at a point out of view of the main battle area. The rest of the project will involve the rehabilitation[1] of the Soldiers National Cemetery and the restoration of the main battle areas to the way they appeared when the armies gathered for battle in early July 1863.

The new visitor center was designed by the New York firm Cooper Robertson and Partners, known for its work at Monticello, the Brooklyn Botanic Garden, and the Lincoln Center in New York City (Figure 6.1). Drawing on the building traditions of the region, the design features a complex of smaller, interconnected pieces and will use local materials of stone, timber, and brick with metal roofs. "The structure should look like it belongs in this beautiful pastoral[2], agricultural landscape," says Jaquelin Taylor Roberson, an architect and principal in the firm (Figure 6.1, Figure 6.2).

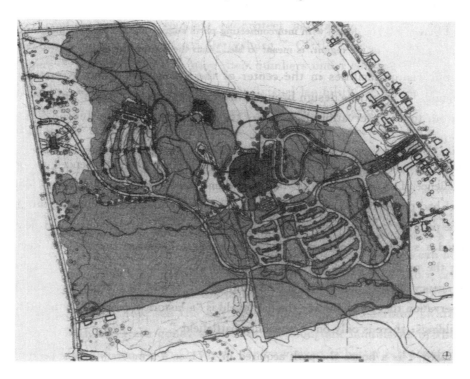

Figure 6.1 The new visitor center site

The 139,000-square-foot visitor center will house a museum, the famous cycloramic painting "The Battle of Gettysburg," and the park's immense collection of Civil War artifacts[3]. But it won't have several other attractions in its original plan: an IMAX film, a for-profit arts and crafts gallery, a National Geographic Store, and a tour center gift shop. These were eliminated after public complaints that the project had become too commercial for this hallowed ground. Another controversy has centered on the plan to demolish the Cyclorama Center, which has held the painting since the building opened in 1962 as the park's visitor center. This modernist building was placed on Cemetery Ridge (where Union forces turned back Pickett's Charge) as part of Mission 66, the crash building program

---

[1] rehabilitation: 恢复，修复，恢复原有地位、正常生活等
[2] pastoral: 田园诗，牧歌
[3] artifact: 人工制品

Figure 6.2 The design with interconnecting parts suggesting the rural architecture of the region, is meant to blend into the pastoral setting

that sought to put visitor facilities in the center of their most important attractions, an idea now viewed as anathema[1]. Neutra's design brought visitors into a large "drum" to view the painting and then took them out to an observation deck where they could examine key terrain[2] features from the artist's perspective.

This building was troubled from the start. Large sliding doors, intended to open an interior rostrum[3] to audience of thousands on a lawn, jammed and never worked properly, and fountains and pools on an upper level had problems and were pulled out. The drum, touted as a climate-controlled wonder that would protect the painting for a century, has been plagued with leaks that have damaged the canvas. Neutra's defenders mounted a successful campaign to have the building listed on the National Register of Historic Places, claiming the problems arose from poor maintenance, which the park service concedes[4] to some degree. But the agency is going ahead with the demolition[5], backed by a report from the Advisory Council on Historic Preservation that rehabilitating the battlefield is "a historic mission of the highest order There are other Neutra buildings; there is only one Gettysburg battlefield."

Also to be razed[6] is a house the park acquired in 1971 as a temporary visitor center under a plan to build a new one away from the town's central business district. Local pressure prevented another move, however, and the house, with 14 additions over the years, remains the visitor center today, holding the park's artifacts collection in a basement with damaging levels of humidity and offering "a disorganized hodgepodge[7] of exhibits that... fail to give visitors to Gettysburg a clear understanding of the significance of the campaign or its place in American history," in the words of one consulting historian.

The overhaul at Gettysburg goes far beyond the physical facilities and involves a completely new

---

[1] anathema: 极令人讨厌之事物
[2] terrain: 地域，地带，（尤指从军事观点而言的）地势，地形
[3] rostrum: 讲台，讲坛
[4] concede: 承认，让与，容许
[5] demolition: 拆除，推翻，毁坏，破坏
[6] raze: 彻底破坏（城市，建筑物），尤指夷为平地
[7] hodgepodge: 杂乱的一堆东西，杂混在一起的东西

approach to this site. The new management plan identifies three nationally significant landscapes at Gettysburg-the landscape of the battle, the Soldiers National Cemetery, and the commemorative landscape, the roughly 1,800 monuments, large and small, erected by veterans' groups over the years-but says the fundamental purpose of the park is to preserve the historic landscape of the battlefield as an 1895 memorial landscape instead of a battle landscape was a public relations and resource management failure.

Previous managers, the report states, also failed to catalogue the topography or natural features of the battlefield and made few attempts to keep boundaries of woods within their historic limits. "This meant that a very significant feature of the battle landscape, the pattern of open and wooded terrain that determined where the armies moved and why, was lost or obscured," the plan states. The park's efforts to make its fields easier to farm under its agricultural program actually helped destroy historic lanes, fence lines, wetlands and water-courses, and boundaries of orchards and farms, "in direct conflict with the mandate to protect natural and cultural features that were significant to the outcome of the Battle of Gettysburg." Cattle and deer also contributed to the damage, and rangers[1] now shoot deer to keep their numbers under control.

To rehabilitate key battle areas to improve their "visual legibility[2]," researchers consulted old maps and photographs taken after the battle to locate fences and old farm lanes that formed important obstacles and avenues of approach for soldiers. For example, in the mile-wide field of Pickett's Charge, the legendary assault by 15,000 Confederates in a final effort to break the Union army in two and march on Washington, fences will be rebuilt to show some of the barriers faced by these soldiers in what today is largely an open expanse that makes it seem easier to cross than it was.

Work has already begun on a demonstration project, the Codori-Trostle Thicket, involving the removal of several acres of woods and the planting of 16,000 shrubs to bring back a thicket that had grown up over the years into a large stand of trees. With the trees gone, visitors can now get the same view as Union General Winfield Scott Han-cock as he threw reinforcements into a critical gap in the northern lines, probably saving his army in the process. This is a difficult part of the battlefield to understand, and the ability to now see the hill over which the Confederates swarmed is just one benefit, rangers say. With the trees gone, several other important landmarks of the battlefield, in particular Cemetery Ridge and the Peach Orchard, can be viewed in better relationship to each other, and that will bring significant changes in the way rangers explain what happened at Gettysburg. "The force driving all this is understanding the battle." Says ranger D. Scott Hartwig.

The park is also continuing its effort to undo, at least in part, more than a century of development on and near the battlefield, where preservation efforts began almost as soon as the dead had been buried. Even then it was recognized that the turning back of Robert E. Lee's final invasion of the North was a momentous event. Private groups assembled pieces of critical terrain-Culp's Hill, Little Roundtop, and Cemetery Ridge-that were absorbed into the national military park established in 1895. But much private land remained within the environs[3] of the park, allowing tourist clutter[4] to

---

[1] ranger: (美) 森林看守人员
[2] legibility: (指字迹、印刷物) 易读的, 清楚的
[3] environs: 郊外, 近郊
[4] clutter: 乱塞, 乱堆

encroach[1] on hallowed ground, including the infamous Gettysburg Tower, a 310-foot observation tower that loomed[2] above Cemetery Ridge until it was demolished in July 2000 after a long court battle. Congress established a new boundary for the 6,000-acre park in 1990 and the government has made nearly 40 property acquisitions at Gettysburg since then.

Other changes will be subtler. In the exhibits at the new visitor center, the traditional theme of the park's interpretive program-that the battle represented the "high-water mark" of the Confederacy as Robert E. Lee's brave band of outnumbered soldiers made one last, heroic effort to overcome the superior numbers and technology of the North-will be downplayed in favor of the "new birth of freedom" embodied in Abraham Lincoln's Gettysburg Address, which he delivered at a spot now within the Soldiers nation Cemetery. That's part of an effort to bring more minorities to Civil War sites, the result of much soul-searching in recent years within the park service-after a prod[3] from Chicago Congressman Jesse Jackson, Jr. —over the way battlefields are presented to the public.

The agency inherited these historic landscapes from the Army in the 1930s among other new responsibilities for historic site preservation, and that tradition of military study has always held sway in exhibits and ranger talks. By sticking to the dry facts of strategy and troop movements, park managers could also avoid explosive topics such as whether slavery was the chief cause of the war. But under legislation sponsored by Jackson, who toured many federal Civil War sites and found them wanting, the park service is including material about slavery, civilians, and other nonmilitary topics in the visitor center displays and movies at its roughly 30 Civil War-related sites. Not to be forgotten are younger generations with notoriously low levels of history knowledge. The park will include material about the causes and consequences of the war in the new visitor center and in ranger talks to better orient the historically illiterate a job that not all rangers are excited about. "You're asking us to do what a college education hasn't done," one said. Traditionalists also worry that military history will get short shrift. But Superintendent Latschar saves battlefields must be made more relevant to more Americans. "We need to get away from 'who shot who where' and more into a discussion of why they were shooting".

Notes

1. Extracts from

Text

Fordney, Chris. New Birth for Gettysburg [J]. *Landscape Architecture*, 2002 (8): 46-49.

Pictures

(1) Figure 6.1: Chris Fordney. New Birth for Gettysburg [J]. *Landscape Architecture*, 2002 (8): 49.

(2) Figure 6.2: Chris Fordney. New Birth for Gettysburg [J]. *Landscape Architecture*, 2002 (8): 49.

2. Background Information

Chris Fordney 克里斯·福特内。

---

[1] encroach：超出正当范围，侵入，侵害

[2] loom：隐约地威胁性地出现

[3] prod：以尖物推或刺

自由撰稿人，英国英格兰南部汉普郡的首邑文契斯特人。

Gettysburg 盖茨堡。

美国宾夕法尼亚州的城镇。1863年南北战争中，罗伯特·李将军率领的南军（Confederates）在此被米德（Meade）所率领的北军（Federals）击败。

3. Sources of Additional Information

http：//www.nps.gov/gett/

http：//www.gettysburg.travel/index.asp

Further Reading C

Place & Project: Moody Historical Gardens Design

By Geoffrey Jellicoe

This is a guide to the designs for the Historical Gardens that will form part of the Moody Gardens at Galveston. The work on the gardens is planned to begin in 1992, giving time for the further preparatory[1] studies that are essential to make them the most scholarly as well as most dramatic of their kind in garden history. The challenge is indeed breathtaking: to compress the experience of a time scale of three thousand years and the space scale of the globe into a time scale of a few hours and a space scale of twenty-five acres. Clearly it cannot be done through realism; so it is being tried through surrealism[2] or the projection of the idea or essence of a culture.

The story of how and why the gardens were conceived is unique.

Galveston Island in the Gulf of Mexico lies two miles off the coast of Texas, some fifty miles south of Houston. Thirty miles long and averaging two miles in width, it acts as a barrier reef[3] to the mainland and provides a remarkable natural harbour[4]. Originally little more than a windswept sand bar[5], with scrub[6] oaks and the like, civilized history began when a Spanish explorer landed here in 1528. In 1685 it was claimed by the French. In 1786 it became Spanish under the governor Bernardo de Galvez, after whom it was named. In 1817 it was Venezuelan, in 1821 Mexican, and in 1836 it became part of the Republic of Texas. Texas joined the Union in 1845 and the Confederacy in 1861. In 1868 slavery was abolished by proclamation[7] and in the following year Colonel William Lewis Moody (1828-1920), a native of Virginia, settled here with his family to help create one of the most prosperous cities in America. Then, on 8th September 1900, came disaster with the virtual destruction of the city by hurricane[8] and inundation[9].

In the worst natural disaster in North American history six thousand persons lost their lives, many of the population of 38,000 fled to the mainland[10] never to return, and the city was left in ruins. Its economy was destroyed and its place as the commercial centre of southern Texas was taken by Houston. Nevertheless, as with so many cities destroyed in war, the ethos of the place was, and

[1] preparatory: 初步的
[2] surrealism: 超写实主义
[3] reef: 礁，暗礁
[4] harbour: 港
[5] bar: 沙洲，沙滩
[6] scrub: 矮树
[7] proclamation: 宣言，公布
[8] hurricane: 飓风
[9] inundation: 泛滥，淹没，洪水
[10] mainland: 大陆

still is, so strong that those of the original inhabitants who survived (BOI-Born on the Island - remains a proud term) set out on the prodigious work of reconstruction. Unlike cities destroyed in war which could not foresee the tactics of future attack, the Galvestonians knew precisely the powers of the enemy and planned accordingly in a manner that is itself an epic[1]. The city on average was raised twelve feet above the existing three feet above sea level, a task involving dredging within the island itself and the total realignment of services. Elsewhere by law all new buildings on the island were to be built on stilts[2] with a minimum of fifteen feet above sea level. The huge sea wall and boulevard[3], now extending ten and a half miles along the south shore facing the Gulf, was virtually completed by 1910, reputedly the most formidable[4] of its kind in the world. Further hurricanes and inundations have caused little damage to building, but Hurricane Alice in August 1983 was particularly destructive of the vegetation that, ever since oleanders[5] were introduced in the early nineteenth century, had been painstakingly developed by individual efforts.

Today the modern city with a population of 62,000, a special interest in conservation of the past and a rapidly growing tourist trade, has an air of prosperity. But it is deficient in vegetation, exceptions proving that trees and shrubs and flowers can flourish given encouragement and the properly sheltered environment. The establishment therefore of these gardens in an area naturally hostile to plants will be both a challenge to the elements and an immense stimulus to the island as a whole.

The location of the gardens lies on what may be described as the soft under belly[6] of the island, on an inlet[7] of the north shore known as Offatts Bayou, a long narrow strip, partly of wet-lands or marshes wedged between an airfield and the sea (Figure 6.3). The wet-lands are strictly wildlife preserves and account for the baroque shape into which the historical gardens will be fitted. The total length of over a mile is divided roughly into three parts: the east includes the already constructed Hope Arena and Conference Hall, and proposed hotel, quay[8] for pleasure steamers, foreshore plage[9], restaurants and other pleasantries leading to the educational

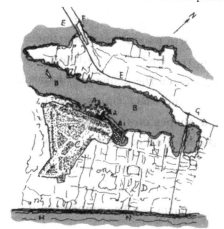

Figure 6.3 Plan of site
A—Moody Gardens; B—Offatts Bayou;
C—Airfield; D—Suburbs; E—Bridge
and approach from Houston from
which outstanding views of the
garden are seen; F—Railway
bridge; G—Broadway to
City centre; H—Sea wall

campus with glass houses and theatre. The centre contains the Historical Gardens described in this guide. To the west are the walkways beyond the dyke[10], seen skipping across the wet-lands to the

---

[1] epic: 描写英雄事迹的诗，史诗，叙事诗
[2] stilt: 高跷
[3] boulevard: 大马路，大道（两旁常植有树木），林荫大道
[4] formidable: 难以克服的，艰难的
[5] oleander: 夹竹桃
[6] belly: 肚子，腹部
[7] inlet: 湾，插入物，镶入物
[8] quay: 码头，横码头
[9] plage: 海滩（尤指在时髦的海滨游乐地）
[10] dyke: 堤

great dyke-protected circle of the nursery.

The most comfortable and briefest way of seeing the Historical Gardens is the one chosen for this guide—the water-bus. This is only intended, however, to whet[1] the appetite and invite a visit on foot along the two miles of paths. A serious study could take two days or more. Although the whole is intended to be a single surrealistic picture, it is also intended to be academically correct in spirit and in detail. It is sound, therefore, that the opening is not for some years, for the research that follows these drawings (made to a scale of eight feet to one inch) will be considerable. Consider plants, for instance, which on the drawings are only shown suggestively[2]. First there must be a study of what plants were likely to be in use at any one period. Then that they are available today and, if not, the selection made of an alternative. After this there must be at least a two year period to see if the species can be acclimatized[3]. On the data thus known there now follows the making of an aesthetic planting plan. Finally comes the planning process itself on previously prepared sites.

By far the most difficult part of this exercise has been the basic choice of subject, disposition, proportions and neighbourliness of all the various cultures. The composition for the whole peculiar site resolved itself without a struggle. The straight line marking the boundary with the airfield suggested the geometry of classicism; the short middle of the site suggested the great divide between west and east; eighteenth century European romanticism, so important to the modern world, filled the shapes on the further side of the classical waters; China and Japan were equally accommodating to the long narrow space in which they had to be contained.

No study of history can be absolute when interpreted through the research of an individual. No two historians will think precisely alike. It is, for instance, this designer's own view that landscape classicism can be compressed into a box (like Shakespeare's Globe), but that romanticism cannot and that the styles which quarreled in history need not do so today.

The concept of the garden is that they are a projection into space, up to the nineteenth century, of my own and my wife's *Landscape of Man*, first published in 1975.

The reader, unlike the voyager in reality, must forgive the presentation in draughtsman's rectangles. These were devised prior to the idea of a written guide and will ultimately be pieced together in a four-metre square on the wall of Reception.

Notes

1. Extracts from

Text

Jellicoe, Geoffrey. *The landscape of Civilisation—Created for the Moody Historical Gardens* [M]. East Sussex: Garden Art Press Ltd., 1989: 17-19.

Picture

Figure 6.3: Geoffrey Jellicoe. *The landscape of Civilisation—Created for the Moody Historical Gardens* [M]. East Sussex: Garden Art Press Ltd., 1989: 18.

---

[1] whet: (喻) 促进, 刺激 (胃口, 欲望)
[2] suggestively: 暗示地, 提醒地, 引起联想地
[3] acclimatize: 使服水土, (喻) 使适应新环境

2. Background Information

（1）Geoffrey Jellicoes 杰弗里·杰里科（1900~1996年）。

建筑师、风景园林师、著名的建筑协会学校的教师、规划师，是英国风景园林师学会的创建者之一，并从1939年起担任了10年学会的主席。1948年还担任了国际风景园林师联合会（IFLA）的首任主席以及历届IFLA的名誉主席，是风景园林规划设计领域的一代先驱。设计生涯几乎跨越70年的时间，其经历与国际风景园林学科的发展紧密相关。从1925年出版第一部著作《文艺复兴时期的意大利花园》起，到1992年规划设计亚特兰大市历史学会庭院，一生出版专著10多部，完成的建筑与风景园林规划设计100余项，其中约有60个是关于风景园林的。

（2）The landscape of Civilisation—Created for the Moody Historical Gardens《文明的景观——穆迪历史花园》。

1989年出版，书中详细阐述了穆迪历史花园的设计思想。穆迪历史花园规划是由历史花园和景观要素构成的。每一个要素并不是真实花园的复制品，而是一个抽象概念，代表着那个时代，以时间为线索，通过景观和花园的历史来描述人类文明的历史。穆迪历史主题花园展示了杰里科在风景园林历史的大舞台上的表演能力。在这里杰里科运用了潜意识，他认为穆迪历史花园的目的就是"转换的一个真实世界"，园林中，水是阴，山是阳，它们统一了可见的世界中不同的事物。藏于这个花园景观表层之下至少有三个潜意识的层次。最上面一层是由这些构思唤起的情感，中间一层是由伊甸园、希腊众神和东方的佛代表的宗教，最深的一层是神秘的山水，在它上面参观者就像一个航海者，穿过三千年及半个地球的浩瀚时空。

本文选自该书场地及项目介绍部分内容。

3. Sources of Additional Information

http://www.moodygardens.com/

Exercises

1. Questions（阅读思考）

认真阅读本单元文章，并讨论以下问题。
（1）谈谈如何理解遗产文化园林的保护政策（保存、保护、复原、重造、重建）。
（2）结合实例谈谈对历史文化园林的理解。
（3）谈谈不同历史文化背景下所形成的风景园林的特色。
（4）联系我国情况，谈谈历史文化园林面临的问题并谈谈针对这些问题有哪些解决的建议。

2. Skill Training（技能训练）

（1）查找并浏览有关历史文化园林英文网页，选择自己感兴趣的某个主题进行深入研究，作专题论文。
（2）每位同学简要介绍自己所作专题论文，并以"历史文化园林"为总的主题，进行交流和讨论。

Tips

风景园林中的历史文化保护

1. 历史景观（Historic Landscape）

历史景观没有明确的定义，通常附属于文化景观体系，如罗伯特 Z. 蒙耐克（Robert Z. Melnick）提出："历史景观可以被定义为文化景观的一个类型，这一类型的景观与特定的人物、事件及特别重要的历史时期相关联"。历史景观涵盖的内容相当广泛，包括居住区环境（Residential Grounds）、纪念性场所（Monument Grounds）、公共建筑环境（Public Building Grounds）、花园（Garden）、小型开放空间（Minor Public Grounds）、战场（Battlefield）、墓地（Cemetery）、街道景观（Streetscape）、公园（Park）、历史性村落（Museum village）、城区（District）、城镇（Town）、史前遗址（Pre-historic Site）及公园体系（Park System）等18个子类型。

2. 文化景观（Cultural Landscape）

文化景观和自然景观是美国国家公园管理局（NPS）负责管理的两类最重要的景观系统。其中关于文化景观，1995年国家公园管理局通过的《内政部历史遗产保护处理措施标准》（The Secretary of the interior's Standards for the Treatment of Historic Properties）有明确的定义：文化景观为一包含文化资源与自然资源的地理区域，包含野生鸟兽或驯养动物在其中。文化景观关联着历史事件、活动、人物或者展示其他文化或美学价值。文化景观可分为4种普遍的形态，彼此间互不排斥：历史地段（historic site）、历史的经过设计的景观（historic designed landscape）、历史的乡土（本土）景观（historic vernacular landscape）和文化人类学景观（ethnographic landscape）。其中，历史地段是指与某一特定的历史事件、活动或人物相联系的景观，如古战场、总统府邸和遗产等；历史的经过设计的景观是指经过风景园林师、花园总设计师、建筑师、工程师、园艺师等根据设计原理进行设计或由业余的园艺师以某一可辨别的风格或传统设计的景观。这一景观可能与风景园林领域内某一重要的人物、景观实践或景观发展方向相联系，或能够说明景观领域在理论或实践上的某一重要发展。美学价值是设计景观的重要特征，例如公园、校园景观和不动产等；历史的乡土（本土）景观是指经过使用者的活动和占有，不断发展演变而形成的景观。历史的乡土景观通过个人、家庭、社区的社会和文化生活态度，反映日常生活的物理、生物、文化特征。功能是乡土景观的重要角色。乡土景观可以是一块地产如农田，也可以是集中的地产如河谷附近的一块历史性农业地带。乡村、工业混合体、农业景观等都属于历史的乡土景观；文化人类学景观是指包含各种与人类相关联的自然资源和文化资源的景观，也被称作遗产资源。例如当代聚落、宗教圣地以及广泛的地理结构体。也包含小型的植物群落、动物、生境等。

文化景观涵盖内容广泛，尺度规模及类型多样。与文物建筑和历史地区一样，文化景观通过它们特有的形式和特征揭示了国家的起源和发展，具有很高的历史价值和文化价值。

3. 历史园林（Historic Garden）

1982年12月15日，国际古迹遗址理事会注册通过的历史园林保护的《佛罗伦萨宪章》，对历史园林的概念界定如下：

第一条 "历史园林应是以其历史性和艺术性被广为关注的营造兼园艺作品"，同时它应被视作历史古迹——本质。

第二条 "历史园林是主要以植物为素材的设计营造作品，因而是有生命的，意即有荣枯盛衰，也有代谢新生"。因此，其表象反映着四季轮转，自然的兴衰与造园家和园艺师力求保持其长盛不衰的努力之间不断的平衡——特征。

第三条 作为古迹，历史园林必须根据《威尼斯宪章》的精神予以保存。然而，既然它是一个"活"的古迹，其保存亦必须遵循特定的规则进行——保护原则。

第四条 历史园林的设计营造体系包括：其平面布局和空间构成；其植物配置，包括品种、面积、配色、间隔以及各自尺度；其景园布局结构特征和景观特征；其映照着天空的水面，动态或静态水景——内涵。

第六条 "历史园林"这一术语同样适用于小型花园和大型园林，不论其是属于几何规整式的还是自然风景式的园林——外延。

前四条分别对历史园林的本质、特征、保护原则和内涵作出了明确的界定。首先，历史园林是指"以往"的"营造兼园艺作品"，属于"历史古迹"的范畴，类同于"古建筑"、"历史地段"、"历史文化城市"等；其次，历史园林又具有鲜明的个性特征：其基本素材—植物所呈现的生命规律，使其具有独特的"动态"特征，即使之成为"活"的古迹，因此，历史园林的保护原则，必须以"古迹"保护原则为基础，同时又有别于建筑等"古迹"的保护。必须制定专门的保护原则。历史园林保护的对象，必须涵盖其设计营造体系所包括的全部专项内容，如"平面布局和空间构成、植物配置、景园布局结构特征和景观特征、水面等所有园林要素"。第六条则进一步明确了"历史园林"的外延。

4. 历史园林保护

国际上对于历史园林保护原则和策略的制定，基本上源于古迹（文物建筑和历史地段）保护的理论和实践。

文物建筑和历史纪念物的保护，就其广泛的意义而言，至少可以追溯到古罗马时代，到了文艺复兴时期，又有了进一步的发展。近代的文物保护，始于18世纪末，英法等国的思想界，首先发起了对文物建筑价值的讨论，继而引起建筑界对于保护和修复文物建筑工作的重视。19世纪中叶起，这项工作开始走向科学化，并在以后的150多年中，逐渐发展和完善了它的一些基本概念、理论和原则。这些国际性的保护理念和原则，主要体现在由联合国教科文组织及有关非政府组织通过的一系列世界文化遗产保护的国际文献，如《雅典宪章》、《威尼斯宪章》、《内罗毕建议》、《华盛顿宪章》和《保护世界文化和自然遗产公约》等。《威尼斯宪章》明确了文物建筑保护的四项基本原则，即最低限度干预原则、可识别性原则、可逆性原则、与环境统一原则。历史地段的保护较之文物建筑保护更加复杂，因为城区和老街等历史地段是"活"的，《华盛顿宪章》中对于历史城市和历史地段的保护原则和目标、方法和手段，都做了原则性规定。

而最近兴起的历史园林保护，则建立在上述国际古迹保护的理念基础之上。1982年，国际古迹遗址理事会注册通过了一项专门的历史园林保护宪章《佛罗伦萨宪章》，作为《威尼斯宪章》的附件，该宪章业已成为国际上共同遵守的历史园林保护准则。

《佛罗伦萨宪章》在讨论历史园林维护、保护、修复、重建方法时，基本上遵循《威尼斯宪章》的规定。例如对于历史园林的修复和重建策略问题，两个宪章有近乎相同的规定，不论是文物建筑还是历史园林，在修复前必须首先进行深入的研究。《威尼斯宪章》第九条规定："修复是一件高度专门化的技术……它必须尊重原始资料和确凿的文献，它不能有丝毫臆测"；《佛罗伦萨宪章》第十五条则规定："在未经详尽的前期研究之情况下，不得对某一历史园林进行修复，特别是不得进行重建"。此外，关于修复时要尊重各个时期的历史信息，对修复、保护工作要有明确的记录等规定，两个宪章也基本相同。

Unit 7

Ecological Planning & Design

- 风景园林生态设计理论
- 设计方法
- 设计实例

These sprays, dusts, and aerosols are now applied almost universally to farms, gardens, forests, and homes nonselective chemicals that have the power to kill every insect, the "good" and the "bad", to still the song of birds and the leaping of fish in the streams, to coat the leaves with a deadly film, and to linger on in soil all this though the intended target may be only a few weeds or insects.

By Rachel Carson

Text

Silent Spring: The Obligation[1] to Endure

By Rachel Carson

> 本文讲述人类对于自然环境的影响，这种影响是巨大的，而且是不利的。本文介绍了人类创造的化学物质是如何对自然环境产生破坏，又是多么地触目惊心。由于篇幅的限制，本文为原文（原书的第二章）的前半部分。

The history of life on earth has been a history of interaction between living things and their surroundings. To a large extent, the physical form and the habits of the earth's vegetation and its animal life have been molded by the environment. Considering the whole span of earthly time, the opposite effect, in which life actually modifies its surroundings, has been relatively slight[2]. Only within the moment of time represented by the present century has one species-man-acquired significant power to alter the nature of his world (Figure 7.1).

Figure 7.1 Silent spring

地球上生命的历史一直是生物及其周围环境相互作用的历史。可以说在很大程度上，地球上植物和动物的自然形态和习性都是由环境塑造成的。就地球时间的整个阶段而言，生命改造环境的反作用实际上一直是相对微小的。仅仅在出现了生命新种——人类之后，生命才具有了改造其周围大自然的异常能力。

During the past quarter century this power has not only increased to one of disturbing[3] magnitude[4] but it has changed in character. The most alarming of all man's assaults upon the environment is the contamination[5] of air, earth, river, and sea with dangerous and even lethal[6] materials. This pollution is for the most part irrecoverable; the chain of evil it initiates not only in the

---

[1] obligation：义务，债务
[2] slight：轻微的
[3] disturbing：烦人的
[4] magnitude：巨大
[5] contamination：污染
[6] lethal：致命的

world that must support life but in living tissues is for the most part irreversible[1]. in this now universal contamination of the environment, chemicals are the sinister[2] and little-recognize partners of radiation in changing the very nature of the world—the very nature of its life. Strontium[3] 90, released through nuclear explosions into the air, comes to earth in rain or drifts down as fallout, lodges in soil, enters into the grass or corn or wheat grown there, and in time takes up its abode in the bones of a human being, there to remain until his death. Similarly, chemicals sprayed on croplands or forests or gardens lie long in soil, entering into living organisms, passing from one to another in a chain of poisoning and death. Or they pass mysteriously[4] by underground streams until they emerge and, through the alchemy[5] of air and sunlight, combine into new forms that kill vegetation, sicken cattle, and work unknown harm on those who drink from once pure wells. As Albert Schweiszer has said, "Man can hardly even recognize the devils of his own creation."

在过去的 1/4 个世纪里，这种力量还没有增长到产生骚扰的程度，但它已导致一定的变化。在人对环境的所有袭击中最令人震惊的是空气、土地、河流及大海受到了危险的、甚至致命物质的污染。这种污染在很大程度上是难以恢复的，它不仅进入了生命赖以生存的世界，而且也进入了生物组织内，这一罪恶的环链在很大程度上是无法改变的。在当前这种环境的普遍污染中，在改变大自然及其生命本性的过程中，化学药品起着有害的作用，它们至少可以与放射性危害相提并论。在核爆炸中所释放出的锶 90，会随着雨水和漂尘争先恐后地降落到地面，居住在土壤里，进入其上生长的草、谷物或小麦里，并不断进入到人类的骨头里，它将一直保留在那儿，直到完全衰亡。同样地，被撒向农田、森林、花园里的化学药品也长期地存在于土壤里，同时进入生物的组织中，并在一个引起中毒和死亡的环链中不断传递迁移。有时它们随着地下水流神秘地转移，等到它们再度显现出来时，它们会在空气和太阳光的作用下结合成为新的形式，这种新物质可以杀伤植物和家畜，使那些曾经长期饮用井水的人们受到不知不觉的伤害。正如阿伯特·斯切维泽所说："人们恰恰很难辨认自己创造出的魔鬼。"

It took hundreds of millions of years to produce the life that now inhabits the earth—eons[6] of time in which that developing and evolving and diversifying[7] life reached a state of adjustment and balance with its surroundings. The environment, rigorously[8] shaping and directing the life it supported, contained elements that were hostile as well as supporting. Certain rocks gave out dangerous radiation; even within the light of the sun, from which all life draws its energy, there were short-wave radiations with power to injure. Given time-time not in years but in millennia—life adjusts and a balance has been reached. For time is the essential ingredient[9]; but in the modern world there is no time.

为了产生现在居住于地球上的生命已用去了千百万年，在这个时间里，不断发展、进化和演变着的生命与其周围环境达到了一个协调和平衡的状态。在有着严格构成和支配生命的环境中，包含着对

---

[1] irreversible：不能逆转的
[2] sinister：险恶的
[3] strontium：锶
[4] mysteriously：神秘地
[5] alchemy：魔术，炼金术
[6] eon：无限长的时代
[7] diversify：使多样化
[8] rigorously：残酷地
[9] ingredient：成分

生命有害和有益的元素。一些岩石放射出危险的射线，甚至在所有生命从中获取能量的太阳光中也包含着具有伤害能力的短波射线。生命要调整它原有的平衡所需要的时间不是以年计而是以千年计。时间是根本的因素；但是现今的世界变化之速已来不及调整。

The rapidity of change and the speed with which new situations are created follow the impetuous[1] and heedless[2] pace of man rather than the deliberate[3] pace of nature. Radiation is no longer merely the background radiation of rocks, the bombardment[4] of cosmic[5] rays. the ultraviolet of the sun that have existed before there was any life on earth; radiation is now the unnatural creation of man's tampering with the atom. The chemicals to which life is asked to make its adjustment are no longer merely the calcium and silica and copper and all the rest of the minerals washed out of the rocks and carried in rivers to the sea; they are the synthetic[6] creations of man's inventive mind, brewed[7] in his laboratories, and having no counterparts in nature.

新情况产生的速度和变化之快产生于人类激烈而轻率的步伐而不是大自然的从容步态。放射性已远远在地球上还没有任何生命以前已经存在于岩石放射性本底、宇宙射线爆炸和太阳紫外线中了；现存的放射性是人们干预原子时的人工创造。生命在本身调整中所遭遇的化学物质再也远远不仅是从岩石里冲刷出来的和由江河带到大海去的钙、硅、铜以及其他的无机物了，它们是人们发达的头脑在实验室里所创造的人工合成物，而这些东西在自然界是没有对应物的。

To adjust to these chemicals would require time on the scale that is nature's; it would require not merely the years of a man's life but the life of generations. And even this, were it by some miracle possible, would be futile, for the new chemicals come from our laboratories in an endless stream; almost 500 annually find their way into actual use in the United States alone. The figure is staggering and its implications are not easily grasped 500 new chemicals to which the bodies of men and animals are required somehow to adapt each year, chemicals totally outside the limits of biologic experience.

在大自然的天平上调整这些化学物质是需要时间的；它不止需要一个人一生的时间，而且甚至需要几代人的时间。即使借助于某些奇迹使这种调整成为可能也是无济于事的，因为新的化学物质象涓涓溪流不断地从我们实验室里涌出，单是在美国，每一年几乎有500种化学合成物在实际应用上找到它们的出路。这些化学物品的形状变幻不定，而且它们的复杂性是不可轻易掌握的——人和动物的身体每年都要千方百计去适应500种这样的化学物质，而这些化学物质完全都是生物未曾经验过的。

Among them are many that are used in man's war against nature, since the mid-1940's over 200 basic chemicals have been created for use in killing insects, weeds, rodents[8], and other organisms described in the modern vernacular[9] as "pests"; and they are sold under several thousand different brand names.

这些化学物质中有许多应用于人对自然的战争中，从20世纪40年代中期以来，200多种基本的

---

[1] impetuous：冲动的，猛烈的
[2] heedless：不注意的
[3] deliberate：深思熟虑的
[4] bombardment：袭击
[5] cosmic：宇宙的
[6] synthetic：合成的
[7] brew：酿造
[8] rodent：啮齿动物
[9] vernacular：本国的

化学物品被创造出来用于杀死昆虫、野草、啮齿动物和其他一些用现代俗语称之为"害虫"的生物。这些化学物品是以几千种不同的商品名称出售的。

These sprays, dusts, and aerosols[1] are now applied almost universally to farms, gardens, forests, and homes-nonselective chemicals that have the power to kill every insect, the "good" and the "bad", to still the song of birds and the leaping[2] of fish in the streams, to coat the leaves with a deadly film, and to linger on in soil-all this though the intended[3] target may be only a few weeds or insects. Can anyone believe it is possible that those poisonous chemicals will do no harm to any life on earth? They should be called "lifecides[4]" instead of "pesticides".

这些喷雾器、药粉和喷撒药水现在几乎已曾遍地应用于农场、果园、森林和家庭,这些没有选择性的化学药品具有杀死每一种"好的"和"坏的"昆虫的力量,它们使得鸟儿的歌唱和鱼儿在河水里的欢跃静息下来,使树叶披上一层致命的薄膜,并长期滞留在土壤里——造成这一切的原来的目的可能仅仅是为了少数杂草和昆虫。谁能相信在地球表面上撒放有毒的烟幕弹怎么可能不给所有生命带来危害呢?它们不应该叫做"杀虫剂",而应称为"杀生剂"。

Notes

1. Extracts from
Text
Carson, Rachel. *Silent Spring* [M]. Boston: Houghton Mifflin Company, 1962: 5-8.
Picture
Figure 7.1: Rachel Carson. *Silent Spring* [M]. Boston: Houghton Mifflin Company, 1962: 1.
2. Background Information
(1) Rachel Carson 蕾切尔·卡逊。
蕾切尔·卡逊是20世纪最具有影响力的环境学家和环保主义者。她关注污染对于环境的毁灭性影响,提倡对污染物的控制。作为一名海洋生物学家,卡逊的其他主要著作有《我们周围的海》和《海的边缘》。
(2) *Silent Spring* 《寂静的春天》。
蕾切尔·卡逊的成名作是一部具有里程碑意义的环境保护书籍,她在书中描述了滥用杀虫剂和其他污染物对于环境的危害,对于公众具有很大影响。美国政府也是在此书的背景下成立的环境保护局。本文为书的第二章的前半部分。
3. Sources of Additional Information
http://www.silentspring.org/
http://www.rachelcarson.org/

---

[1] aerosol:浮尘
[2] leap:跳跃
[3] intended:有意的
[4] —cide:杀

Further Reading A

Bioregional Planning and Ecosystem Protection

By Clair Reiniger

The earth is composed of ecosystems, the borders of which are not represented by political demarcations[1] but follow nature's contours. Areas that are defined by natural boundaries have come to be called bioregions, or life territories, from the *Greek bio* for "life" and Latin *regio* for "territory". The basic root meaning of bioregion has been expanded over the years to mean "part of the earth's surface whose rough boundaries are determined by natural characteristics rather than human dictates[2], distinguishable[3] from other areas by particular attributes of flora, fauna, water, climate, soil, landforms, and by human settlements and cultures those attributes have given rise to."

近年来，生物区域的本义得到延伸，已经用于表示"地球表面的部分，它的大致边界由自然特征而非人类规定而决定，可以根据植物、动物、水、气候、土壤、地形等特征和这些特征所引起的人类居住地和文化来与其他地域进行区别"。

The protection and management of natural resources in the United States is largely a compartmentalized[4] bureaucratic activity. This is evident from the multitude of federal agencies responsible for various aspects of resource management, including the Forest Service, Natural Resources Conservation Service, Bureau of Land Management, Environmental Protection Agency, National Park Service, Fish and Wildlife Service, Bureau of Reclamation, Bureau of Indian Affairs, Geological Survey, Army Corps of Engineers, Federal Energy Regulatory Administration, as well as the state agencies with similar overlapping responsibilities.

Each agency is responsible for the preservation and development of only a portion of our nation's natural resources. Few agencies take a bioregional approach. The result is a lack of unified resource management in the United States as well as competition among federal agencies and their state counterparts. Many countries around the world follow the

---

[1] demarcation: 划分
[2] dictate: 指示
[3] distinguishable: 可区分的
[4] compartmentalize: 划分

American example. Bioregional planning is a way of understanding the complexities of ecosystems as they relate to regional culture; it is an integrated approach to resource management as defined by the ecosystem's characteristics. To understand the ecosystem of a bioregion, a Bioregional Resource Inventory is complied, providing detailed analyses of climate, precipitation[1], vegetation, soil, geology, animal life, air quality, surface hydrology, groundwater depths, surface-water drainage, topography[2], land-use patterns, population density and settlement patterns, metropolitan development trends, economic patterns as well as the extent of natural energy sources of sun, wind, water, and biomass[3]. All of the components just listed, as well as the relationships among them, comprise the structure of a bioregion, or what might be called its anatomy. These elements are dynamic: Energy, water, and materials flow through them, and this gives life to the system, or what might be called its physiology.

The critical concept to comprehend is that a bioregion functions as an ecosystem, which basically captures solar energy through its vegetation and then transforms this energy through the process of growth, maintenance, storage, species reproduction[4], and system regeneration. Vegetation in general but forests in particular play a critical role in the hydrological cycle that brings moisture from the oceans and, through the repeated cycle of rainfall and forest evapotranspiration[5], carries it inland toward the interior of continents. The soils of an ecosystem capture and store water, making it available to plants and releasing it slowly to streams and aquifers[6]. The micro-organisms in soil break down organic materials as well as the minerals in the underlying rock, making these available as plant nutrients. The basic processes by which an ecosystem functions are related to these fundamental flows of energy, waste, and materials through the system.

In general, the more a climate varies, the more surplus energy, water, and materials an ecosystem will store against

---

[1] precipitation：降水
[2] topography：地形学
[3] biomass：生物量
[4] reproduction：繁衍
[5] evapotranspiration：水分蒸发
[6] aquifer：蓄水层

hard times in stressful[1] periods. Plants and animals in temperate climates, for example, store more energy to survive the winter, or more water to withstand the dry season, than do plants and animals of the humid tropics where the climate is fairly even and, consequently, where energy flows are more continuous throughout the ecosystem.

We harvest these ecological surpluses, and in some cases with us for these resources. Our near elimination of the bison and antelope habitat in the Great Plains is a case in point. When we harvest more than the surplus produced by an ecosystem and begin to consume the system itself, we are, in effect, mining the resource. This can trap us in a vicious[2] circle in which the system's yield is never enough for us, so we keep using up the system's capacity to produce and the yield keeps diminishing. This is what has happened to most of our nation's ecosystems, especially during the twentieth century.

这会诱使我们进入一个恶性循环：系统所产生的永远不够我们使用，所以我们不断地穷尽系统的生产能力，从而导致系统的产量不断减少。

Part of the challenge of resource management in the American West, when I live and work, comes from the persistence[3] of the boom-bust mentality[4] that has dominated the region since Anglo-European settlement began. To date, we have mined our forests, rangelands, soils, and aquifers as though there is no tomorrow. Only at the turn of the twentieth century, in Theodore Roosevelt's day, did people distinguish between nonrenewable and renewable[5] resources. The establishment of the Soil Erosion Service (later known as the Soil Conservation Service and now the Natural Resources Conservation Service) put this recognition[6] into practice in 1933, but a view of resource areas as something akin to a stockpile[7] of goods dominates attitudes toward resources to this day. When a forest is viewed as a stockpile of timber and not as part of an ecosystem, there is little incentive[8] to reforest. Just clear-cut and move on to the next stockpile, or just clear-cut and replace it with a monoculture[9] plantation.

[1] stressful：紧迫的
[2] vicious：坏的
[3] persistence：坚持
[4] mentality：心智
[5] renewable：可更新的
[6] recognition：共识
[7] stockpile：存储
[8] incentive：动机
[9] monoculture：单一栽植

The same principle goes for rangelands, agricultural soils, and aquifers. With ecological awareness, people begin to understand that renewable resources are not just stockpiled in nature, but are produced by living systems; to maintain output, one has to maintain the health of the producing ecosystem.

One of the consequences of deforestation[1] and overgrazing[2] is increasingly unreliable water supplies and destructive periodic flooding. The response is to build more dams, levees[3], and reservoirs. But this solves only part of the problem, namely, fluctuating water availability. The primary problem remains the destruction of the original water-capturing ecosystems - soil plus vegetative cove in forests and grasslands. The bioregional viewpoint establishes the absolute necessity of the health maintenance of the natural ecosystems from which we obtain our renewable resources and on which, ultimately, our lives and livelihoods universally depend.

生物区域的观点确立了维持健康自然生态系统的完全必要性。我们从生态系统中源源不断地获取资源，而且我们的生命和生存最终也全要依赖它。

This ecologically oriented view, however, does not mean that we ignore human needs or economies. All but a few of us work to make a living. Bioregionalism views human economies as ecosystems within ecosystems, as subsystems of a larger whole. We, too, use energy to transform material into forms we can use to sustain ourselves. But like other members of the animal kingdom, we depended on the plant kingdom to transform the primary source of energy on earth, which is sunlight, into usable forms of energy - including the fossil fuels on which today's explosive population growth and agricultural, industrial, and technological civilization depend.

但是像动物王国的其他成员一样，我们需要植物王国将地球上最初的能量来源——阳光——转化成可用的能量形式——包括今天人口的爆炸性增长和农业、工业和科技文明发展所依赖的化石燃料。

Our economies are embedded within larger ecosystems. We depend on these ecosystems to provide us with energy, water, and materials; we, also, depend on them for vital services or "foundational resources" such as water storage, moderation of climate extremes, generation of oxygen and absorption of carbon dioxide, and the filtering, neutralizing[4], and recycling of pollutants. And there is that intangible[5] sense of being "at home" when there are rocks, cacti[6], trees,

我们也靠它们提供必要的服务或"基本资源"，像水的储备、极端气候的缓和、氧气的产生和二氧化碳的吸收，以及污染物的过滤、中和再利用。

---

［1］ deforestation：砍伐树木
［2］ overgraze：过度放牧
［3］ levee：防洪堤
［4］ neutralize：抑制
［5］ intangible：无法明了的
［6］ cacti：仙人掌（复数）

gardens, and other aspects of life around us that we miss when we become isolated in unfamiliar surroundings.

In economics, the fundamental concept of cost-benefit[1] analysis carries over into the watershed planning instituted by the Soil Conservation Service. <u>In these plans, we find cost-benefit analyses for proposed structures such as dams, sediment basins, or ditches, and for a wide range of soil and water conservation practices applied to the land.</u> In wildlife and range management, the concept of "carrying capacity" is key to evaluating wildlife habitat needs and stocking rates. The carrying capacity of the land, or more properly of the ecosystem, is the number of animals of a given species supported by a given area over the long term. In other words, carrying capacity is a measure of the sustainability of a wildlife species or of sustainable yield for livestock.

在这些计划中，我们发现了成本—收益分析，以用来分析拟建的建筑物，如大坝、沉积池或沟渠，以及大面积的水土保护。

<u>Bioregional planning brings together both cost-benefit analysis and carrying capacity or sustainability measures in a way that seeks to balance short-term and long-term costs and benefits.</u> The goal is to help us make a worthwhile living today without compromising the lives and livelihoods of our children and their children or sacrificing and welfare of compatriot[2] lifeforms. Accomplishing[3] this goal requires a recognition of the value of healthy, functioning ecosystems as renewable resource systems. The bioregional approach goes beyond these utilitarian[4] values and states that the quality of human life depends on the health of all life; that we are, in actuality, one immune[5] system. Our health mirrors the health of the ecosystem we live in, and vice versa.

生物区域规划将成本—收益分析和支撑容量或持续性估测有机地结合了起来，以寻求短期成本与收益和长期成本与收益之间的平衡。

Notes

1. Extracts from
Text
Reiniger, Clair. Bioregional Planning and Ecosystem Protection [C]. Frederick Steiner, William Thompson. *Ecological Design and Planning*. New York: Wiley & Sons, 1997: 185 - 199.
2. Background Information
(1) Clair Reiniger 克莱尔·瑞尼格（1950～2005 年）。

---

[1] cost-benefit：成本—收益
[2] compatriot：同胞
[3] Accomplish：完成
[4] utilitarian：功利的
[5] immune：免疫的

瑞尼格毕业于哈佛大学设计学院,是新墨西哥州景观师执业委员会的创始成员,新墨西哥州的第一个注册景观师。她专长于使用本土植物进行设计和地下水源的保护。她是生物区域规划与设计方法的提出者。

(2) Bioregional Approach 生物区域方法。

生物区域方法研究生物区域资源的总量变化和管理,对于一个生物区域的各种因素进行详细的分析。这些因素与它们之间的相互关系组成了一个生物区域的结构。它内部的所有的生命系统(人类、动物、水、植物、土壤和大气)都被看作是动态系统,能量、水和物质在它们之间流动。

生物区域方法是生态规划设计的主要理论之一,与伊恩麦克哈格的可持续性分析、约翰莱尔的再生设计过程、卡尔斯坦尼兹的生态设计构架等构成了生态规划设计的主要流派。

(3) 本文。

克莱尔·瑞尼格的这篇文章系统地介绍了生物区域规划对于生态系统保护的意义。由于篇幅的原因,这里选取文章的第一部分,介绍生物区域的定义、美国生态保护的问题、生物区域规划的意义、各种生态元素的影响等。

3. Sources of Additional Information

Approaches to Bioregional Planning

http://www.environment.gov.au/biodiversity/publications/series/paper10/index.html

Bioregional Planning

http://www.ruralfutures.une.edu.au/projects/3.php? nav=Landscape%20Mosaics&page=78

http://www.bioregionalplanning.uidaho.edu/default.aspx? pid=97844

Program: Bioregional Planning/ Utah State University

http://ella.gis.usu.edu/bioregionalplanning/program.htm

Centre for Bioregional Resource Management"

http://www.ruralfutures.une.edu.au/projects/3.php? nav=Landscape%20Mosaics&page=78

Further Reading B

Translating Environmental Values into Landscape Design

By Elizabeth K. Meyer

Over the last quarter century, environmentalism has shifted from a fringe[1] issue to a central theme in American cultural consciousness[2] and political discourse[3]. Environmentalism's evolution from a special interest to a broad-based concern among the general population paralleled a re-centering in the practice of landscape architecture. Motivated[4] by environmental values, landscape architects became increasingly knowledgeable about ecological principles and systems. The associated types of design practices were not monolithic[5], representing a single school of thought, but diverse, ranging from "scientific" restoration ecology to sitespecific[6] "artistic" interventions[7], from projects that simulated nature to those that revealed the act of human creativity and construction. The practices of several landscape architects bridged the "great divide" between ecology and design and between science and art that characterized the profession in the 1970s. In constructing this bridge, a body of work has emerged that not only applies ecological environmental values to a design language, but also suggests a strategy for breaking out of the restrictive[8] tenets[9] of modern art that so marginalized[10] the landscape as a medium and subject.

One thread in the tapestry[11] of the postmodern, postenvironmental movement landscape involves the search for significant forms and spaces that might embody[12], reveal, and express ecological principles while embodying and inculcating[13] environmental values. The focus here is directed toward those works of landscape architecture that represent a new type of practice, one that makes the natural world—its ecological and geological processes, its rapid phenomena, and its invisible substructure—more evident, visibly legible, and meaningful to those who live, work, and play in the landscape. One of the trajectories[14] that connects the disparate[15] work of the last two decades in landscape architecture is the

---

[1] fringe: 边缘
[2] consciousness: 意识，觉悟
[3] discourse: 论题
[4] motivate: 推动
[5] monolithic: 单路
[6] sitespecific: 特定场地的
[7] interventions: 干涉
[8] restrictive: 限制性的
[9] tenet: 原则
[10] marginalize: 忽视，排斥
[11] tapestry: 交织
[12] embody: 体现
[13] inculcate: 劝导
[14] trajectory: 轨道
[15] disparate: 全异的

desire to make palpable[1], physical, and aesthetic the intimate interconnections between humans and the natural world, thus constructing experience and engendering[2] a sense of affiliation[3] between humans and nature. This desire has inserted ecological environmentalism into the design process in many places—in programming, site analysis and interpretation[4] form grammars, and construction techniques. This impulse[5] has also challenged the tenet[6] of modern form as an isolated, bounded form or space experienced by a detached, contemplative[7] observer by focusing on the construction of aesthetic experiences bound to, and enmeshed[8] in, their specific cultural and ecological context. The projects infused[9] by these desires and impulses are environmental experiences, not bounded landscape objects, and they constitute what the cultural critic Andreas Huyssen regards as a critical postmodern reconsideration[10] of modern art and culture as filtered through a new lens—in this case, the lens of ecological environmentalism.

The merit[11] of exploring this genre of work lies in its contribution as a mediating[12] practice between two disparate discourses, each with its own language and principles—the dialectic[13] between science and art and between ecological environmentalism and landscape architectural design. For a designer, one conundrum presented by the environmental movement was the disconnection between site analysis and design expression or, in other words, between environmental values and form generation. After identifying the most ecologically valuable or fragile[14] places, designating them "no build" zones, and ascertaining[15] the most ecologically fit location to site a building or construct a designed landscape, how was one to shape the forms and spaces of that landscape? Was it possible to create places with forms significantly different from those of earlier designers who were not as ecologically literate? Could one make the ecological planning process visible to those who came to the site? Would they be able to decode[16] it? Was it necessary to create places that were recognizably different from existing landscapes for a contemporary public that was considerably more environmentally aware? Once having decided that the design vocabulary, syntax[17], and content should inflect the changing values of both designers and design patrons[18], the landscape architect encountered a second problem in the 1980s. How could one give form to dynamic processes and

[1] palpable：明显的
[2] engender：造成
[3] affiliation：联系
[4] interpretation：阐述
[5] impulse：推动
[6] tenet：原则
[7] contemplative：冥想的
[8] enmesh：陷入
[9] infuse：鼓舞
[10] reconsideration：再考虑
[11] merit：优点
[12] mediate：仲裁
[13] dialectic：辩证的
[14] fragile：虚弱的
[15] ascertain：确定
[16] decode：解译
[17] syntax：句法，语法
[18] patron：资助人

fluctuating systems but not resort to the modern design codes that privileged[1] static, bounded, ideal objects in art and architecture and often relegated[2] landscape to visual scenery, a stripped-down version of the pastoral[3]?

These were among the questions that confronted designers during the quarter century after the first Earth Day, which occurred in April 1970. When landscape architects such as Susan Child, George Hargreaves, Catherine Howett, Anne Whiston Spirn, and Michael Van Valkenburgh began their academic and design practices, two strong models existed. The first, environmental or ecological design, had emerged out of the writings and teachings of educators such as Ian McHarg. Its primary contribution to the design process was to structure the preconceptual[4] design phase according to a more defensible, scientific method. The second model, landscape architecture as art, had emerged from the teachings and practice of educators such as Peter Walker who were concerned that the design process had become so beholden[5] to analyses—ecological, social, and behavioral—that the art of making the landscape visible, beautiful, and memorable[6] had been made subservient[7] to the landscape's function. This model's primary contribution was its application, in the conceptual and design development phase, of the vocabulary and tactics[8] of contemporary art to the making of landscapes. These two models existed in isolated opposition from one another, cognizant[9] of the other but operating in separate worlds, based on separate value systems and vocabularies.

This isolation troubled, even confounded, those landscape architects drawn to the discipline in the aftermath of the 1970s environmental movement. The divide between science and art that was integral to late-modern design theory and practice, such as McHarg's and Walker's, was called into question by such postmodern concerns as environmentalism, a strong cultural undercurrent[10] felt by young landscape architects. From the perspective of this postmodern postenvironmental movement, two key issues were identified and then problemmatized in the works of the next generation. First, the lack of formal inquiry[11] or invention in much environmental planning and design ensured landscape architecture's continued invisibility[12], a legacy of modern urbanism. This invisibility was frequently clothed in the pastoral, a romantic conceit[13] preferred by many modern architects and landscape architects who envisioned their projects surrounded by a background of "natural" scenery. This form of pastoral ecological design, so ably chronicled[14]. by the geographer Denis Cosgrove, perpetuated[15] the visual ideology[16] of the modern landscape that reduced the land to pretty scenery

---

[1] privilege: 给与特权
[2] relegate: 转移，转入
[3] pastoral: 牧歌，田园诗
[4] preconceptual: 预概念的，概念前的
[5] beholden: 对……表示感谢
[6] memorable: 难忘的
[7] subservient: 有用的
[8] tactics: 策略
[9] cognizant: 认知的
[10] undercurrent: 潜流
[11] inquiry: 调查
[12] invisibility: 看不清
[13] conceit: 想法
[14] chronicle: 编入编年史
[15] perpetuate: 使不朽
[16] ideology: 意识形态

devoid[1] of ecological and cultural content. Ecologically planned or not, these landscapes did not look managed or designed to most people. They allowed the public as well as developers and designers to ignore the actual impact of construction and sprawl. Second, while the alternative model was successful in "making the landscape visible" through artistic devices such as gesture, flatness, and objectification[2] and in overcoming the emptiness of much modern urban open space, Walker's vocabulary did not acknowledge the difference between the land's surface and materiality and that of a canvas or gallery floor. As such, Walker's minimalist landscapes perpetuated modern art values and ideals, objecthood[3] and detachment, at the very time such values were being challenged by environmental and conceptual artists.

Landscape designers immersed[4] in the postmodern culture encountered artistic and architectural works and theories that questioned the objecthood of sculpture and buildings. These works explored the site-specific characteristics that conditioned their response, reveled in a "systems aesthetics" that intermingled[5] cultural and natural processes, and valued the regional and the place particular over the universal and ideal. Influenced by such theoretical richness, landscape architects found the venues[6] for exploration between the two models to be rich, varied, and productive. Landscape as a subject, a medium, and an inquiry was no longer marginal, but central to contemporary cultural debates and concerns. What could and did this mean for the discipline whose medium, subject, and canvas was the landscape?

The inquiry that followed was not direct and singular, but made by many in meandering and opportunistic[7] forays[8] that sought clues and inspiration from various sources. These included recent works of landscape architecture, such as those by Lawrence Halprin, that represented a first step toward creating a design vocabulary predicated on landforms created by natural processes, such as erosion and deposition[9] due to water and wind; recent works by conceptual artists, such as Hans Haacke and Alan Sonfist, who were probing[10] the boundaries of art objecthood in their process pieces and performances; contemporary environmental and site artists, such as Robert Smithson, Michael Heizer, Mary Miss, and Robert Irwin, who were making site-specific works outside the gallery; and contemporary critics and artists who were translating the ideas of phenomenologists about bodily experience, duration, immersion, and place making into design and art theories.

To some it may seem odd that landscape architects looked toward art and design theory and practice when seeking direction about folding ecological principles and environmental values into their creative processes. But this simultaneous[11] look to art as well as science and to theories of site

---

[1]　devoid：缺乏的
[2]　objectification：对象化，客观化
[3]　objecthood：客观事物
[4]　immerse：沉浸
[5]　intermingle：混合
[6]　venues：地点
[7]　opportunistic：机会主义的
[8]　foray：袭击
[9]　deposition：沉积
[10]　probe：探查
[11]　simultaneous：同时的

specificity and phenomenology as well as ecology was critical to the successful integration of environmentalism into landscape architectural design. Post-Earth Day environmentalism was more than a movement to solve individual ecological problems. It was an attempt to change the value systems that had created those problems and then to modify the institutions that acted on those values. As such, it is not surprising that some landscape architects saw environmentalism and ecological concerns as cultural as much as scientific concerns. These designers created designed landscapes that operated as focusing lenses for knowing the natural world, that instigated[1] aesthetic experiences that reduced barriers between humans and the natural world, and that functioned as physical catalysts[2] for changing social rituals[3] affecting the natural world.

Notes

1. Extracts from

Text

Elizabeth K. Meyer. The Post-Earth Day Conundrum: Translating Environmental Values into Landscape Design [C]. Michel Conan. *Environmentalism in Landscape Architecture*. Washington D. C.: Dumbarton Qaks Research Library and Collection, 2000: 187-191.

2. Background Information

(1) Elizabeth K. Meyer 伊丽莎白·K. 梅耶尔。

梅耶尔教授1993年进入弗吉尼亚大学建筑学院，并担任风景园林系主任和风景园林研究生院院长。此前曾在哈佛大学和康奈尔大学任教。于1992年、2003年、2004年先后被美国景观设计教育家委员会（CELA）、美国景观设计师协会、美国弗吉尼亚大学授予"杰出学者和教师"称号。

梅耶尔教授是美国风景园林师协会（ASLA）会员，也是一名注册风景园林师，曾任职于EDAW和Hanna/Olin，她的作品Bryant公园获得美国国家设计与规划奖。

她的教学与学术兴趣主要集中于三个方面：现代风景园林理论的恢复与审核，当代风景园林实践评价体系的建立，以及设计思想如现场说明（场地的文化层面与自然过程）。

(2) Translating Environmental Values into Landscape Design。

本文选自美国哈佛大学敦巴顿橡树园研究中心论文集 *Environmentalism in Landscape Architecture*，是文章的第一部分。伊丽莎白·K. 梅耶尔的论文具体介绍如何将环境观念引入景观设计，从概念、理论开始，分别介绍主要领域，并结合实例给人全面的了解。

3. Sources of Additional Information

University of Virginia School of Architecture

　　http://www.arch.virginia.edu/

Environmentalism in Landscape Architecture

　　http://www.doaks.org/Environmentalism/env8.pdf

---

[1] instigate：鼓动

[2] catalysts：催化剂

[3] ritual：典礼

Further Reading C

Sustainable Stormwater Management Program

By Kevin Robert Perry

Introduction

The SW 12th Avenue Green Street project, located adjacent to Portland State University in downtown Portland, is unique to Portland and the United States in the way the pedestrian[1] zone of this street has been transformed to sustainably manage street stormwater runoff[2]. As part of the City of Portland's commitment[3] to promote a more natural approach to urban stormwater management, this "green street" project converts the previously underutilized[4] landscape area between the sidewalk and street curb[5] into a series of landscaped stormwater planters designed to capture, slow, cleanse, and infiltrate[6] street runoff. Built in the summer of 2005, this street retrofit[7] project demonstrates[8] how both new and existing streets in downtown or highly urbanized areas can be designed to provide direct environmental benefits and be aesthetically integrated into the urban streetscape. Though this green street project maintains a strong functional component, it is the ability of the landscaped stormwater planters to be integrated into the urban fabric[9] that has the design community, developers, policy makers, and local citizens excited about the SW 12th Avenue Green Street (Figure 7.2).

How does the 12th Avenue Green Street Work?

The 12th Avenue Green Street project essentially disconnects[10] the street's stormwater runoff from the storm drain system that feeds directly into the Willamette River and manages it on-site using a landscape approach. Stormwater runoff from 8,000 square feet of SW 12th Avenue flows downhill along the existing curb until it reaches the first of four stormwater planters. A 12-inch curb cut[11] channels the street runoff into the first stormwater planter. Once inside the planter, the water is allowed to collect until the water level reaches a depth of 6 inches. The landscape system within each

---

[1] pedestrian: 步行的，行人的
[2] runoff: 径流
[3] commitment: 承诺
[4] underutilized: 未充分使用的
[5] curb: 路缘
[6] infiltrate: 渗透
[7] retrofit: 式样翻新
[8] demonstrate: 阐明
[9] urban fabric: 城市肌理
[10] disconnect: 使分离
[11] curb cut: 路缘缺口

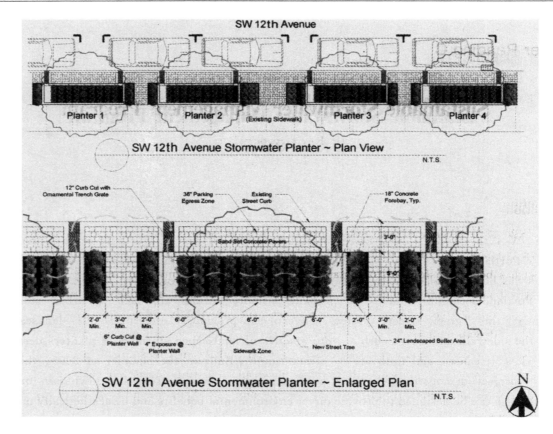

Figure 7.2 SW 12th Avenue Green Street Project

planter allows the water to infiltrate in the soil at a rate of 4 inches per hour. If a rain event is intense[1] enough, water will exit through the planter's second curb cut, flow back out into the street and eventually[2] enter the next downstream stormwater planter. Depending on how intense a particular storm is, runoff will continue its downhill "dance" from planter to planter until all of the stormwater planters are at capacity. Once exceeding capacity, the water exits the last stormwater planter and enters the existing storm drain system. With the new stormwater facilities now in place, nearly all of SW 12th Avenue's annual street runoff, estimated at 180,000 gallons, is managed by its landscape system. In fact, a simulated[3] flow test has shown that stormwater planters at SW 12th Avenue have the ability to reduce the runoff intensity of 25-year storm events by at least 70 percent. Where communities struggle with ever-increasing impervious[4] areas and degraded water quality, these simple landscape approaches can have a measurable positive impact.

Design Solutions for Multiple Uses

The paramount design challenge for retrofitting SW 12th Avenue Green Street was finding sufficient space to locate the stormwater planters while minimizing conflict with other streetscape elements. It was difficult to integrate pedestrians, on-street parking, street trees, landscaping, street

---

[1] intense：剧烈的

[2] eventually：最终

[3] simulate：模拟

[4] impervious：不可渗透的

lighting, signage[1], and stormwater planters within the 8-foot wide zone of space. It was quickly realized that for the stormwater planters to work in tandem[2] with on-street parking that the design must focus on strong pedestrian circulation and connection from the sidewalk to the parking zone. Because of this, multiple design strategies were employed. A 3-foot wide parking egress[3] zone was dedicated for people to access their vehicles without competing with the stormwater planters. Furthermore, perpendicular[4] pathways were located between each stormwater planter so that a pedestrian would not have to walk very far to access their cars or the sidewalk. The design also calls for a 4-inch curb exposure at each planter to help indicate[5] to the pedestrian that there is a drop in grade. Each curb cut that allows the street runoff to enter the stormwater planters has an ADA accessible grate to allow for unencumbered[6] pedestrian flow along the parking egress zone (Figure 7.3).

Figure 7.3 Typical Cross Section

Special attention was also given to the landscape component to the SW 12th Avenue Green Street. The native grooved rush[7] (*Juncus patens*) planted within each stormwater planter is the workhorse[8] for stormwater management. The upright growth structure of *Juncus patens* slows down water flow and captures pollutants, while its deep penetrating[9] roots work well for water absorption. The *Juncus* plants were installed deliberately[10] in rows 18-inches on center to allow

---

[1] signage: 标识
[2] tandem: 串联
[3] egress: 出口
[4] perpendicular: 垂直正交的
[5] indicate: 显示
[6] unencumbered: 不受妨碍的
[7] grooved rush: [植] 灯芯草
[8] workhorse: 驮马
[9] penetrate: 渗透
[10] deliberately: 故意地

space for a leaf rake to easily remove any accumulation of sediment[1] and debris[2] commonly found in urban conditions. The design also boldly locates the project's street trees right in the middle of each stormwater planter. Tupelo[3] Tree (*Nyssa sylvatica*) was chosen because of its tolerance to both wet and dry conditions and its beautiful fall color.

Special Design Considerations and Project Goals

Several design elements used at SW 12th Avenue have helped with the success of this project. One is the elegant and detailed treatment of the curb cuts that allow the stormwater to enter the landscaped planters. Because the focus of this project is on water, it made sense to expend extra detail and design attention by choosing an ornamental trench[4] grate[5] that covers each curb cut. Also, sand-set tumbled concrete unit pavers[6] were utilized in all of the project's pathways so that there was a clear physical and aesthetic separation from the sidewalk zone. Lastly, a landscape buffer was added on the outside of each planter's sidewalls in order to further "soften" the look of the stormwater planters as well as define where the access paths are located.

The design of the SW 12th Avenue Green Street has met three important goals: ①it is low-cost in its design and execution; ②it benefits the environment and embodies community livability[7]; ③it provides a model for other jurisdictions in addressing important national and local stormwater regulations. These stormwater planters are well integrated into the urban streetscape and bring natural hydrologic[8] functions back into the City. The retrofit of SW 12th Avenue Green Street with landscaped stormwater planters cost approximately $30,000 to construct.

Community Partnerships

The success of innovative stormwater projects like the SW 12th Avenue Green Street is dependent on community partnerships. Representatives from Portland State University were active participants in the project's design process. Communication with Portland State University representatives continues to this day to determine the overall success of the project from the neighborhood perspective. In addition, many students and professors within the urban planning and environmental studies departments at Portland State University are excited to use this project in their own research.

In a unique partnership, the city and Portland State University have agreed to share responsibilities in maintaining the new stormwater facilities at SW 12th Avenue. To further engage the community, a small interpretative[9] sign has also been placed at the project site to describe how the stormwater facilities function, as well as how to find more information on sustainable stormwater management practices. Despite the fact that the SW 12th Avenue Green Street has not even reached

---

[1] sediment: 沉积物
[2] debris: 碎片
[3] tupelo: 篮果树
[4] trench: 沟渠
[5] grate: 篦子
[6] paver: 砌块
[7] livability: 宜居性
[8] hydrologic: 水文的
[9] interpretative: 解释的

its first anniversary, the project has caught the attention of visitors from all over the United States, Europe, and Asia. The aesthetic appeal and intrigue[1] of the new stormwater facilities has created a community asset that promotes both environmental stewardship[2] and education at the neighborhood level within the urban core of the city.

Notes

1. Extracts from

Text

Perry, Kevin Robert. Sustainable Stormwater Management Program [EB/OL]. http://asla.org/awards/2006/06winners/341.html, 2007-05-28.

Pictures

(1) Figure 7.2: Kevin Robert Perry. Sustainable Stormwater Management Program [EB/OL]. http://asla.org/awards/2006/06winners/341.html, 2007-05-28.

(2) Figure 7.3: Kevin Robert Perry. Sustainable Stormwater Management Program [EB/OL]. http://asla.org/awards/2006/06winners/341.html, 2007-05-28.

2. Background Information

此项目获得了2006年ASLA常规设计奖，特点在于如何将雨水收集与管理和设计美学相结合。凯文·罗伯特·佩里是将雨洪管理与城市设计结合的领军人物。他设计了十几个波特兰的绿色街道和雨水花园项目。他的工作融合了艺术、教育和生态功能，为他赢得了好几次ASLA奖项。

2007 National ASLA Award of Honor: Mount Tabor Middle School Rain Garden.

2007 National ASLA Award of Honor: NE Siskiyou Green Street.

2006 National ASLA Award of Honor: SW 12th Avenue Green Street.

2006 Oregon ASLA Visionary Honor Award: NE Siskiyou Green Street.

2006 Oregon ASLA Visionary Honor Award: SW 12th Avenue Green Street.

3. Sources of Additional Information

http://asla.org/awards/2006/06winners/

http://www.asla.org/members/awrd_toc.htm

---

[1] intrigue: 谋划，考量
[2] stewardship: 工作

Exercises

1. Questions（阅读思考）

认真阅读本单元文章，并讨论以下问题。
（1）风景园林生态规划与设计的根本宗旨和目标是什么？
（2）风景园林生态规划与设计主要理论有哪些？各自的重点是什么？
（3）在20世纪70年代，环境观念是如何引入风景园林规划设计的？
（4）联系我国情况，谈谈现阶段生态规划与设计在风景园林学科中的地位。

2. Skill Training（技能训练）

（1）查找并阅读国外某个风景园林生态规划与设计项目，记录其背景、内容构成、特色和结果。
（2）用英文简要介绍国外某个风景园林生态规划与设计项目的内容，重点介绍其如何将生态与景观设计进行结合，并以"生态规划与设计理论与实践"为主题，进行交流和讨论。

Tips

风景园林生态规划与生态设计

　　现代风景园林规划与设计专业的产生时候，生态规划与设计就伴随它出现了。现代风景园林的先驱奥姆斯特德及埃里奥特等在城市与区域绿地系统和自然保护区系统的规划，都带有早期生态规划与设计的特征。例如埃里奥特的大波士顿都市区的绿地系统，就是引入生态学的概念来规划设计。

　　然而，此时的生态规划和设计还处于雏形时期，规划与设计者多是具有环境或生态意识，而不具有深刻的生态学知识，因此没有形成完整的生态规划体系。直到麦克哈格提出了一整套的生态规划方法与模式，生态规划与设计专业才真正形成。他所创立的层式叠加的分析法，为规划界所广泛使用。

　　随着环境问题越来越受到关注，生态规划与设计越来越成为风景园林重要的组成部分，也出现很多的流派，比较有名的有四个，一个是麦克哈格的可持续性分析、克莱尔瑞尼格的生态区域、莱尔的再生设计和卡尔斯坦尼兹的生态设计框架。从这些不断发展的理论来看，生态规划与设计越来越强调生态系统与能量流的动态平衡，而不是以往把生态系统作为某一时刻的静态因素来考虑。

　　一些国外的教育机构也越来越在教学与研究中强调生态的作用，例如，美国加州大学伯克利分校的设计学院就名为"环境设计学院"（College of Environment Design），传统意义上的风景园林、城市规划、建筑设计和城市设计专业都被置于其中。

　　在具体的设计上，使用生态的方法来处理景观问题成为热点。例如，人工湿地、雨洪系统的生态化与景观化、废弃地的生态恢复等。很多项目已经取得了良好的视觉效果和生态效益。总体而言，生态规划与设计越来越成为风景园林规划设计的主流，在规划层面上，把景观系统作为生态系统的子集或一部分来进行规划，强调生态系统的动态发展和能量的流动；在设计层面上，模拟自然生态系统的恢复过程，来引导风景园林项目实现的"生长"过程。在这里，景观系统不再是一个静态、不变的系统，而是一个不断生长的体系。

　　目前，对于生态规划与设计仍然存在一些争议，例如生态规划与设计到底是运用生态学概念的风景园林规划和设计，还是艺术化的生态学应用实践；一些生态恢复的手段是否是确实有效的；在对一些生态因素保护的同时会不会造成其他生态因素的毁坏。随着生态学的发展以及风景园林规划设计与其他学科的交流，这些争议不断得以消除，生态规划和设计正在向着更加广阔的空间拓展，并取得更加有效的结果。

Unit 8

Planting Design

● 风景园林种植设计

Considered as materials, all plants have definite potentialities and each plant has an inherent quality which will inevitably express itself. An intelligent landscape design can evolve only from a profound knowledge of, and sensitivity to, materials.

By James C. Rose

Text

Planting Design Through the Ages

By Brian Hackett

> 本文简要归纳了风景园林种植设计历史，讨论了规则式种植、个性配植（意境式种植）、风景式种植、自然式种植四种主要的种植设计方式，并较为详细地阐述了这四种种植设计方式产生的时间、地点、方法及文化内涵。文章主要分以下四个部分。
> （1）规则式种植。
> （2）个性配植（意境式种植）。
> （3）风景式种植。
> （4）自然式种植。

The selection and arrangement of trees, shrubs and other plants in a landscape can be considered as the detailing of the landscape design, although a design philosophy which considers the layout[1] of the landscape together with the plants, hard surfaces and artefacts[2] as one exercise is regarded today as the right way to proceed. It is difficult to separate planting design from landscape design, but there are examples in the past, and indeed today, in which it is clear that the plants were to grow in a manner unlike their natural habitat after the landscape design had been finalized, and thus were selected as a later and separate exercise. The layout of these landscape designs may have survived, but the planting will have changed many times through age and fashion, and because the provision for planting did not favour the regeneration of the plants. Thus, our knowledge of the planting of landscapes laid out more than two or three hundred years ago is limited, and is derived from such varied sources as the written word, old paintings, and the continuation of tradition in remote places. The reconstructed gardens of Colonial Williamsburg[3], Virginia, in the U.S.A. are examples of careful

很难把种植设计从园林设计中独立出来，但不管是过去还是现在都有大量的例子表明：经过园林设计的植物和生长在自然环境中的植物的生长方式是不同的。因此，种植设计通常可以作为园林设计之后的一个专项工作。

---

［1］ layout：设计
［2］ artefacts：人工物
［3］ Colonial Williamsburg：殖民地威廉斯堡，位于美国佛吉尼亚州，是 17 世纪英国殖民者到达美洲后的最初居住地之一

historical research.

Formal Planting

The evidence about planting that is available from ancient times suggests that plants were used to define and highlight formal patterns[1], particularly by the repetition of one species at equal intervals. In the days of Ancient Rome, the art of topiary[2] carried this approach to planting to the stage where landscape design became a form of building design. The Romans did, however, record an interest in plants which produce easily-visible flowers, and if no design principles governed the selection of the species, the plants were contained within the framework of a formal landscape design.

When the Dark Ages[3] gave way to the cultural revival of the Age of Humanism[4] in Europe, there is evidence that the new liberalism[5] was reflected in a less contrived[6] approach to planting. In Italy, during the 15th century, there is evidence of an interest in wild flowers being allowed and encouraged to grow in grassed areas, and of evergreen Holm oak (*Quercus ilex*)[7] woodlands being accepted as part of a "designed" landscape. Soon, however, the planting changed with the selection of a single species to be used for filling a space in a pattern. The species used were often selected for their ability to maintain the pattern, and included evergreen plants and those which would grow with a close texture after clipping. In the 17th century, the pattern, or parterre[8] as it is better known, was more often defined by box (*Buxus sempervirens*) edging. While in France the parterre dominated large areas of the landscape, in Italy it was usually surrounded by woodland with an informal silhouette[9] created by cypresses[10] (*Cupressus* spp.) and pines (*Pinus* spp.).

Some other developments in planting that are associated

---

[1] formal patterns: 规则式图案
[2] topiary: 灌木修剪法
[3] Dark Ages: 欧洲中世纪早期，大约在公元 476～1000 年
[4] Humanism: 人文主义
[5] liberalism: 自由主义
[6] contrived: 人为的
[7] Holm oak: 圣栎（冬青栎）
[8] parterre: 花坛，花圃
[9] silhouette: 外形，轮廓
[10] cypresses: 柏木属植物

• 规则种植

不过，罗马人的确记载了他们对开花植物的偏爱，在没有种植设计原理来决定植物选取的情况下，植物的选取就被列入规则式园林设计的框架之中。

在 17 世纪，规则式图案或者说花圃（这是人们更熟悉的名称）周围总是种着黄杨木，作为花圃的边界。在法国，这种花圃在景区里占地很广，而在意大利，它们由松柏林环绕，形成不规则式图案。

with Italian gardens of the Renaissance[1] were the use of fruit trees for visual effect as well as to produce fruit, and plants in pots and tubs were used in gardens as though they were works of sculpture. The cypress avenue was another form of planting and directed attention to a view beyond the garden or to some feature in the garden, and to a certain extent reduced the strong visual effect of planting arranged to form a pattern.

另一种种植形式是形成柏树林荫道，把游览者的注意力转移到园子以外的视线或者园子中其他特色景点，并且或多或少地削减了排列成图案的植物带来的强烈的视觉效果。

The Individual Plant

• 个体植物

The evidence from the garden landscapes of the Far East[2] discloses an interest in planting design in the visual effect of one plant in relation to another plant, compared with the emphasis on arranging plants to accord with a pattern, which happened in the European civilizations. The preponderance of conifers[3] and broad-leaved evergreens[4] in the gardens of the Far East demonstrated a concern for the Winter effect, and also for the subtlety of the textural effects of the leaves and for the gentle modulation[5] of green colours towards yellow and blue. But the major consideration seems to have been the contrasts between the form of the various plants and the habit or line effects brought out so well in the brushwork of the artists, for example, a twisted and rugged pine tree which had been planted near to a spreading cherry (*Prunus* spp.) or a weeping willow (e.g. *Salix matsudana* "Pendula"). Much the same principles were applied to the smaller plants, such as the water lily (*Nuphar japonicum*)[6], the veronica (*Hebe murorum*)[7], *Pteris aquilinia*, and bamboos (e.g. *A rundinaria angustifolia*), decorative grasses and mosses.

但主要是考虑到把不同植物的形状、习性或者线条感进行对比，就像画家用画笔所表现的那样。例如：在粗犷的松树附近栽种舒展的樱花树或者垂摆的柳树。

Short-lived seasonal effects of strong visual contrast were more sought after in the gardens of the Far East compared with the slower and more gentle changes of the later herbaceous borders and rose gardens of the West. Trees like the lilacs (*Syringa* spp.)[8], laburnum (*Laburnum* spp.)[9], almond

相对于后来西方的草本花坛和玫瑰园缓慢而温和的变化，远东的园林更多地追逐由于季节变迁而带来的强烈的视觉对比。

---

[1] Renaissance：文艺复兴时期，大约在公元 14～16 世纪
[2] Far East：远东地区
[3] conifer：针叶树
[4] broad-leaved evergreen：阔叶常绿树
[5] modulation：韵律的、和谐的运用
[6] *Nuphar japonicum*：日本萍蓬草
[7] veronica (Hebe murorum)：婆婆纳属
[8] lilacs (*Syringa* spp.)：丁香属
[9] laburnum (*Laburnum* spp.)：金链花属

(*Prunus dulcis*)[1] and peach (e.g. *Prunus persica*) were planted for Spring effect, and maples (*Acer palmatum and rufinerve*) for Autumn effect. The Spring effect was further emphasised by groundflora, such as the wild hyacinth (*Scilla* spp.)[2], violets (*Viola* spp.) and primroses (*Primula* spp.)[3].

The association between the landscape designs of the Far East and the natural landscape is familiar to those interested in the history of landscapes, especially in the stylized simulation of mountain and river landscapes, and also the philosophy embodied in this association. On a smaller scale, this philosophy Influenced the planting design. The plant we call Jew's mallow (*Kerria japonica*)[4] and *Lespedeza formosa*[5] when planted near to a stream were meant to convey the idea of a river landscape, and mounds planted with cherries and maples signified a mountain landscape. Clipped evergreen shrubs were planted among rocks to represent islands and hills, There was also an association between certain plants and a particular situation, like *Rhododendron indicum*[6] and the fern[7] (*Polypodium lingua*) at the base of stones.

The Green Landscape

In these historical examples from the Far East and the Moslem world, the interest was in the individual plants as well as in their ability to contribute to the realization of the landscape design, except where grouped effect and function were paramount[8], with topiary and tree planting primarily for shade. With the emergence of the English School of Landscape[9] of the 18th century, plants—in the sense of individuals—were replaced by a very limited range of tree species in groups and belts. Grassed parks with deciduous trees, whose potential monotony[10] was avoided by topographical variety and lakes and streams, were the main elements of the landscape design. This description does not do

远东地区的园林设计和自然风景之间有着密切的联系，对风景史尤其是对山水景观的模仿以及存在于这一联系中的哲学理念有兴趣的人都非常熟悉这种联系。

● 绿色景观

在远东和穆斯林国家的这些历史园林中，人们栽种个体植物不仅因为喜爱，而且还因为它们有助于景观设计的实现，不过，当人们把群体植被效应和功能放在首要位置时或是种植树和灌木主要为了提供绿荫时，人们就不在乎个体植物了。

当时的园林主要是设计有草坪和落叶树的公园，不同的地形以及湖泊、溪流使得这些公园不至于千篇一律。

---

[1] almond (*Prunus dulcis*)：杏树
[2] hyacinth (*Scilla* spp.)：风信子属
[3] primroses (*Primula* spp.)：报春花属
[4] *Kerria japonica*：棣棠
[5] *Lespedeza formosa*：美丽胡枝子
[6] *Rhododendron indicum*：皋月杜鹃
[7] fern：蕨类植物
[8] paramount：最重要的
[9] English School of Landscape：英国自然风景园林流派
[10] monotony：单调

justice to some innovations in planting which were introduced from time to time. The Cedar of Lebanon (*Cedrus libani*)[1] and the evergreen Holm oak are two grand specimen trees that were known in Britain in the middle of the 17th century; towards the end of his life, the 6th Earl of Haddington[2] (died 1735) expressed an interest in flowering trees and shrubs, and in the yellow stemmed willows (*Salix alba* "*Vitellina*"); the Hon. Charles Hamilton at Painshill in Surrey in the middle of the 18th century planted trees and shrubs that were new to Britain—in contrast to Capability Brown's[3] familiar manner of planting as many as 100000 trees on one estate, using mostly one or two species, such as beech (*Fagus sylvatica*)[4], oak (*Quercus* spp.) or pine (Figure 8.1). Henry Hoare at another famous example of the English School of Landscape—Stourhead in Wiltshire—also moved a step away from the bare trees and grass to add cherry laurel[5] (*Prunus laurocerasus*) underplanting[6].

香柏和常绿圣栎是17世纪中期英国人熟知的两大树种。哈迪顿伯爵六世（卒于1735年）晚年偏爱开花的植物、灌木和金丝垂柳；18世纪中期，令人尊敬的查尔斯·汉密尔顿先生在萨里郡的彭斯希尔种了许多不为当时英国人所知的树和灌木——这和可为布朗截然相反。可为布朗通常在一个庄园种上10万株之多的树木，而树种大多只有1~2种，例如山毛榉、橡木或是松木。

Figure 8.1 Blenheim Palace
(the English School of Landscape)

While the practitioners of the English School of Landscape were using a limited range of plant species, a few botanical gardens and, of course, the traditional cottage garden contained a wider range. Certainly by the end of the first quarter of the 18th century, the importation of exotics[7] into Britain from abroad had commenced, especially from North

英国自然风景园林流派的追随者使用的植物种类非常有限，但一些植物园，当然还有传统的别墅花园里的植物种类却相当繁多。

---

[1] *Cedrus libani*：雪松
[2] 6th Earl of Haddington：哈迪顿伯爵六世（托马斯·汉密尔顿，1680~1735年），农学家
[3] Capability Brown：可为布朗，英国园林设计师
[4] beech (*Fagus sylvatica*)：欧洲山毛榉
[5] cherry laurel：桂樱树
[6] underplanting：下木栽植
[7] exotic：异国情调的，外来的

America. In 1765 the collection at Whitton, near Hounslow, listed 342 different plant species, and by 1768 the Royal garden at Kew[1] had over 3000 different species with the precise figure of 5535 in 1789. More interesting to planting design, however, was the record of Dr. John Fothergill's garden laid out in 1762 at Upton House, East Ham, because this included a wild area in which hardy exotics successfully naturalized. Dr. Fothergill also cultivated alpine[2] plants.

The landowners of the 18th century in the rural areas were not alone in the contribution they made to increasing the range of plant species available for planting landscape designs, because the employers, and their employees, of the new manufacturing industries were also active in an interest in plants. In particular, expensive plants like unusual tulip bulbs were associated with the employers, and the hybridization[3] of varieties of auricula (*Primula auricula* var.), carnations[4] (*Dianthus caryophyllus* var.), pinks (*Dianthus plumarius* var.) and polyanthus[5] (*Primula vulgaris elatior* var.) with their employees.

Natural planting

During this period when the range of plants used in gardens increased, the interest of some people was concentrated on a particular family of plants such as *Rosa* species (the catalogue of Rivers and Son of Sawbridgeworth, Hertfordshire, in 1836 lists many species and varieties), and books by James Shirley Hibberd published later in the 19th century included the titles The *Fern Garden* (1869) and *The Ivy* (1872). But this concentration of interest, compared with several plant species grouped in a happy association based upon Nature's patterns or a visual effect, was countered during the latter part of the century as a result of the influence of William Robinson, Gertrude Jekyll and Reginald Farrer, with whom we associate planting expressed in "the wild woodland garden", "the herbaceous border", and "the alpine garden". These particular examples of planting design were far removed from the influence of geometry and building architecture, and allowed the planting to speak for itself.

---

[1] the Royal garden at Kew：丘园，英国皇家植物园
[2] alpine：高山的，阿尔卑斯山的
[3] hybridization：杂交，杂种培植
[4] carnations：荷兰石竹，康乃馨
[5] polyanthus：西洋樱草

Across the Atlantic Jens Jensen, working in the Chicago area of the U. S. A. from 1886, soon began to use native plants in his landscape projects. But particularly from 1905 his design approach embraced the natural prairie[1] landscape of the Mid-West as the model, although he does not seem to have used prairie grasses. Nevertheless, he wrote late in life:

Leave the native growth alone, and fill in with oaks, sugar maple, cherry, ironwood, shad, hawthorn[2], plum and crab apple. Elm[3] and ash[4] for the low places. You may also use tulip, pepperidge, dogwood[5], and sassafras[6]. Sycamore[7] in low places. Hickories[8] are difficult to plant, but they are a good mixture.

Jensen was not alone in the use of native plants in the U. S. A. for W. Miller, O. C. Simonds, and F. L. Olmsted must also be given credit for this important approach to planting design which has come to the fore again with an ecological emphasis from the middle of the present century.

除了延森，还有 W. 米勒，O. C. 西蒙兹及 F. L. 奥姆斯特德都在种植设计中使用了美国本土的植物，而这一重要手法自本世纪中期以来一直被普遍运用，并且强调了生态理念。

A very different use of native plants was evident in the early work of Burle Marx[9], the Brazilian landscape architect, around 1937, when he collected wild plants which would form controlled abstract patterns on the surface of the ground instead of arranging the plants in mixed associations not too dissimilar from associations in natural landscapes. But in the contemporary scene, the less formal and contrived ideas of Robinson, Jensen and the others are relevant to the counter movement away from all that mechanization[10] has brought, and to the economic and social change which has made the gardening operations for maintaining formalized planting so expensive, for example, as described in an account of the restoration of the formal gardens at Herrenhausen in Hanover, West Germany. Thus, the use of wild plants, groundcover plants in place of neatly weeded earth, and the mass planting of

他收集能在地面上形成一定抽象图案的野生植物，而不是将植物排列成不是那么不同于自然群丛的混合群丛。

---

[1] prairie：大草原，牧场
[2] hawthorn：山楂
[3] elm：榆树
[4] ash：岑树
[5] dogwood：山茱萸
[6] sassafras：檫木
[7] sycamore：小无花果树
[8] hickories：山胡桃树
[9] Burle Marx：布雷·马克斯，巴西著名风景园林师
[10] mechanization：机械化

a single species of shrub are typical of planting design today. There is also a more detailed concern for the visual interest of the individual plant and in the way it associates with another plant in both an ecological and visual sense.

Notes

1. Extracts from

Text

Hackett, Brian. *Planting Design* [M]. London: E &F. N. Spon Ltd., 1979: 2-11.

Picture

Figure 8.1: 郦芷若, 朱建宁. 西方园林 [M]. 郑州: 河南科学技术出版社, 2002: 325.

2. Background Information

(1) Brian Hackett 布赖恩·汉克特。

英国纽卡斯尔大学风景园林教授。

(2) *Planting Design*《风景园林种植设计》。

该书在风景园林种植设计方面影响甚广。全书共分15章，讨论15个种植设计方面的问题，深入浅出，文字简练，自出版后深受风景园林学子欢迎。

本文选自该书第一章，有删节。

3. Sources of Additional Information

Colonial Williamsburg

    http://www.history.org/

English School of Landscape

    http://www.apl.ncl.ac.uk/coursework/IThompson/Informal_2.htm

Hamilton Park at Painshill in Surrey

    http://www.painshill.co.uk/

Further Reading A

Plants Dictate Garden Forms

By James C. Rose

It is quite common and almost a boast among some Landscape Architects that they have "never planted a seed" or "don't know one plant from another" —an "art for art's sake" attitude which puts landscape design farther and farther from contemporary life and is therefore the worst possible salesmanship[1]. Can you fancy an architect selling a client on the basis that he knew and cared nothing about brick, wood, and concrete[2]? Or that he was too concerned with beauty to bother about them?

Considered as materials, all plants have definite potentialities[3] and each plant has an inherent quality which will inevitably express itself. An intelligent landscape design can evolve only from a profound knowledge of, and sensitivity to, materials. When we force materials in architecture or sculpture we are sure of at least one thing: that the form, however offensive[4], will be relatively constant. But with plants the struggle is endless and results in victory neither for the plant nor the man who clipped it. If the plant should win the design would be lost; and if the man should win he would succeed only in preserving something false from the beginning.

The 20th century landscape, although hardly touched, even in theory, would of necessity result in the honest use of materials as in the best modern architecture and sculpture. Plants are not applied to a preconceived ground pattern. They dictate form as surely as do use and circulation.

It is only fair to mention that some Beaux Arts[5] landscape designers have used plants for their inherent quality, but under this system they could never be more than decorative elements made to conform to an eclectic[6], ornamental, imposed, geometric pattern with an eye to pictorial composition alone. No striving for the picturesque[7] has resulted in more than a superficial camouflage[8] or facade[9] which obscures rather than expresses the real meaning. Character develops from the actualities which have been solved rather than obscured. Otherwise, we have only the personality of a "great individual" who softens everything with a disguise of ornament to protect hothouse souls from contact with reality.

---

[1] salesmanship：推销术，说服力，销售，推销
[2] concrete：混凝土
[3] potentiality：潜力，可能性
[4] offensive：讨厌的，无礼的，攻击性的
[5] Beaux Arts：巴黎美术学院艺术，也称学院派艺术风格美术
[6] eclectic：折衷的，折衷学派的
[7] picturesque：独特的
[8] camouflage：伪装
[9] facade：正面

It is a vain person who believes that he is free of his own times and can create a detached "thing of beauty." The personal equation exists only as a minor part of the social equation; as society becomes interdependent its expression is social and the "great individual" passes from the scene. It is then that the elements of our own environment become integrated and virtuosity[1] loses meaning. It is then that an expressive style evolves and design ceases to be an eclectic adaptation of ornament to provide a picturesque setting for idle people.

We cannot live in pictures, and therefore a landscape designed as a series of pictures robs us of an opportunity to use that area for animated[2] living. The war cry is often heard that we must combine use and beauty; but by this is meant that we should develop a ground pattern of segregated[3], geometric areas strung along an axis in Beaux Arts relationship and separated by "embellishments[4]" which compose a picture for the "terminal point" of each area. Something to look at!

This may be called exterior decoration, but from the standpoint of twentieth century design, it has no justification. The intrinsic beauty and meaning of a landscape design come from the organic relationship between materials and the division of space in volume to express and satisfy the use for which it is intended. From this viewpoint, the landscape "picture" fades with the "facade" and clears the deck for animated design. Now we can throw away the rubber stamp of Beaux Arts tradition and, although a continuity of style will rightly develop, the solution of each problem will acquire individuality and distinction because it is based on the organic integration of almost inexhaustible[5] material, existing conditions, and the factors of use which could never repeat themselves exactly in all cases.

Ornamentation with plants in landscape design to create "pictures" or picturesque effect means what ornamentation has always meant: the fate call of an outworn[6] system of aesthetics[7]. It has always been the closing chapter of art which had nothing more to say. In one last hasty attempt to propagate, it sings the same old song with a more rasping[8] voice and sordid[9] emphasis. What a pity so few Landscape Architects realize the opportunity they are overlooking in not examining the possibilities of the contemporary approach. It might justify a profession which has, until now, rightly been tagged a useless luxury for the idle and not-too-intelligent rich.

Notes

1. Extracts from
Text
Rose, James C.. Plants Dictate Garden Forms [C]. Marc Treib. *A Critical Review*. London:

[1] virtuosity：艺术鉴别力
[2] animated：活生生的，活泼的，动的，愉快的
[3] segregate：隔离
[4] embellishment：装饰，修饰，润色
[5] inexhaustible：无穷无尽的
[6] outworn：用旧的，废弃的
[7] aesthetics：美学，美术理论，审美学，美的哲学
[8] rasping：锉磨声的，令人焦躁的
[9] sordid：肮脏的

The MIT Press:1993:73-75.

    2. Background Information

James C. Rose 詹姆士·C. 罗斯（1910～1991年）。

美国现代主义景观设计（风景园林）运动的领军人物。本文发表于1938年 *Pencil Points* 杂志，是现代风景园林种植设计重要文献。

    3. Sources of Additional Information

      http://www.jamesrosecenter.org/jamesrose/bibliography/index.html

Further Reading B

The Aesthetics of Planting Design

By Michael Laurie

In order to achieve this level of refinement[1] in landscape design, it is necessary to define the units of the medium (*i.e.*, plants) in terms of their shape or form, their texture (as produced by their leaves and branches), and color (as produced by leaves and flowers). These characteristics are found in plants of all types in the horticultural[2] classification. Just as plants can be ordered into lists on the basis of their usefulness for practical purposes, so they can be ordered in terms of these characteristics.

Form: Size, Shape, Habit, Density

A critical aspect of shrubs and trees in design is their ultimate size: height and spread (often referred to as stature[3]). Where plants are to provide horizontal and vertical definition of space, alone or in combination with architecture, knowledge of size is essential if the desired ultimate scale and proportion[4] relationships are to be achieved. Next comes the question of shape. It is tempting to think of shape as silhouette[5], to say that a young pine tree is triangular or that the form of an oak is round. They are, of course, conical[6] and globular, i.e., three dimensional, with substantial depth, casting shadows which reinforce[7] the reality of their volume. These volumes, which consist chiefly of branches, twigs, and leaves, vary from one another according to their growth pattern, i.e., how the branches originate from the ground, and how they divide and bear the leaves, flowers, and fruits. This is often described as the habit of a plant, the third variable within the category of form. To some extent trees and shrubs with sparse[8] or open foliage[9], or because the foliage is close to the branches, reveal their branching habits more than those densely clothed. Deciduous trees[10] automatically reveal their structure in winter when the growth pattern becomes clearly visible. In any event, plants that have easily identifiable growth pattern, e.g., a horizontally, a drooping, or an ascending quality, can be selected accordingly for a design purpose (contrast, drama, focus, and so on). With time and cultivation, the habit of a tree may change.

---

[1] refinement: 精致，(言谈，举止等的) 文雅，精巧
[2] horticultural: 园艺的
[3] stature: 身高，身材，(精神、道德等的) 高度
[4] proportion: 比例，均衡，面积，部分
[5] silhouette: 侧面影像，轮廓
[6] conical: 圆锥的，圆锥形的
[7] reinforce: 加强，增援，补充，增加……的数量，修补，加固
[8] sparse: 稀少的，稀疏的
[9] foliage: 树叶，植物
[10] Deciduous tree: 落叶树

The organization of the branches, the twigs, and the size of the leaves gives plants yet another quality-relative density. This is a measure of the extent to which light can filter through the structure. This affects the intensity of shade (from dappled to deep) and the degree of visual screen and wind shelter. This quality also varies with time, and (particularly with deciduous plants) with season.

Within the attribute of form, the four characteristics of size, shape, habit, and density are relatively definable.

Color

Color is an attribute of plants provided by all of their parts-leaves, flowers, fruits, twigs, branches, and bark. Frequently the effects are seasonal and therefore, subject to change. These characteristics are further complicated by other variables from the outdoors, weather, light, and shade. These variables modify and constantly change the color of plants and other objects in the landscape. Thus, absolute color theory, which is a quantifiable[1] science for the artist or interior designer, becomes for the landscape architect the basis of a relatively unpredictable[2] phenomenon due to environmental conditions.

Green (and the range of greens from yellow to blue) is essentially a restful color. As a contrast to architecture, it is soothing[3] to the eye. In the landscape the many shades of green, yellow, and brown provide an harmonious matrix[4] for incidents such as buildings, patches of bright color and water, which create focus and contrast, vitalizing[5] the view. Since it is so fundamental a color in the appearance of plants, it is important to have some way of describing the various shades of green. Bracken suggests the selection of a plant with middle green color (*Acer saccharum*[6] is his choice) as a point of reference against which the green of other plants is compared on two scales, dark to light and yellow to blue. Thus, the foliage of any given plant would be described as either darker or lighter, yellower or bluer than the green of the sugar maple (or any other tree chosen as the median). But it is not that easy, since the foliage of deciduous plants frequently changes as the season progresses from spring to fall. The yellow green of newly opened leaves turns to the darker green of summer and before falling may turn bright red or yellow.

In addition to green and the seasonally induced changes, the foliage of some plants is, in fact, a color other than green, or may be green combined with another color on the reverse side of the leaf, or in a variegated[7] form. Thus purple[8] and gray and combinations of white and green must be added to the already complex set of foliage colors and seasonal changes.

Even though we cannot be very scientific in our expectations from the use and combination of colors in the landscape, a basic knowledge of color theory and color phenomena helps to explain the

---

[1] quantifiable：可以计量的
[2] unpredictable：不可预知的
[3] soothing：抚慰的，使人宽心的
[4] matrix：矩阵
[5] vitalize：激发
[6] *Acer saccharum*：糖枫（植物拉丁名）：原产加拿大东部和美国，属落叶乔木
[7] variegated：杂色的，斑驳的，多样化的
[8] purple：紫色的

striking and successful effects identified in the field. But for landscape purposes, and with the uncontrollable variables[1] already discussed, we need to derive color principles that have a realistic chance of noticeable impact in the three dimensional dynamic outdoor environment. It is my judgment that only the crudest theory of harmony and contrast in the use of color (and texture) is worth consideration in landscape design. In more detailed situations, flower borders and particular areas subject to close scrutiny[2], it may be possible to be more sophisticated[3] in the organization of color combinations.

In the same way that green varies according to its lightness or darkness, blueness or yellowness, other colors, found in flowers and building materials, can be described in terms of their relationship to the basic hue[4] (whiter or blacker) and to the adjoining[5] hue in the color spectrum[6] (the tendency of blue to yellow through green or of red to yellow through orange, and so forth). The problem for the landscape architect is to what extent color distinction observed scientifically in the laboratory can be made meaningful for design in the outdoors. It seems possible to identify eight variations of any of the six basic hues and this makes a workable system.

Texture

The practical difficulties of consistency, which we have described in dealing with plant color in the landscape, apply equally to the attribute of texture. Variables in the environment such as light and distance, and in the plants themselves, due to seasons and specific growing conditions, make it difficult to produce a reliable and uniform categorization[7] of plant textures in which minute differences are taken into account. Our interest lies in the broadest differences on a relative scale. As with color, a middle value plant may be selected and the textures of others compared to it on a scale of coarser to finer, with perhaps two major categories in each direction from the median.

Variations between two plants, such as size and form of the leaves, how they are attached to the twigs, and how the twigs are subtended by the branches, are responsible, with light, for differences in texture. Seen close, at a distance of two or three feet, as we see shrubs and flowers, even the surface quality of the leaf or bark[8] (e.g., shiny or matte[9]), is important. The overall effect may be assessed through half-closed eyes or some other device which will subjugate[10] detail for pattern. We must then assign it a value. Fine and coarse are the terms traditionally used. Perhaps the addition of "grain" is helpful, giving us fine grain, and coarse grain. A middle value and perhaps "very fine" and "very coarse" for clearly identifiable extremes, may also be established on a relative scale. But the textural grain of a plant varies with distance. Colvin has suggested three critical distances, other

---

[1] uncontrollable variables：不可控变量
[2] scrutiny：详细审查
[3] sophisticated：诡辩的，久经世故的
[4] hue：色调，样子，颜色，色彩
[5] adjoining：邻接的，隔壁的
[6] spectrum：光，光谱，型谱
[7] categorize：加以类别，分类
[8] bark：树皮
[9] matte：不光滑的
[10] subjugate：使屈服，征服，使服从，克制，抑制

landscape architects, two. The bunching of leaves on shrubs and trees may provide a different pattern of shadows when seen at 50 to 100 feet than it does at close hand. At a still greater distance, say 100 yards to 1/4 mile, plant texture values in relation to others may change again. These greater distance variations apply only to large trees, since shrubs tend to lose their apparent texture at such distances, retaining form to some extent and color. At the most, therefore, we could have a scale of three distances and five values of textural grain. The seasonal variable of greatest importance for texture is the winter appearance of deciduous plants. Thus, branches and twigs give yet another set of texture values to plants differentiating it from its summer appearance.

The clipping of shrubs and trees into hedges or forms[1] (topiary[2]), of course, changes their visual texture. However, hedges made from different plants also vary, but most successful hedges are in the "fine" or "finest" category with possible winter variations if deciduous plants are used.

Notes

 1. Extracts from

 Text

 Laurie, Michael. *An Introduction to Landscape Architecture* [M]. New York: Elsevier Science Publishing Co. Inc, 1986: 74 – 85.

 2. Background Information

 (1) Michael Laurie 迈克尔·劳里（1932～2002 年）。

美国加州伯克利大学风景园林教授。出生于苏格兰，曾在宾夕法尼亚大学和麦克哈格一起工作，以大尺度规划与生态规划方面的研究而著称。20 世纪 70 年代起在加州伯克利大学风景园林系任教，至 1998 年退休，期间曾三次担任系主任。

 (2) *An Introduction to Landscape Architecture* 《风景园林学导论》。

著名风景园林专业教材，本文选自第 2 版 "Plants and Planting Design" 部分。

 3. Sources of Additional Information

 http://www.universityofcalifornia.edu/senate/inmemoriam/MichaelM.Laurie.htm

---

[1] forms：[计] 窗体
[2] topiary：灌木修剪法

Further Reading C

Plant Strata[1], Size and Spatial Issues

By John L. Motloch

Strata refers to the various horizontal layers that constitute a plant community: canopy[2], understory[3], shrub, and ground cover (Figure 8.2). Size usually refers to the height to the top of the plant. In discussing size and strata, we will look at large and intermediate (canopy) trees, small and flowering (understory) trees, large shrubs, intermediate shrubs, small shrub and ground covers.

Large and Intermediate Trees

According to most plant lists, large trees are those generally taller than 40 feet; intermediate trees,

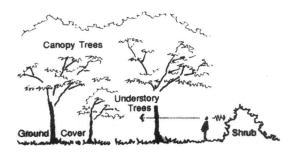

Figure 8.2 Plant strata

grow to 30 or 40 feet. Together, large and intermediate trees form the vegetated canopy. From outside, they create mass; from within, they form canopied space. Their trunks imply but do not enclose space. The spaces they form usually have ceilings but no walls, only columns. They remain fairly open at eye level.

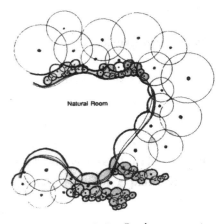

Figure 8.3 Outdoor room

An overhead canopy changes the character of sunlight from hard and glaring to soft and dappled. This dappled[4] quality of light is important in the sensual[5] aspect of being in the woods. When the canopy of large and intermediate trees is fairly continuous, a break in the canopy takes on the character of an outdoor room, open to the sky (Figure 8.3). The hard, direct sunshine penetrating the space contributes to its dynamism[6].

Large and intermediate trees are effective microclimate modifiers. They provide shade from high-and mid-angle sun. If closed at their edge by lower branching material or if great in depth (in the direction of airflow), they can substantially[7] reduce

---

[1] strata: stratum 的复数，[生] 层
[2] canopy: 天篷，遮篷，这里指上层乔木
[3] understory: 林下叶层，这里指中层乔木
[4] dappled: 有斑点的
[5] sensual: 感觉的
[6] dynamism: 活力
[7] substantially: 相当大地

ventilation[1]. On the other hand, if the edge is open and the mass is relatively shallow, they can accelerate airflow because the wind can be compressed and forced below their canopy.

In planting design, large and intermediate trees can provide mass and contribute large scale. If planted to extend the lines or rhythm of architecture into exterior space, tree trunks can give an architectural character to the site.

Small and Flowering (Understory) Trees

Small and flowering trees grow 15 to 20 feet in height. Their form and intensity of flower can differ substantially if they are growing beneath the canopy of large and intermediate trees, or in the open sun. In the open sun, their growth is thicker and rounder and their flowering more intense.

When canopies occur above head height, small and flowering trees imply intimate[2] space. When canopies occur at eye level, they enclose this space (Figure 8.4). These trees are effective in small or intimate courtyards because they provide color and shade without overpowering the space. In such contexts, they are often used as accent[3] plants or focal points.

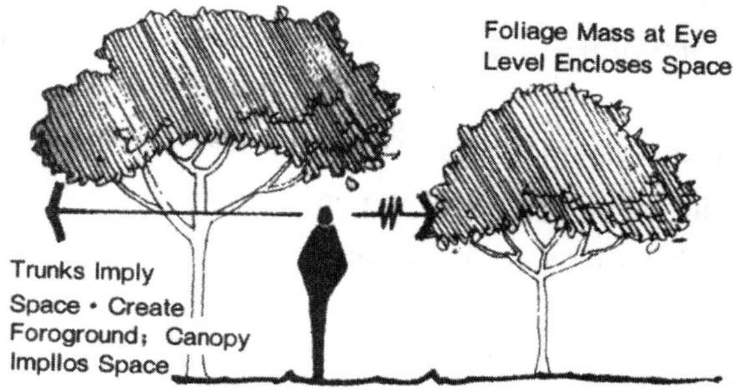

Small and flowering trees are effective at screening mid-to low-angle sun. They are often used on the southwest sides of buildings, or on the west and northwest sides if augmented by low branching shrubs.

Figure 8.4 Small trees and spatial enclosure

Tall Shrubs

Tall shrubs grow to 15 feet in height. They are shorter and lack the canopy of small trees. Their foliage usually extends close to the ground. They provide a strong sense of enclosure and a high degree of privacy. They are effective screens. Tall shrubs can serve as sculptural elements in a large space. They can also be backdrops[4] against which to display smaller plants or sculpture.

Intermediate and Low Shrubs

Intermediate shrubs grow 3 to 6 feet in height, low shrubs, 1 to 3 feet. Low and intermediate shrubs define and physically separate spaces without blocking[5] vision. They serve as constraints to pedestrian movement. Low shrubs provide a weak visual separation; intermediate shrubs, a strong one.

Intermediate shrubs can become disconcerting if their tops occur at eye level. Tension is created as the observer tries to see over the top. Therefore, it is best to avoid materials that top out in

---

[1] ventilation: 通风，流通空气
[2] intimate: 亲密的，私密的
[3] accent: 强调
[4] backdrop: 背景
[5] blocking: 堵塞

between 4 to 6 feet, unless backed by taller materials.

Low shrubs can effectively link groups of larger plants while allowing vision to penetrate between the groups (Figure 8.5). In this way, the shrubs can effectively unify a composition.

Ground Covers

Ground covers define planted areas. Like small shrubs, they can unify groups of larger plants into one composition. They also imply spatial edge and create lines that can lead the eye to focal points, building entries, or other important parts of a design composition. Ground covers also can create lines of visual character and provide detail as they overlap paving or fine textured turf.

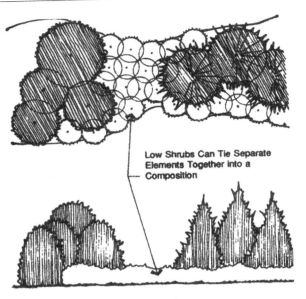

Figure 8.5 Low shrubs linking plant masses

……

Spatial Issues

Spatial issues explored in this section include enclosure, spatial type, spatial depth. enframement, and plant material and landform relationships.

Spatial Enclosure

Enclosure refers to the perceived degree of separation of space. Plant materials that block vision provide enclosure; those that do not only imply enclosure.

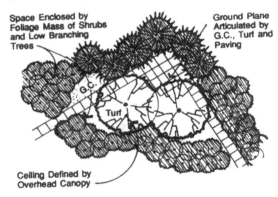

Figure 8.6 Spatial Definition

Plants at eye level enclose space. Solid plant masses provide strong enclosure; porous[1] masses produce only partial enclosure. Overall spatial definition and enclosure are the net visual effect of the various strata working together. As shown in Figure 8.6.

Spatial Type

Spaces vary by type. In scale, they range from intimate to public, in direction from horizontal to vertical[2], and in enclosure from fully enclosed to open and unarticulated.

Spatial Depth

Spatial composition is more effective if it incorporates foreground, middle ground, and

---

[1] porous: 能渗透的，多孔的
[2] vertical: 垂直面，竖向

background. Foreground places the viewer into the space, while middle ground usually serves as subject matter and is displayed in the context of background.

The skilled designer can use a foreground of plants to frame or enclose subject matter, and can provide a high degree of contrast (in color, texture, and so on) between the subject and its background.

The relationship between foreground and background can be designed to accentuate or to mitigate[1] depth. By using coarse-grained plant materials in the foreground and fine-grained ones in the background, the designer can extend visual depth. Conversely, the use of fine-textured materials in the foreground and coarse-grained ones in the background will visually foreshorten space.

Notes

1. Extracts from

Text

Motloch, John L. *Introduction to Landscape Design* [M]. New York: Van Nostrand Reinhold, 1991: 82–86.

Pictures

(1) Figure 8.2: John L. Motloch. *Introduction to Landscape Design* [M]. New York: Van Nostrand Reinhold, 1991: 82.

(2) Figure 8.3: John L. Motloch. *Introduction to Landscape Design* [M]. New York: Van Nostrand Reinhold, 1991: 83.

(3) Figure 8.4: John L. Motloch. *Introduction to Landscape Design* [M]. New York: Van Nostrand Reinhold, 1991: 84.

(4) Figure 8.5: John L. Motloch. *Introduction to Landscape Design* [M]. New York: Van Nostrand Reinhold, 1991: 85.

(5) Figure 8.6: John L. Motloch. *Introduction to Landscape Design* [M]. New York: Van Nostrand Reinhold, 1991: 86.

2. Background Information

(1) John L. Motloch 约翰·L·莫特洛克。

博士，美国注册风景园林师、建筑师、室内设计师，德克萨斯大学（Texas A&M University）风景园林系教授、系主任。

(2) *Introduction to Landscape Design*《风景设计概论》。

德克萨斯大学风景园林系编制教材，被全世界许多院校选为教科书，在国际风景园林界有较大影响，有中译本。本文选自第五章 Resources and Technology 中的 Plant 一节。

3. Sources of Additional Information

http://www.hortworld.com/

---

[1] mitigate：减轻

Exercises

1. Questions（阅读思考）

认真阅读本单元文章，并讨论以下问题。
（1）世界风景园林种植设计有哪几种主要的风格？各有什么特点？
（2）中西方传统园林种植设计的异同和相互影响。
（3）现代风景园林种植设计的特点。
（4）简述风景园林种植设计一般设计原则和步骤。

2. Skill Training 技能训练

（1）每位同学阅读并研究一个英美国家风景园林规划设计案例，分析总结其种植设计特色。
（2）根据研究案例心得体会，虚拟设计一个小庭院，写好种植设计英文说明书。用英语向同学介绍该设计，并开展讨论与交流。

Tips

植物拉丁学名简介

现行用拉丁文为植物命名的体系是由瑞典博物学家林奈（Carolus Linnaeus）于 1735 年出版的《植物种志》(*Species Plantarum*) 中首先提出来的，这个体系称作林奈双名法命名体系（Linnaean binomial system of nomenclature）。世界上的植物种类很多，各国的语言和文字又不相同，因而植物的名称千差万别，为了科学技术的交流，统一使用植物学名是完全必要的。掌握植物的学名，还有助于了解植物的亲缘关系和形态特征。

1867 年 8 月在法国巴黎举行的第一次国际植物学会议制定了《国际植物命名法规》(International Code of Botanical Nomenclature)，称为巴黎法规或巴黎规则。1910 年在比利时布鲁塞尔召开的第三次国际植物学会议，奠定了现行国际植物命名法规的基础。以后在每 6 年召开一次的国际植物学大会上都要对该法规进行修订和补充。国际植物命名法规是各国植物分类学者对植物命名所必须遵循的规章。

1. 植物学名的基本组成

植物学名是用拉丁文或拉丁化的希腊等国文字书写。一个完整的植物学名＝属名＋种加词＋命名人，并规定属名和命名人的第一个拉丁字母必须大写，如火炬松的植物学名是 *Pinus taeda* L.，其中 *Pinus* 为松属的属名，*taeda* 为种加词，L. 是命名人 Linne 的缩写。

值得注意的是，属名、种加词要用斜体或者下划线表示。另外，在多数园林文献中，命名者的名字通常省略，于是火炬松的名字经常写作 *Pinus taeda*。

2. 植物学名中的属名

属名多由名词担任，通常根据植物的特征、特性、产地、纪念人等而命名。

用植物特征命名的，如菊科紫菀属 *Aster*，系指本属植物的头状花序似星，拉丁文的星星就是 aster；菊属 *Chrysanthemum*，希腊文 chrysos 是黄色，anthos 是花；以植物特性命名的，如甘草属 *Glycyrrhiza*（甜的根茎），指本属植物的根茎具有甜味；以人名命名的，如羊蹄甲属 *Bauhinia*，纪念 16 世纪瑞士植物学家 Jean Bauhin（1541～1613 年）及 Caspar Bauhin（1560～1624 年）两兄弟。

3. 植物学名中的种加词

植物学名中的种加词通常使用形容词和名词，也根据植物的特征、特性、产地、纪念人等而命名。使用形容词作种加词，如白皮松 *Pinus bungeana* 的种加词 *bungeana* 含义为白色的，黄山松 *Pinus taiwanensis* 的种加词 *taiwanensis* 含义为台湾的。使用名词作种加词，如用以纪念某一分类学家或某一标本采集者，这种情况在人名后加 i、ii 或 ae，如南洋杉 *Araucaria cunninghamii* Sweet，种加词 *cunninghamii* 是为了纪念植物学家和探险家 Allan Cunningham.

尽管推荐所有的种加词应当小写，但也有些可以例外，在下述情况下种加词的首字母可以大写：①种加词源于一个人的人名；②源于以前的一种属名；③源于一个俗名。于是，人们可以看到大卫枫（David's Maple）可以写作 *Acer Davidii*，因为种加词指 Armand David。

4. 植物学名中的命名人

命名人缩写一般为第一个音节。出现 1 个以上的命名人含义如下：用 ex 连接的两个人名，如珍

珠花 *Spiraea thunbergii* Sieb. ex Bl.，这是表示本种植物由 Sieb. 定了名，但尚未正式发表，以后 Bl. 同意此名称并正式加以发表（Sieb. 和 Bl. 均为人名缩写）。命名人系用 et 连接的两个人名，这表示这一学名系由二人合作命名，如水杉 *Metasequoia glyptostroboides* Hu et Cheng，由胡先骕和郑万钧两人共同定名。

有时在植物学名的种名之后有一括号，括号内为人名或人名的缩写，此表示这一学名经重新组合而成。如青冈栎 *Cyclobalanopsis glauca*（Thunb.）Oerst.，这是由于青冈栎的植物学名由 Thunb. 命名为 *Quercus glauca*，以后经 Oerst. 研究应列入 *Cyclobalanopsis* 属。根据植物命名法规定，需要重新组合（如改订属名、由变种升为种等）时，应保留原种名和原命名人，原命名人加括号。

5. 种下等级的植物学名表示方法

种下等级有亚种（Subspecies）在学名中缩写为 subsp. 或 ssp.，变种（varietas）缩写为 var.，变型（Forma）缩写为 f.，栽培变种（Cultivar），缩写为 cv.。种下等级的学名表示法，为原种名＋种下等级词缩写＋种下等级种加词＋种下等级命名人。如金边大叶黄杨学名 *Euonymus japonicus* Thunb. var. *aureo-marginatus* Nichols.

杂交不同种的植物，通常以"×"来表示。杂种鹅掌楸，是鹅掌楸 *Liriodendron chinense* 和北美鹅掌楸 *Liriodendron tulipifera* 杂交的产物。杂种鹅掌楸的学名是 *Liriodendron tulipifera* × *Liriodendron chinense*，其中"×"表示它是一个杂交种。

6. 植物学拉丁语发音

在《植物学拉丁语》（*Botanical Latin*）一书中，William Stearn 指出："植物学拉丁语本质上是一种书面语言，但植物的学名经常出现在说话当中。如果它们听起来不难听，所有关心它们的人都能够听懂，那么如何发音其实并不太重要。通常可按照古典拉丁语的发音规则，获得它们的发音方法。通常人们把拉丁词与自己语言中的词语类比进行发音"。

Unit 9

The Process of Landscape Design

- 风景园林设计过程解析
- 用地分析
- 方案设计
- 详细设计
- 方案解说

The design process, also sometimes termed a "problem-solving process", includes a series of steps that usually (though not necessarily) follow a sequential order. In general terms, these same steps are also used by architects, industrial designers, engineers, and scientists to solve problems.

By Norman K. Booth

Text

The Design Process

By Norman K. Booth

> 几乎所有的项目规划或设计过程，都遵循着共同的程序和规律，可以分为以下几个连续的阶段。
> （1）接受工程项目阶段。
> （2）研究和分析阶段（包括现场踏查）。
> （3）设计阶段。
> （4）施工图纸绘制阶段。
> （5）施工阶段。
> （6）施工验收阶段。
> （7）维护阶段。

The design process, also sometimes termed a "problem-solving process," includes a series of steps that usually (though not necessarily) follow a sequential order. In general terms, these same steps are also used by architects, industrial designers, engineers, and scientists to solve problems. For site designers, the design process typically includes these steps.

(1) Project acceptance.
(2) Research and analysis (including site visit).
1) Base plan preparation.
2) Site inventory[1] (data collection) and analysis (evaluation[2]).
3) Client interview.
4) Program development.
(3) Design.
1) Ideal functional diagram.
2) Site-related functional diagram.
3) Concept plan.
4) Form composition study.

设计过程，有时也称为"问题解决的过程"，它由一系列的步骤组成，这些步骤通常（并非必须）遵从一定顺序。总的来说，建筑师、工业设计师、工程师和科研工作者也都运用这些步骤来解决各自的问题。

（1）接受工程项目阶段。
（2）研究和分析阶段（包括现场踏查）。

（3）设计阶段。
1）功能分区图。
2）相关环境的功能分区图。
3）设计构思。
4）造型研究。

---

[1] inventory：清查，对能力、资产或资源的评估或调查
[2] evaluation：评价

5) Preliminary design.
6) Schematic design.
7) Master plan.
8) Design development.
(4) Construction drawings.
1) Layout plan.
2) Grading plan.
3) Planting plan.
4) Construction details.
(5) Implementation.
(6) Post-construction evaluation.
(7) Maintenance.

5）初步设计。
6）设计示意图（草图）。
7）总体设计。
8）详细设计。
（4）施工图纸绘制阶段。

（5）施工阶段。
（6）施工验收阶段。
（7）维护阶段。

These steps of the design process represent an ideal sequence of events. Many of the steps overlap one another and blend together so the neat ordering of the outline is less clear and apparent. Furthermore, some of the steps may parallel one another in time and occur simultaneously[1]. For example, client interview and program development may occur while the designer is visiting the site and conducting a site analysis. Or the form composition phase may take place as an integral part of the preliminary design. At other times, the process is not apt to[2] proceed in a neat sequence of phases in which one step is absolutely completed before the next. In many instances it is necessary to move back and forth between the various phases with information gained at one step feeding back to an earlier step. As an illustration, the extent and type of information sought during the site analysis should directly depend on the character and complexity of concepts prepared in a later phase. Likewise, one may find it necessary to revisit the site or talk to the client again once the design phase itself has been started because some item of information was overlooked the first time or one's memory and impression simply need refreshing. And sometimes it helps to revisit the site after starting the design phase because then the designer can look at the site with experience and greater understanding of what limitations or opportunities are present. In other words, no one step of the design process occurs independently of the others.

Several other points also need to be made. First, the application of the design process may vary from one design

在大多设计项目的工作过程中，由于在不同设计阶段搜集的设计资料相互补充，造成前后阶段的划分并不明晰。

---

〔1〕 simultaneously：同时地
〔2〕 be apt to：易于

situation to the next. Each project represents a unique set of circumstances and, therefore, requires a different method for proceeding through the design process. Similarly, the emphasis placed on each of the individual steps may also change. For example, the site may be so barren[1] or nondescript[2] that the site analysis is meaningless. In other cases, the client may not care what is done and have little or no input in the process. In many situations, the entire process may end with the preliminary design. The need to undertake any of the steps of the design process or the emphasis placed on each can be justified only so long as it contributes to the other phases as well as the overall process. In all cases, the limits of a budget dictate how much time can be spent on the various phases of the design process. So the process may be applied slightly differently each time a design is produced.

在所有的设计项目上，项目投资预算的额度决定了设计过程中不同阶段的时间分配。

Preliminary Master Plan

In the preliminary master plan all elements of the design are put together and studied in association with one another in a realistic, semicomplete graphic manner (Figure 9.1). All design elements are considered, some for the first time, as interrelated components of a total environment. Within the framework established previously by the concept plan and the form composition study, the preliminary master plan moves ahead to take into account and study the following.

总体方案设计汇总各设计要素，对其相互关系进行分析研究，以正式或半正式的方式，绘制成图。

（1）The general material of all elements and forms (wood, brick, stone, etc).

（2）Plant materials as masses drawn to approximate mature size. The size, form, color, and texture of the plants are now considered and studied. In this step, plants are described in general terms such as ornamental tree, low evergreen shrub, tall deciduous shrub, and so forth. No plant species are identified.

（2）植物按照成年状态的尺度绘制。分析和研究植物的尺寸、形态、色彩和质感等特征。在这一阶段，植物可以划分为观赏树、低矮常绿灌木、较高落叶灌木等。

（3）The three-dimensional qualities and effects of the design including the location and height of such elements as tree canopies[3], awnings[4], overhead trellises[5], fences, walls, and earth mounds. In other words, the relative height of

---

[1] barren：单调的
[2] nondescript：无特征的
[3] canopy：荫篷
[4] awning：雨篷
[5] trellis：（支架蔓生植物的）棚架，格子架，格子棚

all elements of the design to one another should be considered.

(4) Sketch grading shown by proposed contours drawn at 2 or 5 ft (1 or 2m) contour intervals depending on scale and complexity of the design layout. Major elevation changes between terraced levels as well as tops of walls and fences might also be noted.

The preliminary master plan evolves best as an overlay on top of the form composition study. A number of studies and reiterations[1] on different layers of tracing paper will most likely be undertaken before concluding with a solution with which the designer feels happy. It is also quite possible that initial ideas developed during the form composition study or concept plan (or earlier) will change during the preliminary master plan because new thoughts have come to mind or the designer feels differently about a particular element or form when it is seen in the context of other elements. Having completed the preliminary master plan, the designer should check it against the program to make sure all intentions were implemented. Next the designer should review the plan with the client for feedback. This may be the first time the client has seen the design, so a few days might be allowed for review. Hopefully, the client will accept the solution and have a few suggestions or requested revisions. The designer then makes the necessary revisions and proceeds to the master plan or even to schematics or design development. This depends on how many changes result from the client's review and how elaborate[2] the design proposal is. In some cases the designer's services to the client may end with the presentation of the preliminary master plan and proceed no further.

The preliminary master plan should be drawn freehand and show all the elements of the design in semirealistic, illustrative fashion, as in Figure 9.1. Variations in line weight, textures, and values should be used to make the drawing read clearly. The preliminary master plan should show the following:

(1) Property line.
(2) Existing topography and significant spot grades for design proposal.
(3) Adjoining roads/streets (at least to centerline) and other significant elements such as buildings adjacent to the site.

---

[1] reiteration：重复，反复
[2] elaborate：精心制作，详细阐述

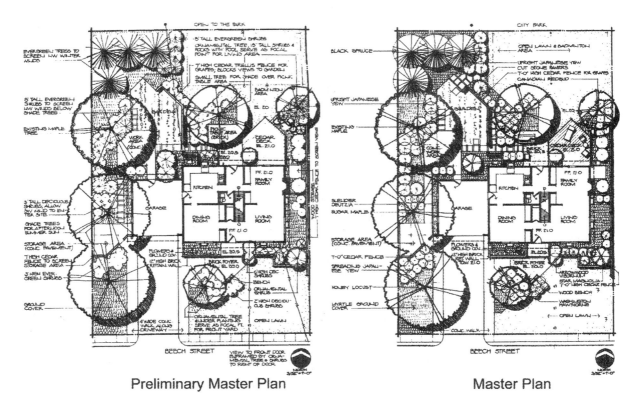

Figure 9.1 Preliminary master plan & Master plan

(4) Outline or "footprint" of all buildings and structures.

(5) All major design elements of the site plan illustrated with their proper graphic texture.

1) Driveway, walks, terrace[1], deck, lawn, etc

2) Roads and parking.

3) Bridges, shelters, docks, etc.

4) Masses of plant materials (both existing that are to be retained and proposed).

5) Walls, fences, etc.

6) Steps, ramps[2], curbs[3].

7) Sketched proposed contours[4].

In addition to graphically showing the above, the preliminary master plan should also identify the following with notes.

(1) Major use areas (examples: lawn, community open space, service area, natural wooded area, amphitheater).

(2) Materials of the design elements and forms.

(4) 所有建筑物或构筑物的平面轮廓线或基础轮廓线。

在总体方案设计中，除利用图示的方式表现上述内容外，还应确定以下几项。

(1) 主要功能分区，如草地、社区开放空间、服务区、自然林地区域、露天剧场等。

(2) 各种设计要素的材料和形态。

---

〔1〕 terrace：台地

〔2〕 ramp：坡道

〔3〕 curb：路缘

〔4〕 contour：等高线

(3) Plant materials by general characteristics (size and type; e. g., deciduous, evergreen, broadleaved evergreen).

(4) Major level changes by the use of spot grades.

(5) Description or justification for special situations.

Master Plan

The master plan, the next step in the design process, is a refinement[1] of the preliminary master plan. After gaining the reactions of the client from the preliminary master plan, the designer may need to revise and restudy certain portions of the proposal. With these changes included, the designer once again draws the site plan in a presentable fashion. One of the primary differences between a preliminary master plan and a master plan, in addition to the necessary design revisions, is the graphic style of each. While the preliminary master plan is drawn in a loose, freehand, yet legible manner, the master plan is typically drawn with more control and refinement, as in Figure 9.1. Rather than drawn entirely freehand, the master plan may have certain parts such as the property line, building outline, and edges of hard structural elements (walls, terrace, walks, decks, etc.) drafted with a triangle and T-square. However, other elements such as plant materials are still drawn freehand. To give the plan a controlled appearance, more time is usually spent drawing the master plan compared with the preliminary master plan.

Because of this additional time required and because the master plan is somewhat redundant[2], many designers elect to prepare a master plan in a graphic style similar to a preliminary master plan to save both time and money. Some designers stop with the preliminary master plan step as noted previously because the cost to the client does not justify the time to draw a highly ornate "pretty picture." Consequently, budget restraints have a direct impact on what is or is not done at the master plan step. A master plan typically shows the same information as the preliminary master plan with similar labeling as well.

Schematic Design

For some projects the design process continues on to

---

[1] refinement：细化

[2] redundant：多余的

schematic design. For smallscale sites such as residence or a vest-pocket park, master plan and schematic plans are synonymous. However, design projects entailing many acres with multiple land uses utilize schematic design to study the proposal in greater detail. Like the master plan, schematic design concerns itself with the entire site.

Graphically, the schematic plan shows the same information as the master plan with the following additional detail.

(1) Building floor plan showing all first-floor rooms doors, and windows.

(2) As appropriate, the plan may show such items as roof overhang (by means of a light, dashed line), down spouts, water spigots, electrical outlets, window wells, air conditioner, etc.

(3) Individual plant materials within masses. Plant materials are identified and labeled according to size, form, color, and texture. If the schematic design is the final step in a design proposal, then species of individual plants may be identified.

(4) Grading shown by proposed contours drawn at 2 or 1ft (1 or 0.5m) contour intervals. In addition, spot grades should be used to represent major high points and low points as well as the tops of walls and fence.

The schematic plan is drawn in a neat, controlled style. Like the master plan, the schematic plan may combine both drafted and freehand lines for the various elements of the plan, though typically the schematic plan tends to be almost entirely drafted.

Design Development

The final step of the design process is design development. In this step of the process, the designer is most concerned about the detail appearance and integration of materials. For example, design development might study the actual pavement pattern, the appearance of a wall or fence, the design of an entrance sign, or the detailing of an overhead trellis. This is often accomplished by studying specific areas of a project (site entrance, arrival court, terrace, pool and deck area, etc.) in plans, sections, and elevations at a detailed scale between 1 in =20 ft and 1/4 in=1 ft. Design development drawings give both the designer and the client a clear idea of what the project will actually look like in critical areas. While quite specific, design development is primarily concerned with the visual quality of the design, not the technical or construction detailing. | 是视觉效果的好坏，而不是具体的技术或构造方面的细节。

Notes

1. Extracts from

Texts

(1) Booth, Norman K. *Basic Elements of Landscape Architectural Design* [M]. New York: Elsevier Science Publishing Co. Inc, 1983: 283 – 304.

(2) 诺曼 K. 布思. 风景园林设计要素 [M]. 北京: 中国林业出版社, 1989: 267 – 287.

Picture

Figure 9.1: Norman K. Booth. *Basic Elements of Landscape Architectural Design* [M]. New York: Elsevier Science Publishing Co. Inc, 1983: 302, 303.

2. Background Information

Norman K. Booth 诺曼 K. 布思。

俄亥俄州大学教授，美国著名风景园林师，以设计住宅园林而著称。

Basic Elements of Landscape Architectural Design《风景园林设计要素》。

一本很著名的园林设计教材，全书共七章，分述地形、植物、建筑等园林要素设计。本文选自第七章，有删节。

3. Sources of Additional Information

http://unjobs.org/authors/norman-k.-booth

Further Reading A

The Site Planning Process

By James B. Root

The first phase of the planning procedure could be captioned[1], "Research Analysis" referring to the process of defining planning objectives and gathering data relative to those objectives. Information obtained from preliminary investigations may in fact suggest alternate goals, or certainly modifications to the original planning concept. For instance, an analysis of a particular population's age groups might lend support to an additional community swimming pool, as opposed to increased housing for the elderly. The growth of residential subdivisions[2] may indicate the desirability of a regional shopping center; on the other hand, the projected traffic volume thus generated by commercial development might exceed the capacity of existing roadway systems. Zoning laws and regulations obviously bear heavily on planning decisions. The feasibility[3], or success potential, of an envisioned development program must be supported by an objective analysis of user demand. Cost factors and construction techniques should be examined, based on the completion of similar projects.

On-site inspections of properties obviously enhance an understanding of the natural environment. Taking inventory of existing conditions is necessary to analyze the development potential of a specific site and to correlate[4] proposed functions properly with unique land characteristics. Pertinent[5] information amassed from observations, soil samplings, and research ultimately comprises the "Site Analysis," sometimes referred to as a "Survey Analysis."

Base maps usually reveal property lines, existing contours, drainage patterns[6]—as suggested by the contours-and bodies of water. Significant trees should be indicated, along with the horizontal scale, contour interval, and the northerly orientation[7]. The results of test borings[8] to determine geological substructures[9] and soil types are noted on a print of the appropriate base map, which serves as an in-field work sheet while preparing the site analysis. Panoramic[10] views, unsightly vistas to be screened, "trouble" spots, indigenous vegetation[11], prevailing winds[12] and seasonal weather

---

[1] caption：加上标题，加上说明
[2] residential subdivision：居住小区
[3] feasibility：可行性
[4] correlate：和……相关
[5] pertinent：有关的
[6] drainage pattern：排水模式
[7] northerly orientation：指北针
[8] test boring：测试钻探
[9] geological substructure：地质基础
[10] panoramic：全景的
[11] indigenous vegetation：乡土植物
[12] prevailing winds：盛行风

patterns, etc., are designated on the site analysis presentation by brief descriptions, charts, and/or bold graphics, usually in color to enhance readability. The recording of access roads, utility easements, and existing land usage lends direction to further development.

The "Land-Use Analysis" phase of site planning refers to the allocation of space for the accommodation of program objectives. "Bubble" diagrams, establishing location priorities for specific activities like housing or recreation, are formulated on tracing paper overlays applied either to the base map, a slope analysis, or the survey analysis. The correlation between the proposed use of an area and the appropriateness of the site manifests the effectiveness of sound land planning practices. Once land areas tentatively[1] have been assigned a particular use, they must be served by vehicular-pedestrian access[2] routes or circulation patterns that are functionally and aesthetically adapted to the terrain.

The arrangement and rearrangement of various land-use overlays[3] produces an organized preliminary plan. At this point the client and the landscape architect discuss relative merits of the "Development Concept," allowing for the incorporation of suggested changes, if any.

The finalized version of the architect's efforts frequently is referred to as the "Comprehensive Master Plan." Detailed construction drawings, known as "working drawings," supplement the master plan, along with additional graphics deemed necessary to convey ideas clearly and precisely. Written specifications[4] usually accompany the working drawings, explaining fully the construction and/or installation procedures to be followed.

The foregoing statements reflect the "methodology" of site planning. Any procedural phase may be emphasized more than others, depending on the nature and scope of specific projects. Also, various steps often are explored concurrently; the findings of one investigation may after the tentative conclusion(s) of another. For instance, the initial idea to construct an eighteen-hole golf course obviously would be influenced by feasibility studies and the availability of suitable land.

The physical factors comprising the "space" of our planet and the systematic approach in developing land planning concepts are summarized in the *site planning outline*.

Ⅰ. Spatial Elements of the Earth

1. Topographical Undulations[5]

(1) Mountains.

(2) Moderate to severe slopes.

2. Vegetation

(1) Trees and shrubs.

1) Evergreen.

2) Deciduous.

(2) Herbaceous[6] materials.

---

[1] tentatively：试验性地，暂时地
[2] vehicular－pedestrian access：车行－人行入口
[3] overlays：叠层
[4] specification：说明书
[5] topographical undulations：地形起伏
[6] herbaceous：草本的

3. Open Space

(1) Meadow grass.

(2) Desert sand.

4. Water

Ⅱ. Land Planning and Design Procedures

1. Research Analysis

(1) Program objectives.

(2) Feasibility studies[1].

1) Need.

2) Location.

3) Support facilities.

a. Existing.

b. Proposed.

4) Budget.

2. Site Analysis

(1) Natural features.

1) Physical characteristics.

a. Terrain.

—Contours.

—Soils.

b. Vegetation.

2) Environmental factors.

a. Air flow.

b. Humidity.

c. Temperature.

d. Solar exposure.

(2) Man-Made features.

1) Architectural structures.

2) Utilities.

3) Roadways.

(3) Cultural aspects.

1) History.

2) Present use.

3) Aesthetic quality.

3. Land-Use Analysis

(1) Proposed functions[2].

(2) Circulation.

1) Vehicular.

2) Pedestrian.

4. Development Concept.

---

[1] feasibility studies: 可行性研究

[2] proposed functions: 设计功能

(1) Design schematics.
(2) Preliminary layout.
5. Comprehensive Master Plan
(1) Working drawings.
1) Plant list.
2) Construction details.
(2) Specifications.

Notes

1. Extracts from

Text

James B. Root. *Fundamentals of Landscaping and Site Planning* [M]. Westport: The Avi Publishing Company, Inc., 1985: 66-68.

2. Background Information

James B. Root(詹姆士·B. 鲁特)。

美国著名风景园林师。

Fundamentals of Landscaping and Site Planning《风景设计和场地规划基础》。

一本以指导风景园林设计实践为中心的专业书，全书共两篇，上篇分述风景园林规划设计中要考虑的种种因素，下篇讨论设计实务。本文选自下篇第一章内容。

Further Reading B

Site Analysis

By Leroy G. Hannebaum

> 本文说明了在园林设计之前进行场地分析的重要性。列举了场地调查和分析的主要内容,如周边环境要素、构筑物情况、气候、噪声、风环境、土壤环境等。

The analysis of a building site includes an assessment of its better features as well as of the problems it presents. Site analysis is most fruitful when it is completed prior to building construction. The buildings then can be designed and placed to take advantage of the best features of the site, while many site-related problems can be prevented. For example, proper building orientation permits passive solar heating in the winter while minimizing exposure to the sun's rays in the summer. Planning for a minimum of northern glass exposure allows the designer to concentrate plantings for further insulation value, rather than for mere protection. Valuable plants and other site features that might otherwise be lost can be retained. Structural materials can be blended with site characteristics to a larger extent when site analysis precedes building design and construction.

在建筑建设之前进行科学的场地分析,可以为后续的建设奠定较好的基础条件。建筑可以利用用地良好的特点进行设计,从而避免很多问题。

Unfortunately, the landscape designer usually is first involved in a design project after buildings have been constructed. The site problems then become building problems as well, and all are considered during the site analysis.

糟糕的是,风景园林设计往往是在建筑建成之后进行。于是用地的问题就转变为建筑的问题。而这一切本应在之前的场地分析环节中得到应有的重视。

A site analysis is recorded on paper while the designer is on the site. This rough diagram is made in a notebook or on other conveniently carried paper. Later, the information may be transferred to tracing paper overlaying a scaled drawing of the site and buildings for a more proportionate view. Of course, the measurements for the scaled drawing are also made during the site analysis. Accurate measurements of the placement of permanent features on the site, notes about all site problems, and the site's assets are all recorded on this diagram (Figure 9.2)

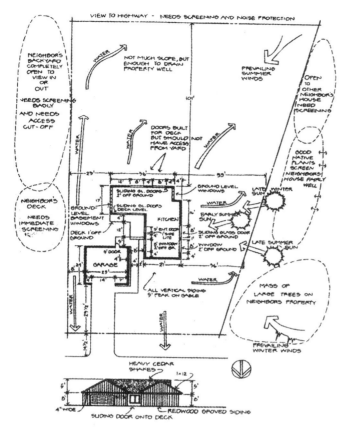

Figure 9.2 A quick, on-the-site diagram provides the designer the measurements and informational notes needed for work at the drawing board

Residential site analysis includes the following steps.

(1) Obtain a plot plan for the building site from city or county offices before visiting the site. Study easements, building setbacks, parking strips, and other legal requirements.

(2) Measure lot dimensions. Locate property stakes to determine boundaries.

(3) Locate the house on the lot. Measure from one corner of the building to two property lines to position the building. Repeat the process at another corner of the house to situate the angle of the house.

(4) Measure and record building features. Locate all windows, doors, and so forth.

(5) Locate water outlets, electric meters and other utility structures on the site and buildings.

(6) Make notations about changes in house siding, trim, shutters, and so forth.

(7) Measure to locate and assess the value and condition of existing plants on the site.

(8) Locate and assess the value of other natural features

住区场地分析包括以下的步骤。

（1）在现场踏查之前，从管理部门索取建造场地的地块规划图。分析场地上的附属构筑物、建筑后退线、停车带，和其他法律法规要求。

（2）测量场地范围，确定界桩。

（3）在场地上确定房屋的准确位置。

（4）测量和记录建筑的形态特征，标注门窗等的位置。

（5）在场地和建筑物上标出出水口、电表及其他设施的位置。

of the site, including rock ledges, boulders, interesting terrain fluctuations, and so forth.

(9) Take the measurements required for elevational drawings. The height to the eaves, distance from eaves to gable peak, height of windows, second story features, and anything else not recorded by previous measurement might be included. Take photographs.

(10) Note the direction of prevailing summer and winter winds.

(11) Study the site's terrain. Plan for a topographical survey if more accurate study is necessary.

(12) Note all good off-property views worth retaining. Measure the distance of that viewing horizon along the property line. Note objectionable views in same manner to prepare for screening.

(13) Move off the property and look into it from all directions. Make notes and measurements in preparation for screening others' views into the property.

(14) Note the need to screen out noise, dust, car headlights, or other possible nuisances.

(15) Make notes about existing macroclimate and microclimate.

(16) Check the soil for depth, rock content, and so forth. Probe the soil and take samples for a soil analysis, if needed.

By the time site analysis is completed, the landscape designer should be prepared to utilize the worthwhile features and deal with the problems presented. This site analysis and measurement process should not be hurried. By allowing plenty of time, a more accurate assessment of the site can be made, and design ideas begin to form. Following a prepared list of site-analysis criteria often prevents costly and unnecessary return trips.

All possible avenues to site study should be explored on the site. Views from each window of a house should be studied from all angles. The location of various rooms and traffic patterns within the house should be noted.

Taking several instant-developing photographs aids recall of site and building features. The same photographs also document improvements made by landscaping efforts for before-and-after remembrances.

（9）绘制立面测绘图，标定屋檐高度、屋檐到山墙顶部的尺度、窗的高度、房屋二层轮廓等之前的测量未包含的内容。

（10）标注夏季和冬季的盛行风向。

（14）记录需要屏蔽的噪声、烟尘、汽车眩光和其他可能的干扰。

（15）记录大的气候环境和小气候环境。

（16）查看土壤的深度、砾石含量等，如有需要，对土壤进行取样分析。

场地分析和测定过程不要过于匆忙。需要充分的时间，获得比较精确的评价结果，从而形成设计思路。在这个过程中，如果事先准备好场地分析标准，并按清单逐项进行调查和分析，可避免不必要的返工。

Notes

1. Extracts from

Text

Hannebaum, Leroy G.. *Landscape Design: A Practical Approach* [M]. New Jersey: Prentice

Hall Career & Technology, 1994: 23 – 26.

Picture

Figure 9.2: Hannebaum, Leroy G.. *Landscape Design: A Practical Approach* [M]. New Jersey: Prentice Hall Career & Technology, 1994: 25.

2. Background Information

Leroy G. Hannebaum 莱若·G. 汉尼鲍姆。

美国风景园林师。

Landscape Design: A Practical Approach《风景园林设计实践方法》。

本文选自该书第七章，有删节。

Further Reading C

Five Rules for Explaining a Project

By Tom Turner

> 本文介绍了规划设计项目汇报和讲解方面的一些基本原则，它们分别是：
> (1) 形成完整的文字说明。
> (2) 注意讲解顺序的科学性。
> (3) 注意在公开场合讲解的语态和姿态。
> (4) 理性地回答方案汇报过程中的疑问和提问。
> (5) 在方案图纸绘制过程中，应用色彩要慎重。

Design organizations live and die by their ability to persuade clients[1] to adopt recommended courses of action, as do non-governmental planning offices[2]. Explaining your scheme to a college jury is a similar. In both cases you have to convince your listeners that you have done a good job and are recommending a sensible course of action, which makes good use of scarce resources that have alternative uses.

Rule 1

The boy scouts were right: "Be prepared". Preparing what to say is an intrinsic aspect of generating a scheme. As initial[3] decisions are usually about what sorts of place to make, express them in words and write them down, instead of just drawing. The words could express qualities, like peace, mystery and adventure. Or they might describe character areas: "market square", "busy social space", "abstract visual space", "quiet retreat", "lush oasis", "outdoor living room", "cottage garden". Adopt the same procedure as you subdivide the spaces: "swamp garden", "fountain court", "bamboo glade", "sheltered haven", "rose walk", The labels will keep changing, and of course you should be inventing new types of

---

[1] client: 客户
[2] offices: 事务所
[3] initial: 开始的，最初的

space. But it must be possible to put the objectives into words: shapes are never sufficient for clients and I doubt if they are sufficient for you either. Instead of producing traditional labelled plans, try making typographic compositions with words. A word plan can be a superb starting point for a design project.

Rule 2

Remember those plans your English and History teachers used to require. They are exactly what you need to coordinate the story-line for a set of drawings into a beginning, a middle and an end. In the beginning, "say what you are going to say". In the middle, "say it". At the end, "say what you have said". Here is an example:

Beginning

"I was asked to design a beach park."

Middle

"As mineral extraction[1] has wrecked[2] the place, we will have to heal the soil (point to the edaphic[3] plan), the water system (point to the hydrological[4] plan) and the vegetation (point to the habitat creation plan). This can be done on the pattern of a sand dune[5] ecosystem, with both exposed and sheltered places (point to the cross-sections and the earthmoving plans)."

End

"This perspective is done from the proposed arrival point. It shows the beach and the car park, which are the two main features, linked by a path, which runs through new habitats."

Producing the best drawings for a project is more like producing a children's storybook than you might imagine. However deep the ramifications[6] of your plans and designs, the story itself should be sweetly simple:

Beginning

"The place is like this..."

Middle

"It should change because..."

　　阐述一个规划项目就像讲述一个完整的故事。开始阶段，讲述"将要讲什么"；中间阶段，讲述"是什么"；结束阶段，总结"讲了什么"。

---

[1] mineral extraction：矿物开采
[2] wreck：破坏
[3] edaphic：土壤的
[4] hydrological：水文学的
[5] sand dune：沙丘
[6] ramification：结果

End

"When the works are complete the place will be like this..."

Once the story plan has been written, it will be easy to judge what drawings you require to illustrate the scheme. Without the story, you may produce a few good pictures but you are most unlikely to produce a project of the type that makes clients reach for their cheque books.

Your English teacher may have asked you to prepare numbered points for an essay, or prompt cards[1] for a speech. It is no bad thing to do likewise for design presentations but, as you can see from the above examples, it is really better if the drawings themselves function as your prompt cards. Please don't read from notes, ever.

Rule 3

Remember all the traditional principles of public speaking:
(1) Maintain eye contact with those you are addressing—talk to the people, not to the drawings, the floor, or the window.
(2) Vary your tone of voice and volume, to keep your listeners awake.
(3) Emphasize the most important points with gestures and dramatic pauses.

Rule 4

When it comes to questions and comment, do not shout, do not abuse[2] your listeners and do not threaten physical violence. If listeners do not understand the scheme, that's your problem. If they do not like your scheme, console[3] yourself that "you can't please all the people all of the time". If they have any useful suggestions, listen carefully and make notes. If you have made a mistake, admit it. If the worst comes to the worst, use bromide[4]: "That's an interesting point-thank you for raising it".

Rule 5

Warning: Colouring can damage your plans!

---

[1] prompt card：提示卡片
[2] abuse：辱骂
[3] console：安慰
[4] bromide：套话

Carefully thought-out well-drawn plans can be damaged, easily, by over-hasty and thoughtless colouring. Five minutes with a felt-tip pen[1] can obliterate[2] five days' work with a technical pen. Colours should be used to enhance the "message" or a plan, not just for prettification[3].

Before reaching for your colour box, stand back and look at the plan, again, from a distance. Try asking yourself the following questions:

(1) Who is this drawing aimed at?

(2) What point should it persuade them of?

(3) What information should it convey?

Depending upon the answers to these questions, you can use colours to achieve some or all of the following:

(1) To bring out the proposed character of the whole place.

(2) To give subsidiary[4] spaces more definition.

(3) To clarify[5] the circulation pattern[6], pedestrian[7] and vehicular[8].

(4) To emphasize the landform[9].

(5) To define the proposed habitat pattern.

(6) To emphasize the planting design.

(7) To show the surface water management proposals.

(8) To explain the materials concept.

<u>With presentation drawings, it is often best not to apply any colour until the end of the drawing-up period. Try to leave a few days for the purpose, and do not begin colouring until you have assigned a role to each drawing.</u> Then remember:

(1) Always do a sample area before starting on the main drawing.

(2) It is better to have too little colour than too much colour.

(3) An overall colour scheme for a set of drawings can have a powerful effect (e. g. cool colours; warm colours; spring

  在设计方案图纸绘制完成之前,最好不要着急上色。先等几天,等你将每张图纸所要交代的内容完全构思好后,再着色也不迟。

---

[1] felt-tip pen:毛毡笔
[2] obliterate:涂去,使湮没
[3] prettification:美化,装饰
[4] subsidiary:次要的,附属的
[5] clarify:使清晰或易懂
[6] circulation pattern:交通模式
[7] pedestrian:步行的
[8] vehicular:车行的
[9] landform:地形

colours; autumn colours; vibrant[1] colours; jungle colours; pastel[2] colours).

(4) Alternatively you can use different-but-related colour schemes for different sections of the project. Interior design books may give you some ideas.

You can also use colour symbolically to represent the character of space, or anything else you wish to symbolize.

Notes

1. Extracts from

Text

Turner, Tom. *City as Landscape: A Post-postmodern View of Design and Planning* [M]. London: E &F. N. Spon Ltd., 1996: 169 - 172.

2. Background Information

Tom Turner 汤姆·特纳。

英国伦敦格林尼治大学教授，是著名的景观设计（风景园林）教师和评论家，在城市规划设计、风景园林设计、花园设计、园林史等方面著作颇丰。代表作品有 *Garden history: philosophy and design 2000 BC—2000 AD*（2005）等。

City as Landscape: A Post-Postmodern View of Design and Plannings 是风景园林学科的重要著作，为麻省理工学院建筑学院必读书之一。本文选自该书第四章"风景设计"（Landscape Design）第二节"工作室技巧"（Studio Craft）。

3. Sources of Additional Information

http://www.gardenvisit.com/garden_history/garden_history_tom_turner.htm

---

[1] vibrant：明快的
[2] pastel：柔和的

Exercises

1. Questions（阅读思考）

认真阅读本单元文章，并讨论以下问题，要求用英文发言。
（1）风景园林规划设计主要分为哪几个阶段？
（2）总体方案设计、总体设计、局部设计、详细设计各要表达哪些内容？
（3）风景园林场地分析包括哪些步骤和内容？
（4）结合 Five rules for explaining a project 内容，谈谈风景园林师如何提高方案讲解能力？

2. Skill Training（技能训练）

（1）案例研究：查阅英美国家风景园林相关资料，并整理成介绍性 PPT 文件，用英语和同学讨论交流。内容可以是某位风景园林师、某个风景园林规划设计理论、或者某个风景园林项目。应包括概况、内容、案例、讨论与评述四方面内容。

（2）学术论文：在案例研究基础上，整理成课程论文。要求用英文书写，格式符合英美学术期刊惯例，字数不少于 3500 字。参考范文：Water & Subconscious in Jellicoe's Works（本书 Part Two-Unit Two-Lesson 3）

Tips

风景园林专业英语论文写作

学术论文包括学术报告、研究论文、学期论文和硕（博）士论文。学术论文的撰写一般分为五个步骤：确立论点，拟定主题，明确目的，构思并完成初稿，修改。值得一提的是，构思阶段不仅仅指作者个人的思考，还可以指和其他研究者进行学术讨论，倾听他们的意见和建议。这个阶段通常被称为"brainstorming"，意即自由讨论，供讨论的论文草案称为 working paper。学术论文旨在展示作者的研究成果或研究发现，语言一般较为正式，语气较为客观，尽量避免使用"我"（I）、"你/你们"（you）"等人称。人称"我们"（we）只有在表示原创性时才可以适当运用。一般情况下，多用"笔者"（the writers），"作者"（the authors），"他们"（they）等人称。为了使论文客观科学，论文中强调资料（数据）和资料分析，因此，句子的主语通常是物而不是人。例如：I carried out an experiment to investigate the effect of sunlight on plant growth. 这句话在学术论文中最好写成：An experiment was carried out to investigate the effect of sunlight on plant growth.

学术论文通常包括以下部分：题目（title），作者/通信地址（author/affiliation），关键词（keywords），摘要（abstract），引言（introduction），正文（body），结论（conclusion），参考文献（references），致谢（acknowledgements）。

1. 题目

题目尽量用名词，名词词组或者动名词，不需要用完整的句子，只要能表达论文的主要内容就行。在拟题目时，要力求精炼简洁；要尽量具体，避免空泛抽象；不要使用问句，如果是疑问句，应该采用疑问词＋不定式的形式，例如：*Steps for Landscape Design: How to Get Prepared*。另外，还要注意使用规范词，不要使用不规范的术语缩写或者符号等。从语法上讲，题目要求使用对称的结构，也就是说名词对应名词，动名词对应动名词。例如：在 *The Redesign and Reconstructing of the Yuanmingyuan Garden* 中，名词和动名词就是不对称了，应该改为：*The Redesign and Reconstruction of the Yuanmingyuan Garden*。至于题目中首字母是否大写，要根据拟投杂志的要求进行改动。

2. 作者及通信地址

（1）姓名的写法。

中国作者的姓名要用拼音形式撰写，单名的作者最好将姓氏全部大写。例如 ZHAO Ying，WANG Fei 等。

（2）写法要求。

如果作者打算投稿或者参加国际会议，则需根据所投杂志或者会议要求的格式。作者一般在四人以下。主要作者的行政职务以及学位称呼不出现在作者姓名之前，如果需要出现的话，则应该放在通信地址处。例如

ZHAO Ying

Professor, School of Landscape Architecture, Zhejiang Forestry University

Lin'an 311300, the People's Republic of China

（3）通信地址的写法。

地址的排列应该从小到大，即从最小的单位开始写，依次往后，直到国家名。例子见上。如果作者是多人的，则根据所投杂志或者会议要求进行书写。要注意的是，有的小机构是某些单位自己设立的，例如："Task-force Team of…"，"Research Office of…"等，这些就没有必要写在通信地址里，以免引起误会。

3. 关键词

关键词是能够反映论文主旨内容的最重要的词，因此关键词的选取非常重要。关键词数量一般不少于2个，不多于8个，最好是4～6个。关键词通常取自摘要中的某些词，建议用名词，而不是动词。例如："调查"一词最好用"investigation"而不是"investigate"。关键词一般位于摘要下方，词和词之间要用逗号或者分号隔开。关键词为缩略词时，需根据ISO标准体系来进行缩写。

4. 摘要

摘要是用最简洁的语句概括论文的主旨内容。小论文的摘要只需50～100字，较长的论文一般需要200字左右的摘要，也可以根据论文长度适当增加字数，但最多不应超过500字。总的来说，摘要应为论文长度的3‰～5‰。

摘要可分为描述型摘要，信息型摘要和描述信息型摘要。描述型摘要主要是给论文定性，即对论文所作研究及分析进行简练描述，相当于主要内容简介。信息型摘要主要提供的是研究数据，而不进行内容的阐释。描述信息型摘要则是二者的结合。

摘要通常包括问题的提出，解决问题的方法以及主要的研究结果。因此，摘要通常由主题句（多为首句），论证句以及结语句组成。主题句主要提出问题，论证句一般由几个句子组成，主要给出解决问题的方法，结语句一般是一个句子，指出研究结果。

5. 引言

引言的作用在于介绍主题，指出研究范围，研究目的及论文的结构。通常，引言以研究背景的介绍作为开头，然后指出现存的问题，接下来，对研究现状进行介绍说明，最后提出论文拟解决的问题以及研究步骤。

6. 正文

正文的展开一般有下列几种方法：根据时间先后，根据研究（设计）过程，从抽象到具体（演绎法），从具体到抽象（归纳法）。

7. 结语

结语部分可以是对某个（些）设计理论进行总结阐释，可以是对某个设计结果进行陈述，也可以是对某个设计作品进行前瞻性预言。

8. 参考文献

参考文献的写法有两种，一种是按照参考文献作者姓氏的字母顺序，另一种是按照引用文献在文中出现的顺序。后者需要在论文中注明上标。

9. 致谢

致谢部分一般有固定的句式，例如：… wishes to express his (her) sincere appreciation to (somebody) for (something or doing something); The author (s) is (are) indebted to… for…; Thanks are due to… for…; Thanks go to… for…; Support by… is gratefully acknowledged;

The above research was made possible by a grant from...

10. 论文投稿

一般学术论文向学术期刊和学术会议投稿。风景园林相关英文学术期刊主要包括 *Landscape & Urban Planning*（SCI 收录期刊），*Environment and Urbanization*（SCI 收录期刊），以及 *Landscape Journal*（ASLA 学术期刊）等。国际学术会议主要有 IFLA 会议、中日韩等区域会议和专题会议等，一般要求先提交摘要进行审核，接到许可通知后提交全文。IFLA 会议通常在每年的 12 月底前交摘要，具体要求和时间要看主办国通知。

PART 2
Major Figures & Their Works

假如我们深入地考察西方现代风景园林历史，就会发现每个时期的作品和设计师正如满天星辰，尽管有着各式各样的光芒却沿着大致相似的轨迹运行——他们从不同角度解决这一时期风景园林师面临的同样的社会、经济和文化问题。而著名风景园林师和作品是满天星辰中最璀璨的几颗，从他们身上可以更明显地看到时代精神和职业发展。

尽管欧洲有着研究和实践风景园林设计的悠久历史，但美国风景园林师奥姆斯特德和他的工作仍然具有里程碑式的意义。奥姆斯特德及其合作者设计了纽约中央公园、波士顿公园系统，并首创了 Landscape Architect 这个名词作为职业名称。奥姆斯特德的两个儿子创立了美国风景园林师联合会（ASLA），哈佛大学现代风景园林教育体系。奥姆斯特德三父子的工作奠定了现代风景园林师职业的基础，对该行业的发展产生了深远的影响。在锐意开拓职业领域的同时，一些风景园林师在传统的花园设计领域大放光彩，比如约翰·托马斯·丘奇（John Thomas Church）。他和以他为首的"加州学派"从形式、功能等方面赋予花园一种新的面貌，开拓了花园及居住环境的现代设计模式。美国风景园林事业风起云涌，其他西方国家也是如此。杰里科在英国，布雷·马科斯在南美都开始了自己的理论和实践。整个20世纪前半叶，虽然因为不同的社会发展水平导致时间上互有参差，但由于整体上面临着现代生活和时代精神的变革，西方国家风景园林普遍形成了新的形式和内容，采取了新的设计方法，一个有悠久历史的职业被赋予了新的内涵——现代风景园林理论和实践体系在西方各国逐渐形成。

第二次世界大战后，世界的艺术和建筑中心从欧洲转移到了美国。1938～1941年间，哈佛大学设计研究生院学生詹姆斯·罗斯、丹·凯利及盖瑞特·埃克博掀起了"哈佛革命"（Harvard Revolution），他们的思想及之后的作品强烈地推动美国的风景园林行业朝向符合时代精神的现代主义方向发展。第二次世界大战以后到20世纪60年代，西方社会和经济状况的稳定和繁荣使现代风景园林蓬勃发展，设计领域不断扩展。1961年，西蒙兹出版了《风景园林学：人类自然环境的形成》一书，在书中西蒙兹写道"风景园林师的终生目标和工作就是帮助人类，使人、建筑、社区、城市及人的生活同地球和谐相处"。20世纪60年代末，风景园林设计思想更加开阔，手法更加多样。劳伦斯·哈普林（Lawrence Halprin）是其中杰出的代表，他的作品类型多样，风格明显。除理论贡献之外，西蒙兹在城镇规划设计方面的实践同样大大拓宽了风景园林师的职业领域。

20世纪70年代，麦克哈格的生态主义思想是整个西方社会环境保护运动在风景园林中的折射。麦克哈格的理论体现了生态学对风景园林学科的渗透，延续并发展了奥姆斯特德以来关注自然和人居环境保护的优良传统。80年代，朱利斯·福布斯（Julius Fabos）开创并领导美国"绿脉"规划的研究，创办了马萨诸塞大学大都市地区景观规划研究室，这是一个由风景园林学家、生态学家、水文学家、土壤学家、社会学家、心理学家和计算机专家等组成的典型的跨学科机构——风景园林规划设计面对新的问题，从相关领域吸取营养，采取革新行动，为现代风景园林的发展注入了新的活力。

从 20 世纪 70 年代末之后，西方风景园林进入一个"后现代主义"的时期。这是一个对现代主义进行反思和重新认识的时期，一些被现代主义忽略的东西被重新认识到它的价值：功能至上的思想受到质疑，艺术、装饰、形式又得到重视，风景园林思想更富有包容性，西方现代风景园林进入了多元化发展的时期。20 世纪 80 年代以来，彼得·沃克（Peter Walker），乔治·哈格里夫斯（George Hargeaves），玛莎·施瓦茨（Martha Schwartz）等相继崭露头角，并至今活跃在风景园林规划设计的舞台上，他们的设计思想，独特的设计手法和设计语言，不断地为现代风景园林添加入新的内容。

今天的西方现代风景园林规划设计正呈现着一种多元化的发展趋势。现代风景园林的概念已极其广阔，从传统的花园、庭院、公园，到城市广场、街道、街头绿地、大学和公司园区，以及国家公园、自然保护区，甚至整个大地都是工作的范围。现代风景园林规划设计比以往更注重多学科的交叉和协作，不断地与其他一些学科进行交流，形成新的分支。从西方现代风景园林的基本构成来看，它并不是一种统一的现象，而是一种组合了许多细流的发展过程。早期现代风景园林的代表——法国现代园林、美国的"加利福尼亚学派"、瑞典的"斯德哥尔摩学派"、英国的杰里科、拉丁美洲的布雷·马克斯等，均是吸取了现代主义的精神，结合当地的特点和各自的美学认识而形成的多样化的集团。丹·凯利、杰里科、布雷·马克斯等人在 20 世纪 80 年代甚至 90 年代初仍然活跃在风景园林的舞台上，使现代风景园林的主流不断前进。风景园林也许是最难描述的行业，随着时代、社会和艺术的发展在不断发展变化之中。

本篇以时间和人物为线索，以三个单元内容介绍 12 位著名的西方现代风景园林师，重点介绍他们的设计思想、设计手法及其代表作品，希望借此使读者进一步了解西方现代风景园林产生和发展的脉络。

Unit 10

- Frederick Law Olmsted
- John Thomas D. Church
- Geoffrey Jellicoe
- Roberto Burle Marx

Frederick Law Olmsted

1. Introduction

弗雷德里克·劳·奥姆斯特德（Frederick Law Olmsted）(1822~1903年)，美国现代风景园林学的奠基人，著名的公园设计师。

奥姆斯特德1822年出生于美国康涅狄格州的哈特福德（Hartford, Connecticut），他的父亲是一名成功的布料商，也是风景爱好者，奥姆斯特德的假日大多花在与家人从新英格兰北部到纽约州北部"寻找美丽风景的旅行"中。1837年当奥姆斯特德即将进入耶鲁大学学习时，受到了严重的漆树中毒，这使得他视力下降，被迫放弃了正常的学业。之后的20年里，他做过水手、商人、新闻工作者等工作。他曾经到英国旅行，对默西塞德郡伯肯黑德公园（Birkenhead Park）留下了深刻的印象，并因此于1852年出版 *Walks and Talks of an American Farmer in England* 一书。

在他的朋友和老师，著名的园林设计师安德鲁·杰克逊·唐宁的介绍下，奥姆斯特德结识了建筑师卡尔弗特·沃克斯（Calvert Vaux）。之后，两人合作于1858年在纽约中央公园的设计竞赛中胜出。就是在这次竞赛中，奥姆斯特德和沃克斯共同署名为 Landscape Architect——从此一个全新的职业名称诞生了。1865年，奥姆斯特德在纽约和沃克斯组建了设计公司，共同完成他们在中央公园的工作，并设计了纽约市布鲁克林区的希望公园（Prospect Park）。之后的30年里，奥姆斯特德完成了一系列典范性的设计，其中包括：大型城市公园、公园大道、城市公园绿地系统、风景保护区、郊外住宅区、私家庭园、公共建筑庭园等，它们证明了风景园林学（Landscape Architecture）能有效地改善人们的生活质量。

奥姆斯特德于1903年逝世，他的两个儿子约翰·查尔斯·奥姆斯特德（John Charles Olmsted）和小奥姆斯特德（Frederick Law Olmsted），Jr. 继承了他的事业。其中他的侄子和养子约翰·查尔斯·奥姆斯特德，在1899年联合其他几个园林设计师创立了美国风景园林师联合会（ASLA）并担任首任主席。他的儿子，毕业于哈佛大学的小奥姆斯特德于1900年回到母校担任教学工作，开创并建立了现代风景园林教育体系。1901年，小奥姆斯特德被老罗斯福总统指定进入麦克米伦委员会（McMillan Commission），负责美国首都华盛顿特区的城市规划和公园设计。

奥姆斯特德和他的公司在他一生中承接了大约500个项目。它们包括约100个公园和娱乐场、200个私人庄园、50个居住社区和小区，还有约40个校园设计。他亲手书写的600份信函和报告得以保留下来，涉及300多个设计项目。在其职业生涯中，奥姆斯特德始终强调风景园林师与其他行业专家进行协作的重要性，尤其是建筑工程师、园艺师和建筑设计师。

奥姆斯特德的朋友和合作者，建筑师丹尼尔·伯罕姆（Daniel Burnham）的一段话比较准确地概括了奥姆斯特德作品的特点："奥姆斯特德是一个画家，他用美丽的湖泊和植物葱郁的山坡作画，他用草地、湖岸和树木茂盛的丘陵作画，他站在起伏的山脉下，以广阔的视野作画"。

2. Major Publications & Design Works

A. Major Publications

- *Walks and Talks of an American Farmer in England*, 1852.
- *A Journey in the Seaboard Slave States in the Years 1853~1854 in two volumes*, 1856.
- *The Back Country*, 1860.
- *Journey through Texas*, 1969.

B. Major Design Works
- Central Park, New York City, 1858.
- Hartford Hospital Institute of the Living, Boston, 1860.
- Prospect Park, New York City, 1866.
- The Emerald Necklace, Boston, 1878.
- The World's Columbian Exhibition, Chicago, 1888.

3. Ideas

A. Landscape Effects

> 选自奥姆斯特德写给费城公园规划委员会主任的设计设想，原文篇幅较长，本文选其中"Landscape Effects"一节。文章主要叙述了费城公园用地中值得关注的景观因素，以及如何因地制宜，营造和谐美观自然景观的方法和途径。

The opportunities that will be offered by the Philadelphia Park for the enjoyment of beautiful natural scenery, will certainly be remarkably extensive; it must necessarily include within its limits a large body of water, with varied and interesting shores, and in whatever direction the additional land to be acquired may be taken, it can hardly fail to be of an agreeable character. It is clear however, that the landscape attractions of such a public pleasure ground as you have in view, must consist of something more than an aggregation[1] of disconnected scraps of scenery.

A City Park to deserve the name, must be a well balanced, artistic composition, and the first aim of the Commissioners should therefore be, to secure additional territory[2] in the direction that will be most efficacious in giving unity and completeness to the various land and water views, already under their control.

Having carefully studied the question of location in this aspect, we have come to the conclusion, that what is most needed to secure to the Philadelphia Park the elements of a very noble work of Landscape Art, is the incorporation within its boundary of a stretch of turf[3], that shall seem to be on the same large scale as the Schuylkill[4] basin, otherwise the land effects will be subordinate[5] in interest to the water effects, which is manifestly undesirable. Of course this aim can only be accomplished to a very measurable extent, but it is therefore the more important, that so far as it can be done, it should be done.

We may here observe, that the question of expensiveness in a City Park, turns less upon the cost of land in the outset, than upon the outlay which will be subsequently required, from year to year, for the maintenance of a suitable finish and character in the grounds when formed. Those parts, which depend for their attractiveness on elegant details of planting or other special decorations, require much constant labor to be kept in satisfactory condition, while every acre of turf may be made

---

[1] aggregation: 总和
[2] territory: 用地
[3] turf: 草地
[4] Schuylkill: 斯古吉尔河
[5] subordinate: 次要的，从属的

at least to pay its own way.

It is also desirable, with the view of avoiding a desultory[1], repetitive general effect, that some culminating point of interest should be determined on, to which other elements of interest may be subordinate[2]; if possible this should be some specially characteristic scene, that may be accepted as the central or apparently central feature of the whole Park.

The last of these desiderata seems to be offered at George's Hill, where a most beautiful view is to be had in the direction of the City, which, from this point, is just far enough off to be entirely picturesque. Fortunately, also, the needed turf stretch lies in the exact line of this view and directly at the base of the hill, from which it gently falls away towards the river, with a simplicity of surface, that is highly suggestive of successful Park treatment.

In our judgment, George's Hill, and the land between it and the river, is the most important territory which it will be in your power to enclose for Park purposes, inasmuch as no other, that came under our observation, will supply these all important requisites, namely a feature of central interest, which is also a point of view, for a large open landscape, the horizon lines and ground is well adapted to a fine, harmonious and thoroughly parklike, artistic development. Moreover, the river, embosomed[3] among the lower hills, and but half disclosed under overhanging trees, is here a delightful, but still, as it should be in the principal view of the Park, a secondary element of the compositon.

The prospect obtained from the Belmont House is of a different character, the City and the River being more prominent, but it is highly interesting, and very desirable to be included in the Park.

We will now however, leave this branch of the subject, as it is not our purpose to enlarge upon the landscape opportunities of the ground, more than is sufficient to explain our views in regard to the general direction that the outline of the Park should take.

B. Report on Prospect Park Design

> 本文选自奥姆斯特德写给布鲁克林公园委员会"希望公园"的设计报告，相当于阶段设计说明书，比较详细地描述了设计理念和具体手法。原文篇幅较长，选入时有部分删节。

January 1, 1874.
To the Brooklyn park Commissioners:
Gentlemen:
The object of the work which has been done on the Brooklyn Park, during the eight years in which we have had the honor to serve your Commission, has been the creation of scenes of natural character as attractive and graceful as the local conditions would allow, and their advantageous presentation.

---

[1] desultory: 杂乱的
[2] subordinate: 次要的，从属的，下级的
[3] embosom: 围绕，围护

With respect to the improvement of the scenery, the work done has, so far, been but preparatory to the greater work asked of nature; the constructions, through the use of which it was to be enjoyed, such as roads, walks, shelters, and places of refreshment, were, on the other hand, to be turned out complete at once. It was therefore inevitable that these constructions should, for a certain period, assume undue importance. In the greater part of the Park that period is already well-nigh passed; special search must be made to find a scene in which nature does not reign supreme, or in which, if artificial objects are to be recognized, they are not relatively unimportant and unobtrusive[1] incidents of convenience.

The general character of the scenery of the Park, even in its present formative condition, is undeniably broad, simple, and quiet, yet the variations of the surface, and the disposition of open woods, thickets, glades, meadows, and of still and running waters, is such that it cannot be deemed monotonous[2]. Its characteristic features in these respects are to be strengthened not only by growth, but also almost equally by timely reduction of whatever will tend to the weakening of distinctive qualities, or to the repression of elements intended to be aggrandized[3].

There were originally two main bodies of natural wood on the site of the Park, connected by a narrow belt at the point where the Long Meadow is now most contracted; the eastern body being broken by bays where the Nethermead, the head of the Lull-water, and the Deer Paddock now are (Figure 10.1).

Figure 10.1 Design for Prospect Park

...

Great care has been taken to secure a natural and picturesque edge to the old woods, both by breaking into them and by planting beyond them the lowest-headed large trees that could be procured. Similar trees have also been used to prevent as far as possible the occurrence of a strong contrast between the old woods and the new plantations. The shape of the ground (its natural features being

---

[1] unobtrusive: 不唐突的，不多嘴的，客气的，谦虚的
[2] monotonous: 单调的，无变化的
[3] aggrandize: 增加，夸大

almost invariably enlarged in the process of grading) has been favorable to the desired result, the outer and more conspicuous parts of the masses being on lower ground than those interior, and the few large trees under these circumstances giving character to the whole.

Most of the plantations, especially on the northern part of the Park, are on very bleak ground, and to lessen the severity of the exposure of the young plants, as well as to provide against and secure greater immediate effects to the eye, many trees have been planted in addition to those intended to remain. That the permanent trees may have the required vigor and not be crowded into ungainly forms, a gradual thinning out of these plantations, a little every year for many years to come, is essential.

...

The charm of the Park will lie chiefly in the contrast of its occasional bodies of low foliage, intricate, obscure, and mysterious, with the more open groves and woods, and of both with its fair expanses of unbroken turf. Its beauty will, therefore, depend on the care and skill with which these respective qualities, each in its appropriate place, are nursed and guarded.

The areas of the Long Meadow and the Nethermead are so large, that it has been deemed unnecessary to maintain the restrictions usually enforced in public grounds in this climate upon walking on the turf. The attractions and the public value of the Park have thus been undoubtedly very much enhanced. But the two dangers which attend this course already begin to be manifest, and it is evident that unless strenuously guarded against, serious evils will sooner or later result. One of these dangers is that of the wearing out of the turf in streaks and patches, the other is that of the destruction of underwood, shrubs, and plants, and the hazardous[1] and inconvenient straggling of visitors across drives and rides, arising from the difficulty of restricting the privilege of walking on the turf within proper limits. The first is to be prevented by so limiting the use of the turf that it will not be trodden upon when in a poachy[2] condition, or when it is excessively dry and inelastic, and by the use of slight guards, frequently shifted from point to point, as patches or streaks of wear become evident. The other may be particularly guarded against by concealed or inconspicuous barriers, and by cautionary signs, but can be permanently kept within tolerable bounds only by special efforts for the purpose made by an active, vigilant, faithful, and numerous body of keepers (Figure 10.2).

Figure 10.2 Lookout Tower

...

The Park gateway should be a handsome architectural structure, with an arcade[3] extending over the walk, so that many persons may with comfort wait for the cars at this point under cover when the weather is showery, and the pavement

---

[1] hazardous: 危险的,冒险的,碰运气的
[2] poachy: (土地)易被牲口踩得泥泞的,湿而松软的
[3] arcade: [建] 拱廊,有拱廊的街道

necessary to carry out this feature of the design has been designed and laid.

A platform for public meetings connects the center of the Plaza with the city, and a large fountain basin is introduced as a central feature.

In the design for the fountain the aim has been to express clearly its special artistic purpose. An artificial flow of water on a liberal scale is prepared for, and has been calculated on—from the outset. So long as the supply to the various jets is inadequate, this design will of course appear to be out of proportion to the result produced, but when the necessary additional forcing power is brought into operation, the stone base, with its bronze corona, will hardly be seen, and therefore will certainly not be considered too large an element in the general design.

It was evident that artificial light should be freely introduced in the Plaza, as it is a public promenade intended for night use; but it was also clear that the lines formed by the play of water and the general artistic effect in the Large Fountain would be much interfered with if a series of lamps elevated on the ordinary high posts should be a part of the design. The lighting has, therefore, been arranged for in connection with the railing for protection that surrounds the Fountain, the intention being, as mentioned in our last report, to have an interior circular line or ring of light below the eye and a few feet only above the water surface, so that the reflection of the globes would form a corresponding line that would be recognized as an element in the design, even by the ordinary observer.

At the north end of the platform, opposite the Lincoln statue, a public rostrum is proposed to be placed, the United States flag being displayed in connection with it. Whenever this feature is added the temporary staffs for the flags of the city and State now erected in the Plaza should be replaced by others of more elaborate and elegant design.

Respectfully,

OLMSTED & VAUX,
Landscape Architects.

C. Olmsted—His Essential Theory

> 选自美国的 *Victoran Society* 杂志，该篇文章记录了奥姆斯特德成为一位景观设计师及形成他自己的设计理念的过程。

Frederick Law Olmsted came to the profession of landscape architecture late in his career. For thirty years after 1837 he served as an administrator-first of New York's Central Park, then of the U. S. Sanitary Commission, and finally of the Mariposa Mining Company in California. Although he was co-designer of Central Park in 1858, it was not until 1865 when he was forty-three, that he made the final decision to devote himself to landscape architecture.

Despite his late commitment to landscape design, Olmsted's ideas on the subject had taken form long before he thought of himself as a landscape designer or attempted anything of the sort in a professional way. His ideas had their basis in the experiences and influences of his youth.

The strongest influence came from his father, who greatly enjoyed natural scenery and devoted most of his leisure time to seeking it out. As soon as young Frederick was old enough, his father set him on a pillow in front of his saddle and took his son through the countryside around their home in Hartford, Connecticut[1]. These short rides expanded to become annual "tours in search of the picturesque" that took Olmsted, by the age of sixteen, through the Connecticut Valley and White Mountains, up the Hudson River, and westward to the Adirondacks, Lake George, and Niagara Falls. As he acknowledged[2] late in his career, "The root of all my good work is an early respect for, regard and enjoyment of scenery? and extraordinary opportunities for cultivating susceptibility to the power of scenery."

The American landscape itself was the source of Olmsted's earliest lessons in aesthetics but that influence was soon supplemented by the writings of late 18th century English landscape gardeners, travelers, and theorists of landscape art. In his youth he read and was influenced thereafter by An Essay on the Picturesque, by Uvedale Price, published in 1794, and Remarks on Forest Scenery, and Other Woodland Views (Related Chiefly to Picturesque Beauty), Illustrated in the Scenes of the New Forest, by William Gilpin, published in 1790. Late in his career, Olmsted described these as "Books of the last century, but which I esteem so much more than any published since, as stimulating the exercise of judgment in matters of my art, that I put them into the hands of my pupils as soon as they come into our office, saying, "You are to read these seriously, as a student of Law would read Blackstone." The professional gardener who most influenced Olmsted was Humphry Repton, whose *Sketches* and *Hints on Landscape Gardening* and *The Theory and Practice of Landscape Gardening* were published in 1795 and 1803, respectively.

The teachings of Price, Gilpin, and Repton and the quiet example of his father provided the basis for Olmsted's aesthetic theories; they also underlay his refusal to follow the gardening fashions of his own time. The horticultural revolution of the early nineteenth century led gardeners to neglect older theories of design and concentrate on displaying the many new species of plants and flowers available to them. The gardening practices fostered by the rapid increase in plant materials led to what Olmsted called "a display of novelty, of fashion, of scientific or virtuoso[3] inclinations and of decoration." Even in England he found that specimen planting and flower-bedding of exotics was intruding on the older designs of landscape gardeners.

Olmsted's dislike of current fashions made him reluctant[4] to accept the term "landscape gardening" for his own art. His feeling that he worked on a larger scale than did gardeners helped to strengthen that reluctance. The term "gardening," he observed, "does not conveniently include exposing great ledges, damming streams, making lakes, tunnels, bridges, terraces and canals." Therefore, he asserted, "Nothing can be written on the subject with profit which extreme care is not taken to discriminate between what is meant in common use of the words garden, gardening, gardener, and the art which I try to pursue."

Olmsted differed from gardeners not only in the style and scale of his work, but also in his concept of the process by which he intended his designs to affect those who viewed them. As a result of his own experience and wide reading, he concluded that the most powerful effect of scenery was one

---

[1] Connecticut：康涅狄格州
[2] acknowledged：公认的
[3] virtuoso：艺术品鉴赏家
[4] reluctant：不顾的，勉强的，难得到的，难处理的

that worked by an unconscious process.

Notes

1. **Extracts from**

Texts

(1) Schuyler, David. & Jane Turner Censer. *The Papers of Frederick Law Olmsted* 1865～1874 [M]. Maryland: The Johns Hopkins University Press, 1992: 234-236.

(2) Schuyler, David. Jane Turner Censer. *The Papers of Frederick Law Olmsted* 1865～1874 [M]. Maryland: The Johns Hopkins University Press, 1992: 664～671.

(3) Beveridge, Charles E. Olmsted—His Essential Theory [J]. *The Journal of the Victoran Society in America*, 2000, 20(2): 32-37.

Pictures

(1) Figure 10.1: David Schuyler, Jane Turner Censer. *The Papers of Frederick Law Olmsted* 1865～1874 [M]. Maryland: The Johns Hopkins University Press, 1992: 664.

(2) Figure 10.2: David Schuyler, Jane Turner Censer. *The Papers of Frederick Law Olmsted* 1865～1874 [M]. Maryland: The Johns Hopkins University Press, 1992: 669.

2. **Sources of Additional Information**

Frederick Law Olmsted

http://www.olmstedsociety.org

http://www.fredericklawolmsted.com

Frederick Law Olmsted National Historic Site (National Park Service)

http://www.nps.gov/frla

Central Park

http://www.centralparknyc.org

Prospect Park

http://www.prospectpark.org

John Thomas D. Church

1. Introduction

约翰·托马斯·D. 丘奇（John Thomas D. Church，1902～1978 年），美国现代园林的开拓者，他从 20 世纪 30 年代后期开始，开创了被称为"加州花园"的美国西海岸现代园林风格。丘奇等加州现代园林设计师群体被称为加利福尼亚学派，其设计思想和手法对今天美国和世界的风景园林设计有深远的影响。

1923 年，丘奇从加州伯克利大学毕业，获得景观设计学士学位。1926 年，他又进入哈佛大学继续深造，并获得景观设计硕士学位。在哈佛期间，丘奇获得哈佛旅行奖学金，得以去欧洲学习意大利和西班牙的园林。1927 年，丘奇回到美国，在俄亥俄州立大学教书。1929 年开始的大萧条使美国经济全面衰退，设计任务急剧减少，此时的丘奇在奥克兰的一家事务所工作了 2 年。1932 年丘奇在旧金山开设了自己的事务所，当时他的设计还主要是传统风格。1937 年他再次游历欧洲，参观了柯布西耶和阿尔瓦阿托的设计作品后深受启发，开始探索如何使用现代材料，非对称构图来满足花园的各项功能与形式的需要。

1939 年金门博览会给了丘奇一次小试牛刀的机会，他所设计的花园运用白水泥，自由钢琴曲线和锯齿状折线，与立体主义风格的画有异曲同工之处。因而被评价为"首位将现代园林设计从研习传统折衷主义风格引向开拓现代设计模式的风景园林师"、"最后一位伟大的传统园林设计师和第一位伟大的现代风景园林师"。花园设计一直是丘奇的主要兴趣所在，他最著名的作品是 1948 年的唐纳花园（Donnel Garden），并也因此而名声鹊起。

第二次世界大战后，美国公共领域的设计迅速增加，设计的尺度和工程的规模相应较大。作为一个声望不断增加的风景园林师，丘奇也参与了一些大尺度的设计。1951 年，丘奇获得 AIA（美国建筑学会）艺术奖章，1976 年获得美国风景园林学会金奖。1955 年，他的著作《以人为本的园林》(*Gardens are for the People*) 出版，总结了他的思想和设计。

丘奇的事务所培养了一系列年轻的风景园林师，如 R. 罗斯坦（R. Royston），D. 贝里斯（D. Baylis），T. 奥斯芒德森（T. Osmundson）和哈普林，他们反过来又对促进"加利福尼亚学派"（California School）的发展作出了贡献。丘奇在 40 年的实践中设计了近 2000 个园林，其晚年仍然孜孜不倦从事园林设计工作，于 1978 年辞世于美国。

2. Major Publications & Design Works

A. Major Publications

- *A Study of Mediterranean Gardens and Their Adaptability to California Conditions*. Master's Thesis, Harvard University, 1927.
- *The Small California Garden*, California Arts and Architecture (May 1933), 16.
- *Gardens are for People*, University of California Press, 1955.

B. Major Design Works

- Aptos Beach House, Northern California, 1948.
- Garden in Sonoma, Northern California, 1948.
- Garden at Woodside, Woodside, California, 1953.
- Stanford Medical Center, Palo Alto, California, 1959.

3. Ideas

A. Design Principles

> 节选自丘奇著作 *Gardens Are for the People* 第三章 设计原则。在本文中,丘奇对花园设计的一些基本准则进行了较为系统的阐述,着重讨论了四个原则。
> (1) 整体观念 (unity)。
> (2) 注重功能 (function)。
> (3) 形式简洁 (simplicity)。
> (4) 尺度适宜 (scale)。

The success of the design will depend largely on these four fundamental[1] principles: unity, which is the consideration of the scheme as a whole, both house and garden; function, which is the relation of the practical service areas to the needs of the household and the relation of decorative areas to the desires and pleasure of those who use it; simplicity, upon which may rest both the economic and aesthetic success of the layout; and scale, which gives us a pleasant relation of parts to one another.

...

Changes in garden design during the last few decades have been enormous. Unless you have had the opportunity to see the best new houses of recent years, you can hardly realize how completely and basically the garden has changed.

Even the term "garden" has changed its meaning. A garden used to have a horticultural meaning-a place where plants were grown to be displayed for mass effects or to be examined individually. It was a place to walk through, to sit in briefly while you contemplated the wonders of nature before you returned to the civilized safety of the indoors. It was generally designed to provide a long vista from some dramatic spot within the house, such as the entry hall, the front steps, or a bay window. It was a place to be looked at rather than a place to be lived in.

The new kind of garden is still supposed to be looked at. But that is no longer its only function. It is designed primarily for living, as an adjunct to the functions of the house. How well it provides for the many types of living that can be carried on outdoors is the new standard by which we judge a garden.

This change in our ideas of what a garden is supposed to do for us was brought about by the force of several circumstances: the shrinking of the size of our houses due to high building costs, the disappearance of gardeners, the coming of power tools, and the increased use of glass. As the house grew smaller, many functions that used to go on inside the house were forced into less expensive space. Smaller rooms set up the need for bigger windows and whole walls of glass-to dissipate the feeling of claustrophobia[2]. So presently it became inevitable that the garden should attach itself to the house, not only in use but structurally and visually. The garden had to go to work for us, solving

---

[1] fundamental: 基本原则,基本原理
[2] claustrophobia: 幽闭恐怖症

our living problems while it also pleased our eyes and our emotional and psychological needs.

So a new trend was bornout of human necessities. No arbitrary whims or designer's caprices created this new kind of garden. It evolved naturally and inevitably from people's requirements. And out of the solutions to these requirements has come a whole new visual aesthetic, contributed to by many landscape designers in many parts of the country. Though this new kind of garden design was born out of the problems of new houses, it has lately been applied to the old as well. For it helps us get more usefulness from our property and gives a new usefulness to old houses.

Peace and ease are the dominant characteristics of the new garden—peace and beauty for the eye and ease of maintenance for the owner. Fewer and simpler lines are being used in the garden, and fewer and simpler materials. All is calculated to give complete restfulness to the eye. If the eye sees too many things, it is confused and the sense of peace is obliterated.

Simplification of lines does not mean eliminating outdoor structure. The closer house and garden are, in use and appearance, the more they begin to interlock visually. More and more the lines and materials of the house are carried outdoors and into the landscape. And more and more the materials of the garden structure penetrate into the house, recalling the outside design and subtly merging indoors and outdoors. The paving of the terrace is penetrating into the house, making the floors more utilitarian in such heavy-duty spots as hallways, stairways, and heavy traffic areas.

Since the garden is being designed more and more to be seen from several parts of the house, the plan of it cannot be rigid and set, with a beginning and an end. The lines of the modern garden must be moving and flowing, so that it is pleasing when seen from all direction, both inside and out.

The eye can be easily fooled. Things can be made to seem longer or wider than they really are. This is a great aid, for we can make a small lot seem bigger, and so create spaciousness, without increasing garden maintenance and real estate taxes. Thus a moving, changing line that creates an asymmetric plan not only pleases the eye but creates a new dimension for the house.

To succeed in making a logical and intelligent plan which will produce the maximum in terms of use and beauty, one must have simplicity of layout, integrity in the use of plant and structural materials, and a sure sense of proportion and pleasing form. Whether your design is "formal" or "informal", curved or straight, symmetrical[1] or free, or a combination of all, the important thing is that you end up with a functional plan and an artistic composition. It must have good proportion and proper scale and plants that have been chosen wisely and cared for affectionately.

Rhythm[2] and movement are essential. You expect them in the pictures you hang on your wall, in the music you listen to, in the poetry you read. In the garden it's the wind in the foliage and the dog running across the lawn. It's the line of the terrace and the repetition of richly textured foliage. The eye is a restless organ.

Symmetry can have motion. It's unimaginative formality that can become static. The eye prefers to move around a garden on lines that are provocative, never lose their interest, never end in dead corners, occasionally provide excitement or surprise, and always leave you interested and contented.

---

[1] symmetrical: 对称的，均匀的
[2] rhythm: 节奏，韵律

Someone may say, "I don't want it formal, laid out on an axis." The truth is your garden is never without at least one axis and probably has two or three. All compositions, however free, are built around them. The great designers of natural gardens may seem to have thrown away their T squares, but the axis is just as strong as in the mirror pool of the Tai Mahal. It's just less obvious.

The axis becomes visual rather than mechanical and needn't be at right angles to the eye. The eye is tolerant. It may be influenced by a view, nudged by a tree, encouraged by a meadow, or seduced by a brook. Don't fret if your garden is never quite perfect. Absolute perfection, like complete consistency, can be dull.

B. In the Hills of Northern Califorina

> 节选自 *Gardens Are for the People* 作品部分。记述了丘奇最著名的作品——1948年设计的唐纳花园（Donnel Garden）。庭院由入口院子、游泳池、餐饮处和大面积的平台所组成。平台的一部分是木铺装，另一部分是混凝土地面。庭院轮廓以锯齿线和曲线相连，肾形泳池流畅的线条以及池中雕塑的曲线，与远处海湾的 S 形线条相呼应。树冠的框景将原野、海湾和旧金山的天际线带入庭院中。

At the end of a long, winding approach through the hilly, oak-studded, grassland landscape of San Francisco's Bay Area, the visitor arrives at a spacious entrance parking court. It is protected from the wind by the surrounding native vegetation, which was retained in the design.

On the south side of the house the rooms open onto a brick terrace and grass area designed around existing large, native live oaks. These extend the architectural overhang of the house and provide wind protection, shade, and foreground[1] for the expansive view beyond. The meeting of garden and natural landscape is clearly marked by a juniper[2] hedge.

The house lies below the topmost knoll. From its terraces a path curves upward and intriguingly[3] disappears behind a rocky outcrop. It turns and arrives at the swimming pool and cabana, situated below the crest.

This recreation area takes advantage of a frame of live oaks offering wind protection and shade, native boulders, and a 30-mile panoramic view of San Francisco Bay (Figure 10.3).

The pool, its shape inspired by the winding creeks of the salt marshes below, was designed to provide adequate space for all water activities. It has a shallow area for children near the recreation room, 60 feet for unobstructed swimming, and a deep section for diving (Figure 10.4).

The concrete terrace around the pool is colored tan to reduce glare. Three people are not lost on it, nor are a hundred crowded.

The cabana has two sides of glass to take every advantage of the view. When the sliding doors are open, it becomes part of the terrace; a long wooden bench follows the stone wall into the room

---

[1] foreground: 前景，最显著的位置
[2] juniper: [植] 刺柏属丛木或树木
[3] intriguing: 迷人的，有迷惑力的，引起兴趣（或好奇心）的

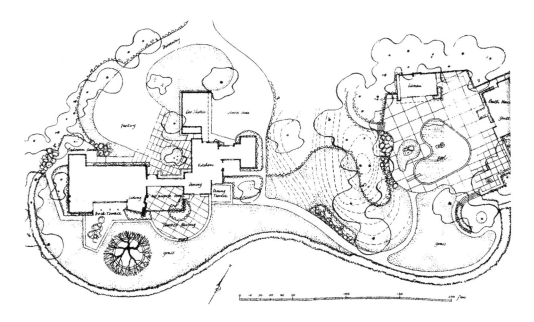

Figure 10.3　Donnel garden

Figure 10.4　The pool of Donnel Garden

to the fireplace. By sliding the side glass door into a slot in the wall, the room may be closed and heated.

C. A Pioneer of California Style

> 节选自著名建筑历史学家 Michael Laurie 为 *Gardens Are for the People* 所做的序言，文章对丘奇的设计风格与哲学思想的形成和发展做了深入阐述。

It is a commonly held belief among garden historians that California was the center of a school of landscape design that broke new ground in the post-World War II years. The garden was the medium through which new concepts were expressed; it reflected aesthetic developments in art and architecture and a new social order. The modern California garden has been described as an informal outdoor living room filled with deck chairs, tables, and swings, more social than horticulture in its intention. Similarities with the relationship of the Islamic house and garden or with the austere[1] restraint and occult balance of a Japanese garden have been suggested. Finally, the California garden has been described as "one of the most significant contributions to landscape design since the Olmstedian tradition of environmental planning in the second half of the 19th Century" (Figure 10.5)

Figure 10.5 One of the Church's Gardens

Until the late 1930s, Church's design could be described as quite conservative[2]; although not replicas of historical models, they were clearly based on traditional principles. After another European trip in 1937 to study the work of Le Corbusier and the Finish architect and designer Alvar Aalto, as well as that of modern painters and sculptors, Church began a period of experimentation with now forms. Two small gardens designed for the 1939 Golden Gate Exposition marked the beginning of this new phase. They demonstrated the possibilities for the evolution of new visual forms in the garden while satisfying all practical criteria. The central axis was abandoned in favor of multiplicity of viewpoints, simple planes, and flowing lines. Texture and color, space and form were manipulated in a manner reminiscent of the cubist painters. A variety of curvilinear shapes, textured surfaces, and walls were combined with a sure sense of proportion, and the gardens incorporated some new materials such as corrugated asbestos[3] and wooden paving blocks. Stylistically these gardens were a very dramatic advance on all previous designs in the United States.

The new house with its small garden, he argued, "must go to work for us, solving our living problems while it also pleases our eyes and our emotional psychological needs." The new kind of garden did not arise from arbitrary whim or designer's caprice; it evolved naturally and inevitably from people's requirements. Out of the solution to these requirements came a whole new visual aesthetic, and "since the garden is being designed more and more to be seen from several parts of the house, the plan of it cannot be rigid, and set with a beginning and an end. The lines of the modern garden must be moving and flowing so that it is pleasing when seen from all directions both inside and out". This notion suggests the influence of cubism on Church's aesthetic thinking about the garden. Similarly Stephen C. Pepper, professor of philosophy and aesthetics at the University of California at Berkeley, suggested in 1984 that the arts had rediscovered space: "The creation of a garden in this new light becomes something halfway between the making of a painting and the making of a house. It is as if the landscape architects were composing an abstract painting for people to live within."

The variety to be found in the approximately[4] 2,000 gardens designed in 49 years of practice

---

[1] austere: 严峻的,严厉的,操行上一丝不苟的,简朴的
[2] conservative: 保守的
[3] asbestos: 石棉
[4] approximately: 近似地,大约

not only substantiates this claim but also illustrates Church's respect for the unique qualities of every situation.

Church had arrived on the landscape scene at a time of transition. He was traditional enough to see value in the old, open enough to consider the new, and sensible enough to know that each, to be good, must be the product of a thorough knowledge of the principles on which it was based. Thus, Garrett Eckbo, a younger and equally distinguished colleague of the California school, has described Church in conversation as "the last great traditional designer and the first great modern designer."

Notes

1. Extracts from

Texts

(1) Church, John Thomas. *Gardens are for People* [M]. Berkeley: University of California Press, 1995: 29 - 34.

(2) Church, John Thomas. *Gardens are for People* [M]. Berkeley: University of California Press, 1995: 182 - 186.

(3) Church, John Thomas. *Gardens are for People* [M]. Berkeley: University of California Press, 1995: 1~9.

Pictures

(1) Figure 10.3: John Thomas Church. *Gardens are for People* [M]. Berkeley: University of California Press, 1995: 182.

(2) Figure 10.4: John Thomas Church. *Gardens are for People* [M]. Berkeley: University of California Press, 1995: 186.

(3) Figure 10.5: John Thomas Church. *Gardens are for People* [M]. Berkeley: University of California Press, 1995: 49.

2. Sources of Additional Information

http://www.ced.berkeley.edu/cedarchives/profiles/church.htm

http://content.cdlib.org/ark:/13030/tf938nb4jf/

http://www.stanfordalumni.org/news/magazine/2003/janfeb/features/church.html

Geoffrey Jellicoe

1. Introduction

杰弗里·艾伦·杰里科（Geoffrey Alan Jellicoe，1900～1996 年），英国现代风景园林学先驱，首届国际风景园林师联合会（IFLA）主席及终身名誉主席。从事风景园林规划设计理论研究与实践的时间长达 70 多年，其经历与国际风景园林学科的发展紧密相关。

杰里科 1900 年出生于英国切尔西，1918～1923 年就读于伦敦著名的建筑学学府 AA，在大学课程结束后的一年里，开始调研意大利花园设计，并在 1925 年出版第一部著作《文艺复兴时期的意大利花园》。20 世纪 20 年代，他和同学谢菲尔德合作开办了设计事务所，承接一些花园和小型建筑的设计。1929 年，杰里科参与创立英国"风景园林学会"（The Institute of Landscape Architects），之后为表示和建筑师的区别，更名为"风景学会"（The Institute of Landscape），从 1939 年到 1949 年，杰里科担任了 10 年的学会主席。1948 年，在杰里科等人的建议下，国际风景园林师联合会在剑桥成立，他担任首任主席。

杰里科一生出版专著 10 多部，完成的建筑与风景园林规划设计百余项。与同是著名建筑师的夫人苏珊（Susan Jelicoe）合著的 *The Landscape of Man: Shaping the Environment from Prehistory to Present day*《人类的景观造园：环境塑造史论》一书是他的代表作，该书自 1975 年初版后，1985 年、1995 年出版了修订本，被译成近 30 国文字，在世界各国规划设计界广为流传。时至今日，仍是风景园林师必读的经典专业书。

2. Major Publications & Design Works

A. Major Publications

- *Italian Gardens of the Renaissance*, 1925.
- *Landscape of Civilization: Created for the Moody Historical Gardens*, Antique Collectors Club, 1989.
- *The Landscape of Man: Shaping the Environment from Prehistory to Present day*, Thames & Hudson, 1995.
- *The Complete Landscape Designs and Gardens of Geoffrey Jellicoe*, Thames & Hudson, 1994.
- *Designing the New Landscape*, Thames & Hudson, 1998.
- *The Oxford Companion to Gardens*, Oxford University Press, 2001.

B. Major Design Works

- Caveman Restaurant, 1934～1936.
- Ditchley Park, 1934～1939.
- Plan for Hemel Hempsted, 1947.
- Harvey's Store, Guildford, 1956.
- Water Gardens, Hemel Hempsted, 1957～1959.
- Kennedy Memorial, 1964～1965.
- Shute House, 1970～1990.
- Hartwell House Garden, 1979～1989.
- Sutton Place, 1980～1986.

- Moody Gardens, 1984.

3. Ideas

A. The Landscape of Man

> 节选自杰里科名著 *The Landscape of Man: Shaping the Environment from Prehistory to Present day* 一书中的后记：走向人文主义的现代风景园林。文章从时空哲学观念开始，讨论人类景观设计风格趋势，着重阐述了现代设计中人文价值观的重要性。本文内容有较强的哲学抽象思辨特点，是反映杰里科思想的重要文献。

The philosophy of landscape design began as belief in myth[1], merged into humanism based on the establishment of fact, and is now grappling with the realization that facts are no more than assumptions[2]. Humanism is passing into another, unknown, phase. It is possible for instance, that the present disruption of the environment can be traced beyond the manifest reasons to one basic cause: the subconscious[3] disorientation[4] now in man's mind concerning time and space and his relation to both.

Artists in the nineteenth century had already sensed not only that all things were in flux (as had the Greeks), but that time and space were not two entities, but one. Now that it has been scientifically proved, the concept is so overwhelming and the break with history so abrupt, that this may be the main reason why today, significantly, time plays little part in the arts. It is the present that matters. The imagination, for example, no longer cares to bridge the gap, peculiar to landscape, between the seedling[5] and the tree: landscape must be instant. Architecture is created for a short life and the discord between old and new is without historical precedent. Such absence of a sense of time is contrary to all previous philosophy, metaphysical[6] or humanistic. It is as though action supersedes[7] contemplation. In extreme contrast, Egypt, ancient India and pre-Columbian America were almost wholly preoccupied with abstract time. China considered buildings to be self-reproducing, like plants; but the new landscapes were to be everlasting. Western civilization has consistently balanced time with space; the Italian philosopher-architect Alberti and the English astronomer-architect Wren held equally that all architecture should be built for eternity[8].

While man's sense of time has diminished, his sense of space seems to have expanded beyond control. He has a command of it, both in microcosm[9] and macrocosm[10], that would have amazed the ancients; but in filling it he is tending to become personally dissociated from it; it is too big and he

---

[1] myth：神话
[2] assumption：假定，设想
[3] subconscious：下意识的
[4] disorientation：方向知觉的丧失，迷惑
[5] seedling：秧苗，树苗
[6] metaphysical：形而上学的，纯粹哲学的，超自然的
[7] supersede：代替，取代
[8] eternity：永远，不朽
[9] microcosm：微观世界
[10] macrocosm：宏观世界

is too small. During the last few hundred years, the mathematical laws of the universe, extracted from outer space by scientists and engineers, have slowly come to dominate the biological laws of the biosphere[1]. Second only to the particular significance of nuclear power lies that of pure mathematics. Civilized life for the human race, as emphasized by J. Bronowski[2] in *The Ascent of Man*, is dependent upon a diversification planned with incredible ingenuity[3] by nature. But mathematics is based on repetition; repetition implies mass production; and this inevitably could lead to the static, efficient and deadly civilization of the bee. Pressure to stamp out individuality is everywhere and is most manifest in state housing or hive; it is not wonder that, under such conditions, the subconscious human instinct for self-expression finds vent in violence and illogical vandalism[4].

Now that we know and can assess the forces battering[5] our planet, can they first be resisted by the defensive mechanism of instinct and then controlled and put to work by the intellect? Balanced and self-renewing ecosystems had already been evolved by past civilizations (notably the eastern), but their scope was limited and their evolution by trial and error slow and laborious. The possibility now before man is the creation, with the services of computer, of an ecosystem that is immediate, comprehensive and based on unlimited recurring energies known to exist in the universe. This can achieve on current theoretical knowledge, but it is not enough. Can we also, as did the simpler past civilizations, turn scientific data into abstract thought and art, thus to sustain and identify ourselves as humans and not as animals in this extraordinary continuum[6]?

The concept of a middle distance, or link between smallness-bigness and immediacy-infinity, is peculiar to the human species. It is primarily concerned with idea: that there is a largeness beyond human comprehension and that this can be approached by an intermediary or stepping-stone. All religions are intermediaries[7], and so is art. In landscape design, the first projection of individual personality has been the complex of home, garden and forest tree; this is the stable foreground from which spring the eternally changing middle distances. In history, the middle distance was almost always metaphysical and abstract, such as the ascending progression of man-sphinx-pyramid-eternity. Although the scene has changed from the metaphysical to the material, the same progression in scale can be experienced today through the enigmatic sculptures and monster structures of Atlanta (Figure 10.6). As the manmade world grows increasingly

Figure 10.6 Sculpture of Atlanta

[1] biosphere: 生物圈
[2] J. Bronowski: Jacob Bronowski (1908~1974年), 英国科学家、作家
[3] ingenuity: 独创性，精巧，灵活性
[4] vandalism: 故意破坏艺术的行为
[5] battering: 用坏，损坏
[6] continuum: 连续统一体
[7] intermediary: 中间物，媒介

superhuman, so the concept of a meaningful middle distance must be extended and deepened.

What abstract form will this middle distance take? Man's new relation to environment is revolutionary and the landscape designer, unlike the artist, is conditioned by many factors that debar[1] immediate experiment. We must therefore turn to the artists for a vision of the future, gaining confidence in the knowledge that the abstract art that lurks behind all art lives a life of its own, independent of time and space. The interpretation of art into landscape is personal to every designer, but a combined study of the aerial view of Urbino[2], the aerial survey of the Philadelphia region, and the painting by Jackson Pollock (Figure 10.7) may suggest the grandeur of a fresh humanistic landscape that will have grown out of history and now lies within our grasp.

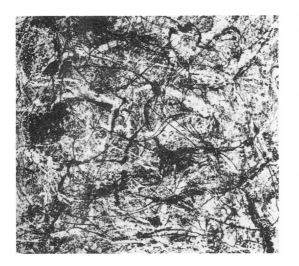

Figure 10.7 Painting by Jackson Pollock

For the first time in history, the shape of the world tat is unfolding expresses collective materialism rather than prescribed religion. In the advanced countries, the individual is evolving his own personal beliefs within his own home. The greatest threat to his existence may not be commercialism, or war, or pollution, or noise, or consumption of capital resources, or even the threat of extinction from without, but rather the blindness that follows sheer lack of appreciation and the consequent destruction of those values in history that together are symbolic[3] of a single great idea.

B. Sutton Place

> 节选自 *Modern Garden Design* 一书中有关杰里科作品的介绍。本文主要介绍了杰里科的著名作品——萨顿花园。

Sutton Place, made between 1980 and 1986 for the wealthy art-lover Stanley Seeger, is perhaps Jellicoe's finest garden, and it was created when he was in his eighties. His comments on it given in the Guelph Lecture at Guelph University, Ontario, Canada, relate to the many layers of history of the house and site, and his reflections on the influences leading to different incidents within the garden. He claimed a debt of gratitude to the Villa Gamberaia, in Tuscany, Italy, whose asymmetry[4] marked a step away from the link between villa and garden in the earlier Italian Renaissance garden. All the features are connected to the underlying theme of Creation, Life, and Aspiration[5]. The

---

[1] debar: 阻止，禁止
[2] Urbino: 乌尔比诺，位于意大利境内
[3] symbolic: 象征的，符号的
[4] asymmetry: 不对称
[5] aspiration: 热望，渴望

garden is a great gathering-place for the new and the old in Jellicoe's repertoire[1]. This time, he was more certain of his theme, and he may have wanted the climax of the journey to be related to a work by Ben Nicholson before he began his plan. Jellicoe found that Seeger was a fellow admirer of Nicholson and Henry Moore[2]. Not all of the original design was implemented- the Avenue of Fountains with its cascades[3] and grotto[4] were not made, the music theatre became a green theatre, and the moss garden (now the plane tree garden) to which Jellicoe made much reference in the Guelph Lecture, was unsuccessful, as the moss would not grow. That said, the garden throughout is marked by diversions and incidents to colour the journey.

The beginning, Creation, belongs to a fish-shaped lake beyond the entrance to the sixteenth-century house. By its shores are two hillocks, one representing a father, the second a mother, and between them was to be the site for a Moore sculpture reminiscent[5] of a foetus[6], called *Divided Oval*; the obvious analogy is of civilization emerging from the water. Moving back towards the house, Jellicoe made additions to the original garden which help to balance the rectangle which encloses the house and west walled garden. To the east, he designed two interlocking gardens within a wall. The first, or paradise garden, lies alongside the house, and is divided from it by a moat[7] with stepping stones; they represent the hazards which await those aspiring to reach the ultimate goal. Within the paradise garden are bubbling fountains framed by airy metal arches; the scent of flowers mingles with the murmuring of the fountains. From the paradise garden, a darker, more contemplative space under the shade of the plane tree in its centre creates a change of mood not envisaged[8] by the original plan for the moss garden. A belvedere[9] at the corner of the east gardens gives a view across to trees and the lawn across the back of the house. On the west side of the mansion, entry to the original walled garden leads to the pool garden. This, with its Tudor[10] pavilions, has a setting of blue, silver and grey planting, which was originally devised by Susan Jellicoe. In the centre of the pool were stepping stones and a rounded, floating raft, designed by Jellicoe to resemble a painting by Joan Mirb; on top of the sun raft were a table and chair, which reminded Jellicoe of *The Beautiful Truths*, a painting of a table on an apple suspended above water, by Rene Magritte.

Figure 10.8 Ben Nicholson Wall

---

[1] repertoire：节目，全部技能
[2] Henry Moore：亨利·摩尔，著名现代雕塑艺术家
[3] cascade：小瀑布，喷流
[4] grotto：洞穴，岩穴，人工洞室
[5] reminiscent：回忆往事的
[6] foetus：胎儿
[7] moat：护城河，城壕
[8] envisage：正视
[9] belvedere：望景楼
[10] Tudor：英国都铎式建筑式样的

From the kitchen gardens next to the pool, a more defined route leads, as at Runnymede and Sutton Place, to a place of disorientation[1] and disquiet. First, the traveller finds himself in an enclosed corridor between a high brick wall and a hedge, with no exit at one end- the Magritte garden. Along the wall are large urns[2] which originated at Mentmore, Buckinghamshire; they have been placed so that their shapes do not diminish into the distance in a manner akin to a surrealist painting. From the Magritte garden an exit is discovered through a dark, winding, evergreen tunnel; the release from uncertainty is found across the grass, where suddenly through an entrance to a hedged enclosure, a large, white relief created by Ben Nicholson appears to hang over a pool (Figure 10.8). The Carrara marble, with its simple shapes reflecting in the pool, give the perfect resolution to the earlier discord.

C. Water and Subconscious[3] in Jellicoe's Works

> 本文是英国格林威治大学（Greenwich/Hadlow College）"花园设计"课程论文。文章介绍了杰里科的生平，并从"水"和"潜意识"两个角度分析杰里科的作品，并讨论其设计思想和创作手法，是一篇研究设计师及作品的较好范文。

Introduction

Sir Geoffrey Jellicoe was probably the foremost British landscape architect of this century and his career spanned almost seventy years. From a background of the classics and a day to day exposure to modern art and architecture he was able effortlessly to combine both traditions.

Born in 1900 Jellicoe had a traditional classical education which inspired in him a lasting love of the Greek and Roman poets and philosophers. From 1919 he studied at the Architects Association School and for his final year thesis he visited Italy and wrote a definitive book on Italian renaissance gardens. This early experience gave him a wealth of detail to draw on throughout his career. Works, from his modifications to Ditchley Park in 1935, through to his designs for Sutton Place in 1980 showed this early influence.

Throughout the 1920's and 1930's Jellicoe remained associated with the Architects Association, as a lecturer and, later, president. During this period the Association was a center of modernism and its influence was shown in one of his first architectural commissions. His design for the restaurant at Goughs Caves in the Cheddar Gorge combined contemporary European lines with a renaissance feel for setting—the long low lines of the building emphasising the vertical grain and scale of the rocks above. This design also introduced one of Jellicoe's main themes in his exciting use of a pool and fountains to bring the terrace to life. Water remained one of his paramount elements, whether in actuality as at Shute, Sutton Place, and many others, or implied as at Horsted Place.

Jellicoe became increasingly involved in landscape design from the mid-1930's and was a founder member of the Institute of Landscape Architects. During and after the war he became more involved in

---

[1] disorientation: 方向知觉的丧失，迷惑
[2] urn: 瓮，缸
[3] subconscious: 下意识的，潜意识的

industrial and urban landscaping schemes and started to address the problems of maintaining human scale within large developments. He also developed his ideas about the relationship between painting and landscape design, bringing them together in a series of lectures which formed the basis for his "Studies in Landscape Design".

After his formal retirement he took on a new lease of life. Having time to think and reflect brought to fruition his ideas on the link between the conscious and the sub-conscious in the appreciation of design. In his eighties he embarked on a remarkable series of projects, culminating in his monumental design for the Moody Historical Gardens in Galveston, Texas.

It is obviously impossible to do justice to such a rich career in a short essay but it is interesting to reflect on two, perhaps overlapping, aspects of his work, viz. the influence of water and of the sub-conscious in his work.

Water

Water reoccurs[1] in many of Jellicoe's designs. In places it is still and reflective, in others tumbling down steep slopes adding an extra, auditory[2], dimension to the design. Indeed, at Horsted Place there is no water. Instead the lawns flow out towards the woods like lakes and planters seem to float on their surface. In his later schemes it is also possible to interpret his use of water in an allegorical[3] sense, or as an attempt to engage the subconscious.

The gardens at Shute, developed from 1969 onwards, seem to have provided Jellicoe with one of his most satisfy projects. Not only was there water in abundance but Jellicoe and his clients, Michael and Anne Tree, developed a superb[4] working relationship. Jellicoe was so impressed by the way that the empathy grew between them that he developed a simple diagram, based on Jung's methods, to explain it.

Shute was blessed with a number of natural springs, one of which rose at the highest point of the garden and fed into an old stone canal. Jellicoe used this as the source of a new stream which ran in a straight line down the hillside in a series of rills, linked by falls. This feature works superbly[5], and on several levels:

—from the top of the fall the waterway acts like a classical "long walk" leading the eye to a distant vista, in this case a small statue. This statue, heroic in style but small in size, helps to increase the apparent length of the vista. From this end of the stream the water appears smooth and tranquil.

—from the base of the rills the view is completely different. Each fall provides constant movement and reflection. The band of dense planting on either side emphasises the slope and heightens the sounds.

—on another plane the movement of the water and its accompanying planting can be seen as an

---

[1] reoccur: 重复出现
[2] auditory: 耳的，听觉的
[3] allegorical: 寓言的
[4] superb: 庄重的，华丽的
[5] superbly: 雄伟地，壮丽地

allusion to human life. At the top is the old canal, the source, or the creation. Initially the water runs through exuberant planting and the first waterfall is designed to give a treble trill as it flows. Progressive drops also drop in note through alto, tenor, and bass whilst the planting becomes more restrained. At its foot the water runs slowly through lawns before returning to water in a river.

At Sutton Place (1980) Jellicoe designed in several areas of still water. In the Paradise Garden a large formal pond bars direct entry to the garden, its tranquil surface hardly disturbed by spouting gargoyles[1]. Access to "paradise" is over a series of stepping stones representing the hazards of human life. Interestingly, Jellicoe here uses a feature which he first formally referred to in his 1960 book "Studies in Landscape Design". In his chapter, "Scale, Diversity, and Space" he gives a critique of Bellini's "The Earthly Paradise" and praises the way the balustrade[2] "interlocks the foreground and middle distance". Here, at the edge of the pond, we have a series of Bellini balustrades.

Further on the swimming pool, known as the Miro Mirror, features an elegant trick. A wooden deck is reached via a series of stepping stones but, unlike the stones, the deck is floating - merely being restrained in place by chains. Stepping onto it can produce an alarming tilt, hinting, perhaps, at the impermanence of life.

At the other end of the series of gardens at Sutton Place, themselves an allegory on human life, stands a monumental sculpture by Ben Nicholson. Apparently floating in space, the effect of this magnificently simple piece is magnified by its reflection in the still pool before it.

The heroic use of water continued right to the end of Jellicoe's career. His last major design, the Moody Historical Gardens, encountered water on a grand scale. On one side of a 5 metre high wall the ocean was to expend its energy—on the other visitors were to be taken through a series of canals by water bus.

The Subconscious

In his later work Jellicoe came to feel that the contribution of the subconscious in the overall appreciation of a design was being underrated[3] or ignored. Earlier in his career he had been closely associated with the work of modern artists as they struggled to suggest a third and fourth dimension in their work. Cubism[4] presented an attempt to show the passing of time on a flat canvas but artists such as Victor Pasmore, moving from representational drawing to abstracted forms, grew to feel that landscape design was a far better medium for their work. Landscapes already existed in three dimensions so the suggestion of the fourth, time, was far more readily achievable. The use of moving elements such as water and, more importantly, changing aspects as people moved through the design supplied the extra dimension. Although most of Pasmore's work was conceived on paper or canvass[5] he felt a strong spatial relationship to each aspect of it. His paintings invite imaginary journeys.

Jellicoe felt drawn to "abstract" work. One particularly rewarding friendship was with Ben

---

[1] gargoyle：怪兽状滴水嘴，（突出的）怪兽饰
[2] balustrade：栏杆
[3] underrate：低估，看轻
[4] Cubism：立体派
[5] canvass：讨论，游说

Nicholson. Not only did Nicholson provide the culminating sculpture for Sutton Place but his abstract, "Painted Relief" provided inspiration for Jellicoe's landscaping around Oldbury-on-Severn power station. Another influence was Paul Klee whose closely detailed works echoed the comparatively new science of aerial photography. Although architects such as Corbusier believed that this new technique could provide an extra dimension to their work the scope for its appreciation by the general public was limited. However this was not to mean that the aerial view was to be shunned. Many of Jellicoe's designs, such as the fish shaped lake for Sutton Place, look stunning in plan view but what gave them an added resonance[1] over a simpler or more "hackneyed"[2] design? Their shape couldn't be fully appreciated from most normal viewpoints so was there another sense at work?

Carl Jung proposed the idea that our subconscious responses to sensations are governed by the interplay between the "personal unconscious" and the "collective unconscious". Jung proposed that a person's response to any situation will be conditioned by "race, nation, family, spirit of the age, and personal experience". Thus a Japanese garden may attract respect and excitement from a British audience but not the empathy[3] that a native would feel. We miss the nuances[4], the references to tradition. We recognise the artistry but lack the "collective memory" necessary to truly appreciate it.

In a similar vein the gardens left by the Raj in India are looking increasingly bereft. Although until recently there has been little money to support any public gardens those based on Moghul and earlier Islamic designs have fared best. Although it can be claimed that the neglect of British style gardens is a conscious rejection of colonialism even the cemeteries around Christian and Anglophile[5] enclaves, such as that close to Old Delhi Station, are decaying. Despite the years of British rule and the care that been put into them by both races the gardens are still, on a more or less sub-conscious level, alien.

Jellicoe's triumph was to use his eclectic[6] knowledge of current and past practice to relate to the "collective consciousness" of the target audience. Thus he could move from a sympathetic restoration at St. Paul's Walden Bury to the water garden at Hemel Hempstead to the landscaping of the centre of Modena.

Jellicoe became increasingly absorbed by the role of the subconscious in landscape and garden design and Jung's theses. At Shute House he was delighted by the empathy between himself and the Trees. He felt that their relationship went beyond the formal architect/client partnership and that they were starting to correspond on a subconscious level. To explain the process he devised a diagram based on a circle, with a triangle within—its base across the diameter and its apex at the lowest point on the circumference. The thoughts of the client and the designer start at opposite sides of the circle and flow backwards and forwards around the rim as their conscious and unconscious thoughts converge. Both will also draw from a deep subconscious at the bottom of the circle. The final stage should be when agreement is reached at the centre of the circle, midway between client and designer.

---

[1] resonance：共鸣，反响，中介
[2] hackneyed：陈腐的，常见的
[3] empathy：移情作用
[4] nuance：细微差别
[5] Anglophile：亲英派的人
[6] eclectic：折衷的，折衷学派的

This diagram epitomised his relationship with the Trees (and with Stanley Seeger at Sutton Place where his initial plans were approved within 10 minutes) but also suggests the importance of the "personal" and "collective unconscious". If client and designer have no shared culture there is less likely to be a successful collaboration.

Conclusion

Sir Geoffrey Jellicoe holds a unique position in modern British landscape design. Coming into practice in the period after "The Great War" when the old nostrums were naturally reviled he combined an appreciation of the classical with a love of the modern. Through a long career he skillfully combined the two and interwove them with a regard for human scale and feelings.

In his work he can be said to have invented the idea of the professional landscape designer, taking the treatment of a building's exterior[1] beyond being just an adjunct to the structural design. His work confirmed the artistry[2] in landscape and garden design.

Notes

1. Extracts from

Texts

(1) Jellicoe, Geoffrey. & Susan. *The Landscape of Man: Shaping the environment from prehistory to present day* [M]. London: Thames & Hudson, 1988: 155-157.

(2) Waymark, Janet. *Modern Garden Design* [M]. London: Thames & Hudson, 2003: 185-189.

(3) Brackenbury, Martin. Sir Geoffrey Jellicoe [EB/OL]. http://users.eggconnect.net/mandvbrackenbury/Geoffrey%20Jellicoe.html, 2008-03-20.

Pictures

(1) Figure 10.6: Geoffrey and Susan Jellicoe 著, 刘滨谊主译. 图解人类景观——环境塑造史论[M]. 上海: 同济大学出版社. 2006: 330.

(2) Figure 10.7 Geoffrey and Susan Jellicoe 著, 刘滨谊主译. 图解人类景观——环境塑造史论[M]. 上海: 同济大学出版社. 2006: 343.

(3) Figure 10.8 Janet Waymark. Modern Garden Design [M]. London: Thames & Hudson, 2003: 189.

2. Source of Additional Information

http://www.greatbritishgardens.co.uk/

http://www.npg.org.uk/

http://users.eggconnect.net/mandvbrackenbury/Geoffrey%20Jellicoe.html

http://eng.archinform.net/arch/2815.htm?scrwdt=1024

---

[1] exterior: 外部，表面
[2] artistry: 艺术之性质

Roberto Burle Marx

1. Introduction

罗伯特·布雷·马科斯（Roberto Burle Marx，1909~1994 年），20 世纪杰出的艺术家和园林设计巨匠，他运用现代艺术语言与巴西当地植物材料，创造出风格独特的园林作品，对现代园林的发展产生了深远影响。

1909 年 8 月马科斯出生于巴西圣保罗市，5 岁时全家迁往里约热内卢，父亲是一位德国犹太裔商人，热衷于文化事业，母亲是音乐和园艺爱好者，受家庭环境影响和熏陶，从小马科斯就对音乐和观赏植物产生浓厚兴趣，1928 年随父亲来到德国学习艺术，在那里他接触了欧洲现代艺术，并在参观柏林达雷姆植物园时，对植物造景产生兴趣，欲把巴西丰富的热带植物资源应用于现代园林设计中。

1930 年马科斯进入里约热内卢国立美术学校学习，师从于著名建筑师和城市规划师 L. 科斯塔（L. Costa）先生，在那里完成其第一个作品施瓦茨居住区景观设计。经过几年的学习和锻炼，马科斯已经将绘画艺术和植物造景相结合，创造出一个又一个优美的环境景观作品。

1936 年是马科斯一生中比较难忘的一年，那一年里他有幸和著名建筑设计大师柯布西耶合作。柯布西耶受巴西政府邀请参与了教育卫生部大楼主体设计，这座标志性建筑的环境景观设计正是马科斯和波尔蒂奈里合作完成的，受柯布西耶影响马科斯在环境景观设计中思路更加清晰，思维也更加缜密，为今后的诸多优秀设计奠定基础。

20 世纪 50 年代，是马科斯大量大型设计的开始阶段，这段时间他参与和设计了里约热内卢的桑托斯杜蒙特机场环境景观设计、加巴那斯岛的加勒翁机场景观设计、巴西大学新址环境景观设计和美国驻巴西大使馆环境景观设计等，并利用 20 年时间完成了里约热内卢滨海开发环境景观设计。1954 年马科斯受聘为巴西大学城市与建筑系教授，并于同年在美国举行了个人园林作品展，1955 年在伦敦举行了个人设计展，1958 年在布鲁塞尔世博会上设计了巴西展馆。50 年代末 60 年代初参与了大量的巴西新首都巴西利亚的环境景观设计。

马科斯一生勤于工作，晚年仍坚持设计。在 50 多年的设计生涯中，共设计了上千件作品，他用艺术手法赋予园林设计新的形式语言，他善于利用当地植物材料设计出具有浓郁地方风情的园林作品。一生中除了在园林设计中的杰出贡献外，其在绘画、壁画、植物学等领域也取得了一定成绩。为巴西园林设计甚至是世界园林设计做出了杰出的贡献。

2. Major Publications & Design Works

A. Major Publications
- *The Gardens of Roberto Burle Marx*, Sagapress, Inc. /Timber Press, Inc., 1991.
- *Burle Marx: the Lyrical Landscape*, Thames & Hudson, 2001.

B. Major Design Works
- The Sitio, 1949~1989.
- Hospital Da Lagoa, formerly Sul-America, Rio, 1955.
- Ministry of Foreign Affairs, Brasilia, 1965~1988.
- Bank of Brazil, DF, 1973.
- Chaim Weizmann Square, Rio, 1983.
- Fazenda Vargem Grande, Areias, 1984~1989.

3. Ideas

A. Finding a Better way of Living

> 本文是马科斯为 *The Gardens of Roberto Burle Marx* 一书撰写的前言。马科斯认为设计花园是一种了不起的艺术,植物、动物和人在花园中和谐相处,好的花园能塑造好的生活方式。

To create gardens is marvelous art—possibly one of the oldest manifestations of art. The Bible records and describes a paradise in which there was a balance between the plants, the animals and man (Figure 10.9). Unfortunately, mankind sought to dominate nature and lost their paradise. With the knowledge we have today of ecology and the importance of our association with trees and plants, we are now trying to regain that lost paradise, and rectify[1] the mistakes of past generations.

Figure 10.9 Burle Marx's drawing for the Praca Arthur Oscar

Sima's love for plants, her knowledge and understanding of their growth; the color, the rhythm, the juxtaposition[2] and the relationship between the sizes of small, medium and large volumes—her respect for the association or harmony of plants and their climatic conditions; her worldwide journeys and her own constant work with those actors in gardens, including her own, has given her a whole and complete vision of the balance in space created by man for man.

We are living at a time in which the destruction of nature is so great that it has become a pre-occupation of thoughtless and ambitious people. In our struggle against the destruction of a legacy, we need to understand that we live in a world where plants exist, not only for material reasons, but also because they depict[3] birth, growth and death, emphasizing the instability of nature. We should always try to understand the mutations[4] and variations in nature and the light, sounds and perfumes that stir our emotions. Through her understanding of these mutations, together with her constant association with plants and gardens, and, as importantly, by her curiosity[5] in discovering plants and their usage in gardens, Sima's life, through her books, succeeds in her objective of revealing to others the beauty and functions that a garden ought to have.

I am very proud and honoured to have Sima write this book about my work. I trust that my work illustrates my intentions as an artist in composing what I believe good gardens ought to be- spaces in which people would like to express or renew their faith or belief in finding a better way of living.

---

[1] rectify: 矫正,调整
[2] juxtaposition: 毗邻,并置
[3] depict: 描述,描写
[4] mutation: 变化,突变
[5] curiosity: 好奇心

B. Hospital Da Lagoa & Chaim Weizmann Square

> 本文节选自 *The Gardens of Roberto Burle Marx* 一书，主要介绍了马科斯二个重要作品：The garden of Hospital Da Lagoa 和 Chaim Weizmann Square，这两个作品比较典型地反映了马科斯的设计手法。

Hospital Da Lagoa, Formerly Sul-America, Rio

The Hospital Da Lagoa was designed by the architect Oscar Niemeyer, while the garden was laid out in 1955 by Roberto Burle Marx, and despite a few changes over the years, provides a luxuriant[1] oasis in the heart of Rio on a long narrow piece of ground before the hospital building (Figure 10.10). The manner in which Roberto designed this garden, leading the viewer's eye along its entire length to end in a splendid focal point, displays a brilliant solution to a plot of difficult proportions.

The entrance at one end takes the visitor along a stepping-stone pathway, which winds between the lawn and a broad bed along the boundary, separating the two. This bed is thickly planted with groups of evergreens, unfolding as one walks, including *Scindapsis*; masses of white-flowered *Agapanthus*; and a creeping *Ficus montana*, with attractive oak-leaved foliage, covering the ground in a low carpet. A line of lofty Royal Palms on the boundary partially screens the city buildings and provides shade for the perennials[2] without depriving the narrow garden of sunshine. A huge clump of *Philodendron bipinnatifidum*[3], emerging from a carpet of pink-flowered *Heterocentron elegans*[4], emphasizes the end of this footpath, as it broadens into a wider path.

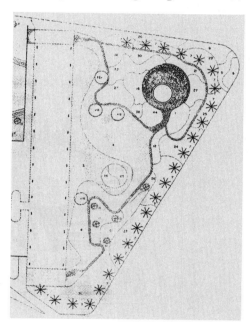

Figure 10.10 The garden of Hospital Da Lagoa

This broad path, composed of round stones, curves into a circular paved area which forms an open-air enclosure for convalescing patients. It is partially rimmed with a low, circular bench and half-enclosed by a tiled wall, backed by shrubs and trees. A small round pool is placed off-center and contains a thicket of upright canes of *Montrichardia arborescens* and a clump of Papyrus Grass, *Cyperus papyrus*[5]. The entire area around the paving and near the building is planted with perennials, creating a verdant atmosphere, but the plants are grouped separately and not overcrowded.

---

[1] luxuriant：丰产的，丰富的，肥沃的，奢华的
[2] perennial：四季不断的，终年的，长期的，永久的，（植物）多年生的
[3] *Philodendron bipinnatifidum*：[植] 春羽
[4] *Heterocentron elegans*：[植] 蔓性野牡丹
[5] *Cyperus papyrus*：[植] 纸莎草

Looking down from above, the long path appears to culminate in a graceful flourish, forming an abstract design of pleasing proportions. The effective simplicity of this design reveals the artist's ability to make a bold statement and integrate it with plants and a variety of building materials and stonework to form a balanced and harmonious composition.

Chaim Weizmann Square, Rio

The library of Botafogo, Rio, is situated in the center of a triangular[1] plot named Chaim Weizmann Square (Praca Chaim Weizmann), bounded by Rua Barao de Itambi, Rua Pinheiro Machado and Rua Farani, immediately opposite the Centro Empresarial building and garden (Figure 10.11). The urban landscape around the library was designed by Roberto Burle Marx in 1973. Its features include all the elements of Roberto's treatment for city squares, with places for people to sit and rest in the shade amid attractive flowering trees, lawns and palms.

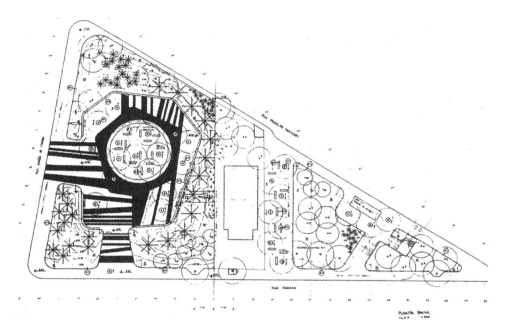

Figure 10.11 Chaim Weizmann Square, Rio

The area before the library has been covered with a black, white and brown mosaic in an abstract pattern comprising lines and circles. The brown circular portion at the center marks the main shade area. Circles containing Hong Kong Orchid Trees, *Bauhinia blakeana*[2], have been let into the design to allow for a group of these small trees to shade the center. These are underplanted with the yellow-flowered groundcover, *Wedelia trilobata*[3]. Short low benches under the trees, placed in such a way that people face in different directions, invite strollers to rest awhile. A running bench encircles the mosaic paving, backed by ramps covered with a tropical grass, *Axonopus obtusifolius*. These ramps are mounded at the center to form a berm masking the road and provide a sense of enclosure to the mosaic area. They also provide added height for the shrubs planted on it.

---

[1] triangular：三角形的

[2] *Bauhinia blakeana*：[植] 红花羊蹄甲

[3] *Wedelia trilobata*：[植] 南美蟛蜞菊

Palms arranged in groups ornament[1] the lawn. One corner holds a cluster of 11 triangular Palms, *Neodypsis decaryi*[2], and the opposite corner, 17 Assai Palms, *Euterpe oleracea*[3]. A dozen Royal Palms, *Roystonea regia*, curve in a line around the front of the building, screening some of the high-rise buildings in the vicinity. Clusters of frilly *Philodendron mello-barretoanum* placed beneath them on the street side assist in the screening with their luxuriant foliage. Red Gingers, *Alpinia purpurata*, cluster in the shade alongside the library building. Tall, yellow-flowered Sibipiruna trees, *Caesalpinia peltophoroides*, are planted in two groups of three in the far corner, with the double pur-pose of acting as the apex[4] of the design and screening the square. Near the library, a group of 7 small Rosewood trees, *Physocalymma scaberrima*, produce their pretty, reddish purple, crinkled flowers in early summer.

Behind the library, two beds mounded into grassy berms, separated by flights of steps, hold groups of trees. *Bauhinia blakeana* and *Physocalymma scaberrima* are repeated on the larger berm, together with a cluster of *Yucca gloriosa*, while the other features a street tree that thrives in Rio, *Moquilia tomentosa*.

The combination of art and design with skilful use of varying plant materials, from low to high, has seldom been demonstrated with greater effect, transforming a small, difficult piece of ground into a delightful city square.

C. The Lyrical Landscape

> 本文是著名当代园林设计师玛沙·斯瓦茨（Martha Schwartz）为马科斯 *The Lyrical Landscape* 一书所作的序言。斯瓦茨认为马科斯较好地在园林创作中融合了艺术与科学，并以其独到的手法表达了设计师的个性和灵魂。

Burle Marx has always been and will remain for me a shining light in the somewhat dim[5] universe of landscape art and architecture. Like my other hero and model, Isamu Noguchi[6], the strength of Burle Marx's work is that it is a direct expression of his spirit. Burle Marx's bold use of form and colour, combined with his painterly approach, express the force of his personality and sensibility, which, to me, is the necessary core to all art. This force, his art, transmits and connects to everyone around him (Figure 10.12).

Figure 10.12 Flamengo park, the museum of modern art and copacabana

---

[1] ornament：装饰物，装饰
[2] *Neodypsis decaryi*：[植] 三角椰子
[3] *Euterpe oleracea*：[植] 阿萨依野棕
[4] apex：顶点
[5] dim：暗淡的，悲观的
[6] Isamu Noguchi：野口勇，当代著名设计师，雕塑家

A great artist in whose mind and heart were no boundaries, Burle Marx was a fluid thinker unconcerned by living within limits. Whether designing a landscape or painting, eating, playing music or throwing a great party, he lived and created wholeheartedly and with gusto[1]. And because his life was as creative as his gardens, he showed us that it is just as important to invent what to do with one's life as *how* to live it.

Burle Marx was not only a prominent landscape architect with an international reputation. In Brazil[2] he was a national hero. Through his passion for the native landscape and his artistic skill in depicting[3] this love, he helped an entire nation define itself. In his fascination with natural processes, with the artifacts[4] of nature that he lovingly collected and nurtured, and the love of our human (and in particular, Brazilian) nature, he balanced humanism with naturalism. I find his connection to his culture as significant as his enthusiasm for his native landscape, and it is precisely[5] this essential connection between humans and their environments that he expressed and celebrated in his work.

Through understanding the great range of Bude Marx's works, interests and the influences on him, we professionals should take heart that it is possible to integrate art and science and to believe that it is only with our personal intuition, spirit and sensibilities that we can bring meaning and form to our environment, wherever that may be.

Notes

1. Extracts from

Texts

(1) Elivson, Sima. *The Gardens of Roberto Burle Marx* [M]. Portland: Sagapress Inc. / Timber Press, Inc., 1991: 7.

(2) Elivson, Sima. *The Gardens of Roberto Burle Marx* [M]. Portland: Sagapress Inc. / Timber Press, Inc., 1991: 119 – 122.

(3) Elivson, Sima. *The Gardens of Roberto Burle Marx* [M]. Portland: Sagapress Inc. / Timber Press, Inc., 1991: 179 – 180.

(4) Montero, Marta Iris. *Roberto Burle Marx: the Lyrical Landscape* [M]. London: Thames & Hudson, 2001: 1.

Pictures

(1) Figure 10.9 Marta Iris Montero. Roberto Burle Marx: the Lyrical Landscape [M]. London: Thames & Hudson, 2001: 58.

(2) Figure 10.10 Elivson, Sima. *The Gardens of Roberto Burle Marx* [M]. Portland: Sagapress Inc. /Timber Press, Inc., 1991: 121.

(3) Figure 10.11 Elivson, Sima. *The Gardens of Roberto Burle Marx* [M]. Portland: Sagapress Inc. /Timber Press, Inc., 1991: 180.

(4) Figure 10.12 王向荣，林箐. 西方现代景观设计的理论与实践 [M]. 北京：中国建筑工业

---

[1] gusto：爱好，趣味
[2] Brazil：巴西
[3] depict：描述，描写
[4] artifacts：史前古器物
[5] precisely：正好

出版社，2002：113.
  2. Sources of Additional Information
    http：//www. burlemarx. com. br/
    http：//www. maria-brazil. org/sitio _ roberto _ burle _ marx. htm
    http：//www. ceramicanorio. com/conhecernorio/sitioburlemarx/SITIOBURLEMARX. html

Unit 11

- John Ormsbee Simonds

- Lawrence Halprin

- Dan kiley

- Ian Lennox McHarg

John Ormsbee Simonds

1. Introduction

约翰·奥姆斯比·西蒙兹（John Ormsbee Simonds，1913～2005 年），美国现代风景园林先驱，著名的风景园林师、规划师、生态学家。曾任美国风景园林师协会（ASLA）主席，英国皇家设计研究院研究员，美国总统资源与环境特别工作组成员，美洲社区规划组织顾问等职。一生在著作、设计实践、教育三方面均取得瞩目的成就。

西蒙兹 1913 年生于美国北达科他州，父亲是基督教长老会牧师。1930 年入密歇根州立大学学习风景园林，1936 年到哈佛大学设计研究生院学习，三年后获硕士学位。1955 年至 1967 年，西蒙兹应邀在卡内基——梅隆大学讲课；1961 年，西蒙兹整理讲课内容，出版了《风景园林学：人类自然环境的形成》一书。此书成为美国现代风景园林史上里程碑式的著作，在书中他写道"风景园林师的终生目标和工作就是帮助人类，使人、建筑、社区、城市及人的生活同地球和谐相处。"

1963～1965 年西蒙兹任美国风景园林师协会主席，1966～1968 年任协会基金会主席；任职期间，他高举改善人类生活环境大旗，开拓一个又一个新的领域，不遗余力地推动风景园林事业的发展。1970 年，西蒙兹将他的公司改名为"环境规划设计公司"（The Environmental Planning and Design Partnership）即 EPD，标志着他的研究与实践转向一个更为广阔的天地。此后，EPD 公司完成了美国佛罗里达州渔民岛总体规划等一大批优秀作品，范围包括新社区规划、公路环境、州立公园、区域资源总体规划、滨水景观等广阔的领域。1978 年，在总结六七十年代工作基础上，发表了《大地景观：环境规划指南》一书，系统全面地阐述了环境规划的内容。

1983 年，西蒙兹从 EPD 公司退休，但仍活跃在风景园林论坛。1994 年出版第三部著作《21 世纪花园城市：创造适宜居住的城市环境》。1999 年，美国风景园林学会授予他"世纪主席奖"——以嘉奖"他在 20 世纪为美国风景园林及风景园林学会作出的不可比拟的贡献"，同年出版第四本专著《启迪》。2005 年 5 月 26 日以 92 岁高龄在匹兹堡家中逝世。

2. Major Publications & Design Works

A. Major Publications

- *Landscape Architecture：The shaping of Man's Natural Environment*，New York 1961.
- *Revised Landscape Architecture：A Manual of Site Planning and Design*，New York 1983，1997.
- *Earthscape：A Manual of Environmental Planning*，New York 1978.
- *Garden City* 21：*Creating a Livable Urban Environment*，New York 1994.
- *Lessons*，Washington，D.C. 1999.

B. Major Design Works

- Mellon Square，Pittsburgh，1953.
- Miami Lakes New Town，1960.
- Master Plan for Chicago Botanic Garden，Chicago，1964.
- Environmental Action Plan for Chattanooga and Hamilton County，Tennessee，1968.
- Comprehensive Plan for St. George Island，Franklin County，Florida，1976.

3. Ideas

A. The Great Capital

> 节选自西蒙兹著作 Lessons《启迪》。该书为笔记体回忆录,记录了西蒙兹在学习和探索风景园林规划设计过程中的思考和体会。
>
> 1939年,西蒙兹从哈佛大学设计研究生院硕士毕业,但他觉得对景观设计(风景园林)还缺乏本质的认识。于是与同窗好友勒斯特·A. 柯林斯(Lester A. Collings)结伴到东方学习考察,足迹遍及日本、韩国、中国、印度等地。历时一年的游历对他后来设计思想的形成产生了深刻的影响。本文记述的是西蒙兹和柯林斯在北平拜访一位皇家建筑师家族的后裔——李先生,李先生向他们介绍了中国"大都"城的规划和建造过程。

"Peking, as you know, was built—largely in its present form—by Emperor Kublai Khan[1], as his capital. He had extended the conquests of his grandfather, Ghenghis Khan[2], until now he was master of most of the known world. His new city, Ta-Tu, "The great Capital", must reflect his unbounded power, wealth, and prestige."

"As was his custom, he first called his advisors together to discuss the purpose and nature of the city to be built. This would be no mere ritual[3]. The success of the Khans lay in large part to wisdom derived from such councils—as to battle strategy, governance, or, as in this case, the planning of a new city. All conceivable problems, possibilities, and alternatives would be thoughtfully explored. There would be, I am sure, long discussions of other cities taken, or made tribute in his conquests. There would be, too, detailed consideration of the merits and faults of other classic Chinese cities dating from earliest times."

"But back to the Emperor, Kublai Khan. After long consultation with his respected advisors, he had put into writing what might be called his proclamation of intent as to what the city would be. I have here a transcription."

Then Architect Li, descendent of the long line of Imperial architects from the time of Ta-Tu's founding, slid back a section of screen and extracted[4] from a lacquered[5] tube a yellowed scroll which he unrolled with care. His translation follows, as best it can now be remembered.

"Here on this well-watered plain, is to be built a great city in which man will find himself in harmony with God, with nature, and his fellow man."

"The decree goes on to describe the purpose and nature of each city component. First, those of defense: the water-filled moats, double-gated walls and drumtowers, the military roads, encampments and parade grounds. Then, centered on the broad avenues, the walled inner 'Forbidden

---

[1] Kublai Khan:忽必烈
[2] Ghenghis Khan:成吉思汗
[3] ritual:仪式,典礼
[4] extract:选录,摘取
[5] lacquer:漆器

City' of the Imperial household, courts and palace from whence the Khan, on his throne of jade, would govern his far-flung empire."

"Throughout the capital were to be located reservoirs[1] in the form of lakes and lagoons, the soil from their excavations[2] to be shaped into enfolding hills, planted with trees and flowering shrubs collected from the farthest reaches of his dominion. Crowning the hills were to be built the temples and civic buildings, that all might look up to see their curving roofs of gold-glazed tile ablaze against the sky."

"The courtyard homes of his nobles would be reached by roads that wound along the lakeside and between the fronting hills. To his trusted officers and officials would be given that most precious gift—the gift of privacy."

Neither their gates nor their homes were to be entered except by those invited or by one bearing the seal of the Khan. Temple compounds, governmental buildings, work areas and market places were to be carefully planned together to provide support and ease of movement from place to place.

"And finally, the matter of parks and open spaces. The Khan decreed that no separate parks were to be set aside. Rather, the whole of Ta-Tu would be planned as one great inter-related garden-park, with palaces, temples, public buildings, homes, and market places beautifully interspersed[3]."

If one would plan a great city, first learn to plan a garden. The principles are the same.

B. The Work of the Landscape Architecture

> 节选自美国 *Modern Architects* 一书中介绍西蒙兹的章节，本节内容是西蒙兹自撰个人简介的一部分。在这段文字中，西蒙兹阐述了风景园林师的工作内容。

The work of the landscape architect, or "architect of the landscape", is the design of functional and expressive out of door spaces and places. These are created within the context of the natural and man-build environment, and range in scope and complexity from a child's playlot to a recreational park, university campus, or the conceptual planning of a community or new town.

On larger commissions the landscape architect often serves as a member of a closely coordinated professional team, which includes architects, engineers, planners, and scientist-advisers. A generalist[4], the landscape architect brings to the planning-design process specialized training in the

Figure 11.1 The parkway in Pelican Bay

[1] reservoirs：水库
[2] excavations：挖掘
[3] interspersed：点缀，散布
[4] generalist：通才

physical sciences—such as physiography[1], geology, hydrology[2], biology, and ecology—and a feeling for the land, human relationships and design.

The past several decades have witnessed a remarkable evolution in the practice of landscape architecture. Within this brief span of time the emphasis has shifted from the design of large estates and sumptuous[3] resorts for the wealthy to subdivisions, public works projects, public housing, urban renewal, military installations, freeway improvements, national, state and urban parks—and more recently to river basin studies and comprehensive resource planning (Figure 11.1). It has been the privilege of the members of our firm to have had an active role in this dynamic transformation.

C. Mystical Pragmatist[4]

> 节选自美国 *Landscape Architecture* 杂志介绍西蒙兹的人物访谈文章。本文介绍了西蒙兹的设计哲学和思想。

He concluded that he was responding to the stimulus of nature. And that's the crux[5] of his career: orienting his new towns and plazas to stir the senses—"planning experience", he calls it. The gophers[6] on the prairie[7] of his native North Dakota, who dug dens in the sun, away from the breeze, safe from the hawks, further inspired his concepts of ecological planning for towns up to 10,000 acres.

If that sounds like the musing[8] of a would-be Druid[9], bear in mind that Simonds' projects are ruthlessly practical. They also make money. Many of the 80 communities he has master-planned are lavish[10] Florida wintering grounds built by such developers as Westinghouse and the Grahams of Washington Post fame. A common-sense mystic[11], he is as devoted to raising subdivisions and freeways as saving pelican[12] habitats.

"Preservation, conservation and development—PCD," is how Simonds initializes his philosophy. Preserve an area's vital ecological features, conserve its wetlands and water supplies and develop intensively on less crucial lands. The ideal result: an ecological new town, densely developed with neighborhood shopping, open space, verdant[13] parkways and bike paths. He can gauge[14] the results of his theories from the window of his studio in Naples, Florida. "The things we've been

---

[1] physiography: 地貌学
[2] hydrology: 水文学
[3] sumptuous: 华丽的
[4] Pragmatist: 实践主义者
[5] crux: 症结
[6] gophers: 土拨鼠
[7] prairie: 牧场
[8] musing: 冥想的
[9] Druid: 德鲁伊教团员
[10] lavish: 丰富的
[11] mystic: 神秘主义者
[12] pelican: 塘鹅
[13] verdant: 青翠的
[14] gauge: 测量

seeing in the past three weeks are miracles. The whole bay just surging with minnows[1]. And you see the pelicans, diving in and raking it up. We saved a sanctuary[2] with 200 nests (at Pelican Bay, one mile away). That's PCD—P for preservation, P for pelican..." (Figure 11.2)

The tale begins in the 1930s, when landscape architects were still decorating country estates. Simonds was part of a movement that nudged[3] the discipline into a mature phase that embraced urbanism and ecological planning. "We were lost sheep," says Simonds. "Now I think we're the second most important profession, after teaching".

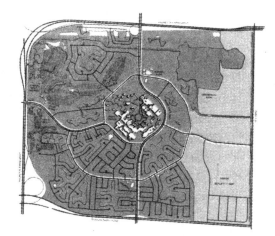

Figure 11.2 Master Plan: Miami Lakes New Town

"John Simonds is one of the last of a generation of modernists who changed the field," confirms Neil Porterfield, FASLA, head of the landscape architecture department at Pennsylvania State University. "Simonds, James Rose, Garrett Eckbo that group, most of whom went to Harvard, really put landscape architecture on the map".

However, Simonds broke free of the abstraction that so bewitched[4] his Harvard colleagues. It colored their work, he believes, with a falseness based on the pursuit of raw geometry. "I was doing geometry, like everyone else," he says of his days at Harvard. "I'd do a grid and twist it and add a circle. It was pattern. Pattern and geometry, instead of responding to the sun".

Some philosophical underpinnings[5] emerge in his writing on the Pittsburgh Aviary-Conservatory[6], the first free-form aviary, built in 1954: "Actual jungle—in India, Borneo, Malaya—is dark, dank, raw and cruel. The stylized jungle of the Aviary has a cultivated, breathtaking beauty resembling[7] no jungle but expressing the highest quality of them all".

In Zen-like pursuit of the "highest qualities" of rocks, water, steel and landforms themselves, the aviary was but one of many Simonds innovation. In Florida he pioneered the planned unit development, a system of clustered building designed to preserve open space. His Chicago Botanic Garden rescued fouled swamps and gravel pits to provide an archetype[8] of urban reclamation. Even his office broke new ground. Long before industrial chic[9], he converted an old, wood-frame warehouse[10] into headquarters for 30 designers.

---

[1] minnow：鲦鱼（一种小淡水鱼）
[2] sanctuary：动物保护区
[3] nudged：推动
[4] bewitched：迷惑的
[5] underpinnings：基础
[6] Aviary-Conservatory：大型鸟舍温室
[7] resembling：类似
[8] archetype：原型
[9] chic：时髦
[10] warehouse：仓库

Notes

1. Extracts from

Texts

(1) Simonds, John Ormsbee. *Lessons* [M]. Washington, D. C. : ASLA Press, 1999: 37 - 39.

(2) Simonds, John Ormsbee. Simonds, John Ormsbee. *Modern Architects* [M]. 编者王欣从西蒙兹处获得的复印资料.

(3) Leccese, Michael. Mystical Pragmatist [J]. *Landscape Architecture*, 1990 (3): 79 - 83.

Pictures

(1) Figure 11.1: Michael Leccese. Mystical Pragmatist [J]. *Landscape Architecture*, 1990 (3): 83.

(2) Figure 11.2: John O. Simonds. Miami Lakes New Town [J]. *Parks & Recreation*, 1970 (10): 17.

2. Sources of Additional Information

A Guide to the John Ormsbee Simonds Collection

http://www.uflib.ufl.edu/spec/manuscript/guides/Simonds.htm

Chicago Botanic Garden

http://www.chicagobotanic.org

Lawrence Halprin

1. Introduction

劳伦斯·哈普林（Lawrence Halprin，1916 年～），美国现代风景园林规划设计第二代的代表人物，美国最著名的风景园林师之一，曾任全美艺术学会会长，历史文物财产保护局委员等职；他所涉足的领域包括城市规划、城市设计、风景园林规划设计及其方法论。

哈普林 1916 年生于美国纽约的布朗克斯区，父亲是一家科学仪器公司的老板，母亲是全国性犹太组织的领导人。他 10 岁开始学习绘画，在上大学之前随双亲移居以色列，在以色列的一个集体农庄生活了三年，这段和自然世界极其亲和的经历，毫无疑问影响了他以后一生的职业兴趣和设计取向。20 世纪 30 年代中后期，哈普林在美国康乃尔大学和威斯康辛大学学习植物学，获植物学学士和园艺学硕士。1942～1944 年进入哈佛大学，在格罗皮乌斯、布劳耶和唐纳德等人指导下学习风景园林。

第二次世界大战期间，哈普林在美国海军驱逐舰服役，因受伤而回到旧金山，后来他在 T. Church（T. 丘奇）的事务所工作了四年，1948 年哈普林参加了 T. Church 最著名的作品 Donnell Garden（唐纳花园）的设计；1949 年哈普林在旧金山开设了自己的事务所。1949～1959 年是哈普林创作生涯的探索阶段；1960 年开始，哈普林设计理念和手法逐渐成熟，代表作相继问世，并成为了美国现代风景园林规划设计第二代的代表人物。

2. Major Publications & Design Works

A. Major Publications

- *Cities*，1963.
- *Freeways*，1966.
- *The RSVP Cycles: Creative Processes in the Human Environment*，1969.
- *Lawrence Halprin Notebooks* 1959 - 1971，1972.
- *The Franklin Delano Roosevelt Memorial* (*Special FDR Memorial Edition*)，2000.

B. Major Design Works

- Meintyer Garden，1958.
- Lovejoy Fountain Plaza, Pettygrove park, Auditorium Forecourt Plaza (Ira C. Keller Fountain Garden), at Portland, Oregon, 1960.
- Chirardelli Square, at San Francisco, California, 1963.
- Freeway Park, at Seattle, Washington, 1972～1976.
- The FDR (Franklin Delano Roosevelt) Memorial, at Washington, 1974～1997.

3. Ideas

A. The RSVP Cycles

> 节选自哈普林专著 *The RSVP Cycles: Creative Processes in the Human Environment* 一书。本文描述了一种"现状——方案——预测分析——执行结果"的不断循环的规划设计理论和方法。批评了以目标为导向的规划设计方法，强调要重视规划设计的过程。

R *Resources* which are what you have to work with. These include human and physical

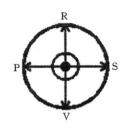

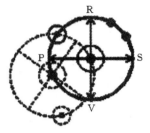

Figure 11.3 The RSVP Cycles

resources and their motivation and aims.

S *Scores* which describe the process leading to the performance.

V *Valuation*[1] which analyzes the results of action and possible selectivity and decisions. The term "valuation" is one coined to suggest the action-oriented as well as the decision-oriented aspects of V in the cycle.

P *Performance* which is the resultant of scores and is the "style" of the process.

Together I feel that these describe all the procedures inherent in the creative process. They must feed back all along the way, each to the other, and thus make communication possible. In a process-oriented society they must all be visible continuously, in order to work so as to avoid secrecy and the manipulation of people.

Together they form what I have called the RSVP cycles (Figure 11.3).

...

One of the gravest dangers that we experience is the danger of becoming goal-oriented. It is a tendency that crops up on every hand and in every field of endeavor. It is a trap which goes like this: things are going poorly (in the realm of politics or religion or building a city or the world community or a personal relationship or whatever). As thinking people we must try to solve this problem that faces us. Let us set ourselves a "goal" upon which we can all agree (most goals after all are quite clearly moralistically based and incontrovertibly "good ideas"). Having set ourselves this goal we can then proceed posthaste to achieve it by the most direct method possible. Everyone can put his shoulder to the wheel, and systems engineering, technology, and our leader (or whatever) will get us to the agreed goal.

It doesn't work! The results of this oversimplified approach, now coming into general vogue[2], are all around us in the chaos of our cities and the confusion of our politics. It generates tension in personal relationships by burying the real problems; it avoids the central issue of education, which is why today's young people are dropping out; it is destroying the resources and physical beauty of our planet; and it avoids the basic issue.

...

We don't really want to be involved in goal-making or goal-solving. Fritz Perls says, "scores face the possible where goals face the impossible." What we want, what we desperately need, is a feeling of close and creative involvement in processes. It is the *doing* that we all enjoy and which is meaningful to us. That is what is needed in education, in the ghetto, and in the young and the downtrodden who feel that they are excluded from the process of decision-making in our communities; certainly it is needed in personal relations. It is ongoingness, the process that will build and develop great cities and regions and a world community on this planet Earth. By involvement in process we all

---

[1] valuation: 评价

[2] vogue: 流行

interact, our input is significant, visible, meaningful, useful, and no one point of view can hold us in thralldom. Scores are not goal-oriented; they are hope-oriented.

This is why "scores," which describe process, seem to me so significant. It is through them that we can involve ourselves creatively in "doing," from which, in fact, structure emerges—the form of anything is latent[1] in the process. The score is the mechanism which allows us all to become involved, to make our presence felt. Scores are process-oriented, not thing-oriented. In dance and theatre this works through open scoring, which establishes "lines of action" to which each person contributes and from which a final performance then emerges. In personal relations scoring allows a constant interaction devoid of the moralisms and shoulds and shouldn'ts which inhibit growth and deep contacts and involvements. In the planning of communities a score visible to all the people allows each one of us to respond, to find our own input, to influence *before* the performance is fixed, before decisions are made. Scoring makes the process *visible*. For that reason scores seem to me the key link in the entire RSVP cycles—though only one link, still at the core of the whole procedure.

The RSVP cycles is a balanced scheme in which all the parts are mutually related and constantly interacting. It functions best when all parts are operating. Its purpose is to make procedures and processes visible, to allow for constant communication and ultimately to insure the diversity and pluralism[2] necessary for change and growth.

B. Freeway Park: Past, Present and Future

> 选自美国文化景观基金会（the Cultural Landscape Foundation）网页，作者是 Brice Maryman（西雅图市风景园林教师）和 Liz Birkholz（黑格风景园林设计事务所设计师）。本文主要描述了哈普林设计的西雅图市大道绿地（Freeway Park）的设计过程、景观特点、现状和拟采取的保护措施。

Freeway Park, executed by Lawrence Halprin's office under the design direction of Angela Danadjieva, is one of the most compelling treatises on post-War landscape architecture that survives today. Perched above Interstate 5 in downtown Seattle and using 5.5 acres of interstate air rights, the linked spaces of the park evocatively and imaginatively engage the three major preoccupations of post-War landscape design as described by Elizabeth K. Meyer: the car, the garden and the growing awareness of ecology. Yet like so many other modernist works, Freeway Park is threatened by poor maintenance, shrinking budgets and the threats attendant with community revitalization.

After Halprin's publication of the book *Freeways* in 1966 and his work with the Federal Highway Administration's Urban Advisors group, the Seattle Parks Commission sought his assistance in designing a park along the edge of the new interstate gorge[3]. Rather than confining himself to the proposed plot of land, Halprin pushed the ideas in his book into the cityscape by proposing an extensive landscape that scaled down the impact of the freeway for both driver and pedestrian[4] by

---

[1] latent：潜在的
[2] pluralism：多重性
[3] gorge：障碍物
[4] pedestrian：步行者

building right over it. Rather than balking at this audacious[1] plan, the proposal was bundled into the countywide open space bond measure called Forward Thrust, and, in 1969 approved local funds were combined with state, federal and private monies to allow the park plan to move forward.

Figure 11.4 Freeway Park

The space is defined by a series of linked plazas that are intertwined and enclosed by board-formed concrete planting containers and walls. These major spaces are known as the Central Plaza, East Plaza, and West Plaza. Consistency and cohesion between these spaces is developed through a shared materials palette of concrete, broadleaf evergreen plantings and site furnishings. The spaces are differentiated through the dynamism of the water features that occupy the spaces and the attendant differentiation of moods (Figure 11.4).

A roiling precipice of water in freefall dominates the Central Plaza, tumbling over an assertive assemblage of overhangs and outcrops. The design effect of 28,000 gallons per minute tumbling over 30-foot tall formed concrete blocks is at once rugged and decidedly urban: a space that is consciously of the city yet inspired by the morphological[2] capriciousness[3] of the Cascade and Olympic Mountains. By placing the water feature over the canyon of the freeway, the "natural" gorge was able to drown out—or at least ameliorate[4]—the roaring sound of the artificial, automotive flows below.

Like an idyllic mountain stream, the fountain was filled with children and parents when it opened on July 4, 1976 as part of Seattle's bicentennial celebrations. Though there are no guardrails protecting visitors from the water; Halprin's design intent heightened an explicit sense of danger so that people are confronted by risk prima facie and are therefore cautious. Near the base of the canyon, a heavy-gauge glass window allows visitors to see cars driving by, creating a dynamic visual dialogue between nature (water) and the city (the cars of the freeway).

The framework for these original elements still exists, but the experience of the canyon today is significantly degraded. A steel screen now covers the canyon's window, obfuscating[5] the connection to the freeway, and the falls themselves are tragically under served. While there were three pumps that originally fed water to the canyon (using two at a time, cycling through the third), today only two pumps remain and with only one pumping water at a time. The capacity of this one pump has since been reduced by 30 percent to where the 28,000 gallons per minute of the original design is now reduced to a relative trickle near 9,500 gallons per minute when running. Most of the time, however,

---

[1] audacious：勇敢的
[2] morphological：形态学的
[3] capriciousness：多变性
[4] ameliorate：改良
[5] obfuscating：模糊

the canyon water feature is not even active. Again due to increased safety standards and reduced maintenance budgets, parks officials are not easily able to access all of the basins and traps within the fountain to clean out debris[1] before starting the pumps.

Throughout the park, the role of vegetation is not limited to aesthetic or architectonic purposes, rather plants were also chosen for their ability to reduce pollution and baffle sound coming from the freeway below. As in Halprin's Portland open space sequence, the original planting plans reveal a placement strategy that develops an analog to the larger landscape surrounding Seattle. Lower levels are heavy with Rhododendron and Alnus; higher levels are dominated by *Kinnikinick*, *Cedrus deodora* and other upland tree species. Although the park appeared sparsely planted at its onset, the vegetation has grown dense and required limbing up for maintenance and security reasons. Similar to the controversy that swirled around San Francisco's U. N. Plaza and Denver's Skyline Park, the park is often viewed as blighted and undesirable as transients and drug users emerge from the vegetated planters above the freeway.

...

Right now, the City of Seattle has allocated funds to conduct a study on how to revitalize Freeway Park. Seattle's Mayor, Greg Nickels, has said recently, "We must not turn our back on the park again." At a recent public meeting, city of Seattle staff unveiled draft recommendations that were submitted as part of a visioning and activation report prepared by the New York City-based Project for Public Spaces (PPS). Propelled by neighborhood interest groups who want to see a beloved park returned to its full glory, PPS made several insightful programming recommendations that could serve to make the park more safe and habitable throughout the day. Yet noticeably lacking from the presentation was any language referring to the precedent-setting status of the park or its ingenious[2] design team. Some design recommendations included significant reworking of the original Halprin plan including demolition[3] of some of the concrete retaining walls, redesigning or removing at least two of the original fountains, and installing a series of exercise stations.

While the implementation of many of these efforts can be done sensitively, there is currently no agency overseeing the preservation of this historic park. A Seattle Landmarks nomination is currently being prepared and will be submitted to the city in the near future. With design monies allocated for 2005, landmark designation will provide an additional level of review before the integrity of yet another Halprin masterwork is lost forever.

C. The Power of Place

> 选自美国 ASLA 期刊 *landscape architecture*。本文介绍了哈普林的重要作品——美国华盛顿市罗斯福纪念园（FDR Memorial）的设计背景、设计构思和营建过程，展示了哈普林在风景园林设计中的主张和观点。

The Flanklin Delano Roosevelt Memorial, dedicated May 2 by President Clinton, represents a

---

[1] debris：残骸
[2] ingenious：设计精巧的
[3] demolition：破坏

historic first—a memorial to an American president that is a landscape, breaking with the tradition that the great are memorialized only by structures. The highest-profile piece of landscape architecture in many a year, it was conceptually designed in 1974 by a collaborative team led by Lawrence Halprin, FASLA, and, following decades of bureaucratic[1] delays over appropriations, finally built over a period of thirty months starting in 1994; the total cost: $48 million. It contains 4,000 blocks of South Dakota carnelian[2] granite[3], the heaviest of which weighs thirty-nine tons—in all, enough to construct an eighty-story building. Yet the memorial is so well fitted into the existing landscape of Washington's Tidal Basin that it is virtually invisible from the basin's opposite shore. Its footprint, however, is big: covering seven and a half acres, it is as long as three football fields.

Figure 11.5 The statue of Roosevelt in FDR

The unprecedented size of the memorial was necessitated by the decision to tell the whole story of the FDR era in a progression of four outdoor granite rooms—one for each of FDR's terms in office—separated by gardens. As visitors stroll from one room to the next the experience is designed to build in intensity "like a symphony in four movements," says Halprin. Statuary and bas-reliefs present carefully chosen images from the FDR era together with numerous inscriptions from FDR's speeches, thereby conveying the great themes of FDR's presidency much like illustrations in a book (Figure 11.5). The "abstract sculpture" of Halprin's rusticated walls was intended to provide a framework for the figurative sculpture of his four collaborators.

One's first impression of the memorial is how disparate it is in style from the classicism of most other memorials in Washington's monumental core. If the majority of these make a bow, at the very least, to Greek and Roman traditions, the FDR Memorial's giant rusticated blocks of carnelian granite seem to hark back to monumental stonework of far greater antiquity[4]. Halprin's intent, he says, was a "timeless primordial quality."

The memorial also harks back to Halprin's own earlier projects—the 1965 Ira Keller Fountain in Portland, Oregon, for example, or the 1970 Freeway Park in Seattle—projects in which water cascades from massive concrete escarpments, bringing an "experiential equivalent" of mountain crags and waterfalls into the heart of a city. To the extent that stylistic trends exist in landscape design the FDR Memorial may seem a period piece for some, but this is a hazard of putting a design on bureaucratic hold for more than two decades. Halprin himself does not seem to have chafed at the delay.

"We've been at this for twenty-three years, and it's a great joy to have it accomplished," Halprin told a press conference the day before Clinton was to dedicate the memorial. "I'm very glad it took as

[1] bureaucratic: 官僚
[2] carnelian: 玛瑙
[3] granite: 花岗岩
[4] antiquity: 上古

long as it did because the design got better and better. It never diverged from the basic idea I had at the beginning; the details just got more and more refined. I'm glad it's over, but I wouldn't give you a nickel[1] to have shortened the process."

Notes

1. Extracts from

Texts

(1) Halprin, Lawrence. *The RSVP Cycles: Creative Processes in the Human Environment* [M]. New York: George Braziller, 1969: 1-5.

(2) Maryman, Brice. & Liz Birkholz. Freeway Park: Past, Present and Future [EB/OL]. http://www.tclf.org/features/freeway/index.htm, 2008-03-20.

(3) Thompson, J. William. The Power of Place [J]. *Landscape Architecture*, 1997 (7): 63-70.

Pictures

(1) Figure 11.3: Lawrence Halprin. *The RSVP Cycles: Creative Processes in the Human Environment* [M]. New York: George Braziller, 1969: 4.

(2) Figure 11.4: Maryman, Brice. & Liz Birkholz. Freeway Park: Past, Present and Future [EB/OL]. http://www.tclf.org/features/freeway/index.htm, 2008-03-20.

(3) Figure 11.5: Thompson, J. William. The Power of Place [J]. *Landscape Architecture*, 1997 (7): 67.

2. Sources of Additional Information

FDR Memorial

   http://www.nps.gov/fdrm/memorial/archart.htm

A Legacy in Stone

   http://www.buildingstonemagazine.com/fall-06/halprin.html

Lawrence Halprin: Changing Places

   http://eng.archinform.net/arch/6621.htm

   http://www.tclf.org/features/freeway/index.htm

   http://www.cityofseattle.net/parks/

---

[1] nickel: 镍

Dan Kiley

1. Introduction

丹·凯利（Dan Kiley，1912～2004年）是世界著名的美国风景园林大师。1992年，他被哈佛大学授予"杰出终生成就奖"。他的作品曾在纽约现代艺术馆、华盛顿国家图书馆、哈佛大学等地巡回展出。1997年被授予美国国家艺术勋章，是首位获得此荣誉的园林设计师。

1912年，凯利出生于马萨诸塞州波士顿市的Roxbury Highlands，1930年高中毕业。1932年，凯利开始了他的第一份工作，在沃伦·曼宁（Warren Manning）的公司做学徒。外向、机敏的他经常被派去施工现场监督，到苗圃挑选植物还有移栽树木。他喜欢这些户外的工作，也正是这些现场经验为他将来的成功奠定了基础。他以植物选择大胆而富有创意而闻名，植物材料是影响凯利设计作品最重要的元素。

1936年他作为一名特殊学生，进入了哈佛大学设计研究生院学习。1939～1941年发表了《城市环境中的景观设计》、《乡村环境中的景观设计》、《原始环境中的景观设计》等一系列文章。1939年春，经人推荐到华盛顿特区美国财政部公共建筑局作助理风景园林师，到了夏天又被调到美国住宅局（USHA）作助理城市规划师。1940年凯利离开住宅局，在华盛顿特区和佛吉尼亚州的Middleburg开设了丹·凯利事务所。1942～1945年的二战期间，凯利在工程兵团的战略设施办公室服役，并于战争结束后的1945年被派往德国，负责重建纽伦堡的正宫作为审判战犯的法庭。这项工作使凯利有机会周游西欧，实地考察欧洲的古典园林。这次欧洲之行对他的影响极为深刻。

凯利近半个世纪的从业过程中创造了不计其数的作品，他的作品并不局限在某种特定的设计语言。他选择生活在森林和农田之间，为的是保持一个开放的思想，"以孩子般清澈的目光去看世界"，设计是生活本身。

凯利本身也是一名建筑师，他在做设计的时候经常从建筑出发，将建筑的空间延伸到周围环境中，因而得到许多建筑师的欣赏，并选择凯利作为自己的合作伙伴。凯利与美国众多一流建筑师有过良好的合作，如路易斯·康、小沙里宁、贝聿铭、凯文·罗奇和SOM等。

2. Major Publications & Design Works

A. Major Publications

- *Kiley, Dan. Dan Kiley - The Complete Works of America's Master Landscape Architecture. Boston*，1999.
- *Daniel Urban Kiley：The Early Gardens*，1999.

B. Major Design Works

- Miller House, Columbus, Indiana, 1955.
- Chicago Art Institutes South Garden, Chicago, Illinois, 1962.
- Dalle Centrale, LLa Defense, Paris, France, 1978.
- Fountain Place, Dallas, Texas, 1983.
- National Gallery of Art, Washington, D.C., 1989.
- The Agnes R. Katz Plaza, Pittsburgh, Pennsylvania, 1998.

3. Ideas

A. Philosophy and Process

> 节选自 *Dan Kiley in His Own Words Americ's Master Landscape Architect*，凯利写出了他的设计哲学和他对设计过程的理解。凯利认为设计并不是千篇一律的模式，设计就是生活本身。而且对于凯利来说，生活不是墨守成规的，而是永远充满各种可能性和乐趣的。在凯利看来设计没有一定的模式，设计总是来源于场地。场地蕴涵的问题、机会和挑战，还有场地的需要是设计的出发点。

Philosophy

The process of addition and expansion, renovation and reinvention is never done. To build is to keep imagining fresh possibilities; if we stopped, the place would be static. Then, we would move out to new, unexplored quarters, like a butterfly emerging from its chrysalis[1]—not because the insect has grown too large for its larval[2] envelop, but because to stay in that envelope would be to seal off new experiences. I am always searching for the purest connection to that which holds us all together - we can call it spirit or mystery; it can be embodied by descriptions of the universe or of religion; it takes the form of sacred geometries and infinitesimal[3] ecologies. There is an evolving, ever-changing, many-faceted order that binds everything into harmonious parts of the greater whole.

Sometimes, the pervading order is violated. Often, it is unseen, unknown, disguised or ignored. Anne and I live as simply as possible on the land, within the forest and farmland of northern New England. We chose to live here, in this way, to keep our minds open and our senses attuned to the organizations and evolutions of nature's order. The challenge is to preserve and feed an open mind, so that nothing is shut out by walls of preconception. The purpose is to see with the clear eyes of a child, so that all is new, so that the intrinsic[4] solution to each problem is apparent. When people do not allow themselves to perceive and intuit[5] the harmonies of form and proportion that surround them, and instead arbitrarily[6] invent and inject disjunctive[7] ideas, they stray from pure design (which is integral with the universal orders) and end up imitating or decorating. The greatest contribution a designer can make is to link the human and the natural in such a way as to recall our fundamental place in the scheme of things.

What is design? It is not something that sits on a shelf, waiting to be taken down. Neither is nature. Both are ever-present manifestations of a greater unity. Design is the same, basically, in all fields, although the tools and language are different. Whether it's music, writing, architecture or dance, the most important aspect is the energy of life itself. The thing that's important is not

---

[1] chrysalis：蛹，茧
[2] larval：幼虫
[3] infinitesimal：微小的
[4] intrinsic：本质的
[5] intuit：凭直觉知道
[6] arbitrarily：武断的
[7] disjunctive：造成分离的

something called design; it's how you live, it's life itself. Design really comes from that. You cannot separate what you do from your life.

This truth can be seen in the work of primitive cultures. They did not set out to "design"; rather, the shelters they built, the crops they planted, the tools they crafted and the modes of communication they employed were direct responses to the climate, topography[1] and resources that defined their communities. Today, we put their pottery[2] in museums and call it art; we study their strategies of protection and call it architecture; we find inspiration in their agricultural practices for form and pattern in landscape design. The intimate assimilation[3] of given conditions produced works of balance, rhythm and scale that please our contemporary senses because they reveal the governing efficiency and orders of nature. The reliance on functionality produced the purity perceived as art - beauty is the result, not a preconception. Today's technology can also be brought into the production of powerful experiences if one can see the determining factors clearly and respond with unadorned honesty.

Yet it is dangerous to be a slave to rationality[4]. Life is fun and should be celebrated. One has to be open to the unexpected and the unserious. Although it seems obvious, many ignore the inspirational fortitude[5] of love - love of family, love of place. When you are passionate, something is bound to happen. If you do something simply for money or to make a big splash[6], that's no basis from which to move forwards. I am motivated by the adventure inherent in seeking out the possibilities of life, and that includes a sense of lightness and spontaneity[7] that frees one from sterile rules and procedures. The strongest artists and designers, such as the great poet Rilke, search for the mystery of who we are; they call it being. The best work comes from that search. The mystical dimension joins with our faculties to prepare us for further growth.

Process

Each time you walk in nature, it is a fresh and original experience. Whether you squeeze through a small opening amongst maple trees, or pick your way across a rushing stream, or climb a hill to discover an open meadow, everything is always moving and changing spatially - towards the infinite. It's a continuing kind of pull. Instead of copying the end-result of an underlaying process, I try to tap into the essence of Nature; the process is evolution; things are moving and growing in a related, organic way; that's what is exciting, this sense of space and release and movement. As Ralph Waldo Emerson put it, "Nature who abhors mannerisms[8] has set her heart on breaking up all style and tricks." Instead, one must go right to the heart and source: the interplay of forms and volumes that, when arranged dynamically, release a continuum that connects outwards. Should not the role of design be to reconnect human beings with their space on their land?

---

[1] topography：地形学
[2] pottery：陶器
[3] assimilation：同化，同化作用
[4] rationality：理性观点，理性实践
[5] fortitude：坚韧
[6] splash：一种强烈但是短暂的印象
[7] spontaneity：自发性
[8] mannerism：（言语、写作中的）特殊习惯，怪癖

The design we are looking for is always in the nature of the problem itself; it's something to be revealed and discovered. You might say that you are searching for the design latent[1] in all conditions. That's how it all starts, really; I am excited first to get a diagram. If the diagram isn't right, no matter how much you "design", you can never solve the problem. The diagram has to be correct in its relationships, just like the human body. You must find out how it unfolds organically, from the overall structure down to the fingernails; don't start with the fingernails. It is intrinsic[2]; it's all there, waiting for you to release it. A site is almost never a big, bland slate waiting for your creative genius; it is a set of conditions and problems for which one seeks the highest solution. I always start from a functional base. I want to protect from the elements, like a farmer or a caveman. For example, the first day at our farmhouse, which is laid out north - south, we planted a row of sugar - maples along the west side to provide shade. That's landscape design: putting that row of trees in is a master stroke of design, it's the start of the structure for the site. Then arriving at and moving in and out of the house, then needing certain things that get planted in relation to the existing environment - it all adds up to compose the diagram. It doesn't come out of the air or your head; it comes straight from what the site tells you. But you must see it with an eye for balance and proportion - that part is intuitive.

One must begin with the site itself and take in all of its attributes with an open mind. When I go onto a site, often I have an immediate understanding of, and reaction to, it. It speaks to me right away. Once you see the physical aspects, the second consideration is the programme. How will the land be used? When? By whom? The programme is an outgrowth of the third, all-important ingredient in the design process: the client. A designer never works alone. Whether the client is a person or a city or a museum, there is immediately a kind of dynamic interchange going on. There is input, counter-input, explanation, encouragement. A wonderful client can make a wonderful project; a bad client should be dropped immediately.

I feel that my current work is reaching a new level of integration and fluid[3] order. We are reaching for a reduction and amplification of elements, relationships and materials to achieve the purest connection to outer orders. My intent is to achieve the efficiency of form that leads to harmonic balance, the result of tapping into the source of life out of which Nature operates. Emerson knew this truth when he wrote, "It is the very elegance, integration, and proportioned harmony between opposing tensions that make our world a manifestation[4] of ordered beauty. Nature is economical and embodies its organic harmonies in the fit and graceful patterns. Nature takes the shortest route and is a fertile balance of tensions."

When one of my sons was young, he would tag along behind the older generation and ask, "When are things going to be real?" This question stirs me even today, as it grasps at the core of life: there is never a point when one can stop and say, "Now I am done; this is the way it is and will be." To maintain a connection to the ever-changing, growing network that is life, we must always be moving, ready to see and respond and evolve. Design is truly a process of discovery. It is an exciting dialogue

---

[1] latent：潜在的
[2] intrinsic：实在的，本质的
[3] fluid：可以改变的
[4] manifestation：显示，表现

that draws upon all of one's knowledge, intuitions, values and inspirations. Luis Barragán[1] once said that "... beauty speaks like an oracle[2], and man has always heeded[3] its message in an infinite number of ways... a garden must combine the poetic and the mysterious with serenity and joy." It is the mystic and the beautiful that we seek to attain through revealing our place in the order of Nature.

B. Fountain Place

> 凯利的代表作之一（1985年）。该场地位于是美国的达拉斯，这是一个工业城市，城市热岛效应严重。炎热干燥的场地让凯利萌生了一个非比寻常的想法——营造城市湿地。在大胆的设想和缜密的工程配合下，最终形成的场地确实给人耳目一新的感觉，真正实现了凯利在第一次造访现场时向建筑师说的话——"你们将会行走在水上"。

I may have startled architect Harry Cobb of Pei, Cobb, Freed when I announced on my first visit to the Bank Tower site: "You will walk on water!" Rarely before had I been struck so soundly and quickly by a vision for a project. My first impression was a reaction to everything the place was not.

Figure 11.6 Fountain Place

Yes, this arid[4] urban plaza in the heart of Dallas could be transformed into acres of cascading, cooling water. It was just what the city needed: an urban swamp[5]. Not a replication of nature, but a compacted experience of nature so intense that it would be almost super-natural. A place so unusual and so refreshing that it would draw visitors from outside the city to come in for picnics; office workers would venture out of air-conditioned hallways to be enveloped in the clamour[6] of falling water and the particles of moisture carried on the breeze; people would stay downtown at night to attend parties amongst the illuminated cypress[7] trees and glowing fountains (Figure 11.6).

Fortunately, the owners and architects agreed that their sixty-storey glass tower would benefit from such a surrounding. We knew what the end result should be a weaving together of water, trees and pavement into an animated composition that could change by the hour and by the season, a place in direct contrast with the hard constancy of the city. We then had to work backwards and examine exactly how the concept would fit the site. The six-acre lot is bounded by busy Ross Avenue and Field

---

[1] Luis Barragán：路易斯·巴拉干
[2] oracle：神谕
[3] heed：注意
[4] arid：干旱的，贫瘠的
[5] swamp：湿地
[6] clamour：喧闹
[7] cypress：落羽杉

Street and is adjacent to Dallas's growing Arts District. The grade drops 12 feet between the 2 streets; this had to be accommodated in a manner that would not disrupt heavy pedestrian traffic through the site. From this condition came the realization that not only would the site be all water, but it would actually be a giant waterfall, or a series of smaller drops, to negotiate the grade change.

...

At Fountain Place, we hoped to achieve a complete transformation of urban space. With the most basic materials-water, trees and concrete-and skilled engineers, a place of wonder was created. The plaza is outside the parameters[1] of conventional urban design; in fact, it is said to be the most extensive water garden built since the Renaissance[2] (this despite the fact that just half of the original plan has been implemented thus far). The architectural critic David Dillon described Fountain Place as "a unique combination of precise geometry and luxuriant[3] nature, reason and sensation intertwined[4]". I see it as a chance to connect people with the essential ingredients of life.

C. The Third Block of Independence Mall

> 本文介绍了凯利1963年所做的美国宾西法尼亚州费城独立大道景观设计，是凯利的代表作之一。在该设计中，设计师用几何树阵创造了一个个充满活力的绿色空间。

With geometric elements, closely spaced monoculture[5] plantings and maintenance plans that specify strict clipping and pleaching of trees, my designs might not be the first that spring to mind to receive accolades from the American National Forestry Board. Yet our scheme for the Third Block of Independence Mall, with its 700 honey-locusts, did just that. The intent was to create an architectural forest within the heart of the city. The difficult selling point, and undoubtedly the most crucial, was to place the trees closely enough to achieve a tangible unity across the entire block. The continuous homogeneity[6] of trunks forms a spatial mass out of which fountain squares are cut, resulting in the division of the whole into smaller, equal parts-much like the grid of streets and blocks that comprise an urban plan. The Third Block's construct of many individuals comprising an integrated body is a nested reflection of its context; it is the nature of the city (Figure 11.7).

The forest is bisected[7] by a wide corridor, on axis with the Liberty Bell and Independence Hall. I feel that it was wrong to extend the central axis out so far from the Hall-it should have been just one block, keeping in scale with the historic building-but to respect the design of the first and second

---

[1] parameters：参数
[2] Renaissance：文艺复兴时期
[3] luxuriant：富饶的
[4] intertwine：（使）交织
[5] monoculture：单作
[6] homogeneity：同质性
[7] bisected：平分

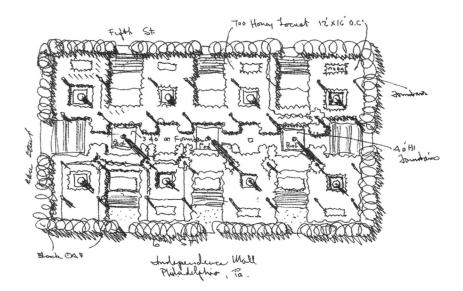

Figure 11.7 The Third Block of Independence Mall

blocks, both bilaterally[1]symmertrical, we used the axis as a starting-point. From there, the design was a decision about the most effective modular scheme. Three square basins with tall jets of water sit on the central-corridor axis. A satellite pool with low fountains is set at the cornor of each basin to form a quincunx. These pools, in turn, were to have a single jet at each of their corners (not included in the final scheme). This recurring composition symbolizes the five squares in William Penn's plan for Philadelphia (laid out by surveyor Thomas Holme). With it, we attempted to evoke a collective memory of the park's historical context, not by emulation of period materials or patterns, but via reference to the original intent of the city's founders. At the same time, we were playing with the idea of unending dilation[2]and contraction; that from one point of reference, one can move outwards into the cosmos or just as easily inwards, into increasingly intimate models of physical or spiritual experience.

Notes

1. **Extracts from**

Texts

(1) Kiley, Dan. & Jane Amidon. *Dan Kiley-The Complete Works of America's Master Landscape Architecture* [M]. Boston: Bulfinch Press, 1999: 8 - 11.

(2) Kiley, Dan. & Jane Amidon. *Dan Kiley-The Complete Works of America's Master Landscape Architecture* [M]. Boston: Bulfinch Press, 1999: 98 - 103.

(3) Kiley, Dan. & Jane Amidon. *Dan Kiley-The Complete Works of America's Master Landscape Architecture* [M]. Boston: Bulfinch Press, 1999: 42 - 43.

Pictures

(1) Figure 11.6: Dan Kiley. *Dan Kiley: The Complete Works of America's Master Landscape Architecture* [M]. Boston: Bulfinch Press, 1999: 98.

(2) Figure 11.7: Dan Kiley. *Dan Kiley: The Complete Works of America's Master Landscape*

---

[1] bilaterally: 双边

[2] dilation: 扩张

Architecture. Boston, 1999: 42.

2. Sources of Additional Information

Kiley Gardens

    http://www.kileygardens.org/kiley.html

Pioneers of American Landscape Design

    http://www.tclf.org/pioneers/kiley.htm

Ian Lennox McHarg

1. Introduction

伊恩·伦诺克斯·麦克哈格（Ian Lennox McHarg，1920～2001 年）被誉为 20 世纪最有影响力的风景园林师之一。

麦克哈格出身于苏格兰，格拉斯哥城外十英里的地方。在他长大的地方，麦克哈格既看到了丑陋的工业城市也看到了美丽的乡村风景，于是成为设计师后的他毫不犹豫地选择了自然。1936 年，16 岁的麦克哈格进入格拉斯哥艺术学院。第二次世界大战使他不得不中断学业。从 1939 年到 1946 年，麦克哈格成为英国陆军的一名志愿兵。

1950 年麦克哈格进入哈佛大学，他取得了风景园林的学士学位、城市规划硕士学位。1954 年麦克哈格受邀前往宾夕法尼亚大学创建风景园林系，在他的领导下，宾夕法尼亚大学风景园林系已经成为美国风景园林方面交叉学科发展最完善的学校之一。

20 世纪 60 年代初，麦克哈格制作并主持了一个名为"我们生活的居所（The House We Live In）"的电视节目，这个节目探讨的是人类与环境的关系。1969 年，他编导制作了美国公共广播公司记录片——"Multiply and Subdue the Earth"。

麦克哈格之所以能够成为杰出的学者、电视人和保护环境倡导者，不仅仅是因为他具有风景园林的天赋并接受了专业训练，还因为他有卓越的写作和演讲能力。

1969 年，麦克哈格的代表作《设计结合自然》出版，这是一部具有里程碑意义的专著。该书在很大程度上扩展了传统"规划"与"设计"的研究范围，将其提升至生态科学的高度，使之真正向着综合性学科的方向发展。

麦克哈格不仅是作家和教育家，还是华莱士、麦克哈格、罗伯特和托德公司（Wallace, McHarg Roberts & Todd）的创始人。他是韩国、台湾和日本的政府顾问。他做过的项目遍及世界各地，并获得了不计其数的奖项。

1984 年，麦克哈格得到了 ASLA 的最高荣誉——ASLA 奖章，1992 年布什总统还为他颁发了全美艺术奖章（National Medal of Art）。2000 年他得到了日本城市规划奖（Japan Prize in City Planning）。2001 年 3 月 5 日麦克哈格在宾夕法尼亚切斯特去世，享年 80 岁。

2. Major Publications & Design Works

A. Major Publications
- *To Heal the Earth: Selected Writings of Ian L. McHarg* 1998 ISBN 1-55963-573-8.
- *A Quest for Life: An Autobiography Ian L. McHarg* 1996 ISBN 0-471-08628-2.
- *Design with Nature Ian L. McHarg* 1969 ISBN 0-471-11460-X.

B. Major Design Works
- The Valleys in Baltimore County, Maryland, 1964.
- Lower Manhattan Study, New York, 1966.
- The Woodlands, Texas, 1972.
- Denver Regional Transportation District Study, Colorado, 1972.

3. Ideas

A. Design with Nature

> 节选自麦克哈格名著 *Design with Nature*《设计结合自然》第一章 *City and Countryside*。本文向人们解释着自然的可爱和不可或缺的重要作用，人如果想要成为生物界的管理员，就必须了解自然并在设计中结合自然。

We need nature as much in the city as in the countryside (Figure 11.8). In order to endure we must maintain the bounty of that great cornucopia[1] which is our inheritance. It is clear that we must look deep to the values which we hold. These must be transformed if we are to reap the bounty[2] and create that fine visage for the home of the brave and the land of the free. We need, not only a better view of man and nature, but a working method by which the least of us can ensure that the product of his works is not more despoliation[3].

Figure 11.8 The countryside

It is not a choice of either the city or the countryside: both are essential, but today it is nature, beleaguered[4] in the country, too scarce in the city which has become precious. I sit at home overlooking the lovely Cresheim Valley, the heart of the city only twenty minutes away, alert to see a deer, familiar with the red-tailed hawk[5] who rules the scene, enamored[6] of the red squirrels[7], the titmouse[8] and chickadees[9], the purple finches[10], nuthatches[11] and cardinals. Yet each year, responding to a deeper need, I leave this urban idyll[12] for the remoter lands of lake and forest to be found in northern Canada or the other wilderness of the sea, rocks and beaches where the osprey[13] patrols.

This book is a personal testament to the power and importance of sun, moon, and stars, the changing seasons, seedtime and harvest, clouds, rain and rivers, the oceans and the forests, the

---

[1] cornucopia：希腊神话里象征丰富的富饶羊角
[2] bounty：恩赐
[3] despoliation：掠夺性
[4] beleaguered：受到侵害
[5] red-tailed hawk：红尾鹰
[6] enamored：使人迷恋的
[7] squirrels：松鼠
[8] titmouse：长尾山雀
[9] chickadees：黑头山雀
[10] purple finches：紫雀
[11] nuthatches：五子雀
[12] idyll：田园景色
[13] osprey：鱼鹰

creatures and the herbs. They are with us now, co-tenants[1] of the phenomenal universe, participating in that timeless yearning that is evolution, vivid expression of time past, essential partners in survival and with us now involved in the creation of the future.

Our eyes do not divide us from the world, but unite us with it. Let this be known to be true. Let us then abandon the simplicity of separation and give unity its due. Let us abandon the self-mutilation[2] which has been our way and give expression to the potential harmony of man-nature. The world is abundant; we require only a defense born of understanding to fulfill man's promise. Man is that uniquely conscious creature who can perceive and express. He must become the steward[3] of the biosphere. To do this he must design with nature.

B. Ecology and Design

> 节选自 *Ecological Design and Planning*。在文中麦克哈格提醒设计师们：如果无视生态，他们的设计将对环境改善毫无意义的，甚至带有破坏性的。麦克哈格肯定了艺术对社会和文化的贡献，但是对于那些只注重艺术而忽略生态的设计持反对态度。麦克哈格还分析了建筑和风景园林专业往往忽视生态的原因所在，并提出在风景园林专业培训中加入自然和社会科学的重要性。最后，麦克哈格激励风景园林专业人员要融合科学与艺术为一体，成就"恢复地球"的重要责任。

Ecological design follows planning and introduces the subject of form. There should be an intrinsically suitable location, processes with appropriate materials, and forms. Design requires an informed designer with a visual imagination, as well as graphic and creative skills. It selects for creative fitting revealed in intrinsic and expressive form.

The deterioration[4] of the global environment, at every scale, has reinforced my advocacy of ecological design and planning. Degradation has reached such proportions that I now conclude that non-ecological design and planning is likely to be trivial[5] and irrelevant and a desperate deprivation. I suggest that to ignore natural processes is to be ignorant to exclude life threatening hazards - volcanism, earthquakes, floods, and pervasive environmental destructions - is either idiocy or criminal negligence. Avoiding ecological considerations will not enhance the profession of landscape architecture. In contrast, it will erode the modest but significant advances that ecology has contributed to landscape architecture and planning since the 1960s.

Yet, you ask, what of art? I have no doubt on this subject either. The giving of meaningful form is crucial; indeed, this might well be the most precious skill of all. It is rare in society, yet it is clearly identifiable where it exists. Art is indispensable for society and culture.

Does art exclude science? Does art reject knowledge? Would a lobotomy improve human competence, or is the brain the indispensable organ?

---

[1] co-tenants：共同的居住者
[2] self-mutilation：自取毁灭的（生活习惯）
[3] steward：管理员
[4] deterioration：退化
[5] trivial：微不足道的

There is a new tendency by some landscape architects to reject ecology, to emphasize art exclusively. This I deplore[1] and reject. Such an approach is tragically ironic[2] when so many world leaders are calling for sustainable development, when architects are issuing green manifestos[3], and professional associations in architecture and engineering are refocusing their attention on the environment.

...

A major obstruction to ecological design is the architecturally derived mode of representation by drawings. This is paper-oriented, two-dimensional, and orthogonal[4]. In contrast nature is multi-dimensional, living, growing, moving with forms that tend to be amorphous[5] or amoebic[6]. They can grow, expand, interact, and alternate. Field design would be a marvelous improvement over designing on paper removed from the site. Yet new representations must be developed to supersede[7] the limitations of paper-oriented, orthogonal investigation with their limited formal solutions. We should be committed in our work to designing living landscapes in urban, rural, and wild settings. Yet there are infinite opportunities afforded to those who would study natural systems, their components, rules, succession, and, not least, their forms. This should be the basis for an emerging ecological design.

...

In 1992 I received a signal honor from President George Bush, the National Medal of Art. The noteworthy aspect of this act was the inclusion of landscape architecture as a category eligible[8] for this high honor. As preface to the award President Bush stated, "It is my hope that the art of the twenty-first century will be devoted to restoring the earth."

This will require a fusion[9] of science and art. There can be no finer challenge. Will the profession of landscape architecture elevate itself to contribute to this incredible opportunity? Let us hope so. The future of our planet—and the quest for a better life—may depend on it. So let us resolve to green the earth, to restore the earth, and to heal the earth with the greatest expression of science and art we can muster. We are running out of time and opportunities.

C. The River Basin: Optimum Multiple Land Uses

> 节选自麦克哈格名著 *Design with Nature*《设计结合自然》，本文是波托马克河流域研究中的一部分。在收集用地中关于地形、水资源、动植物等情况之后，考虑土地利用的各种要求，本文讨论了"最适宜的多种土地利用"问题，文章描述了利用"矩阵"来辅助决策的生态规划方法。

The preceding studies of intrinsic suitabilities for agriculture, forestry, recreation and

---

[1] deplore: 悔恨
[2] ironic: 讽刺的
[3] manifesto: 声明
[4] orthogonal: 直角的
[5] amorphous: 无定形的
[6] amoebic: 阿米巴性的
[7] supersede: 取代
[8] eligible: 合适的
[9] fusion: 融合

urbanization reveal the relative values for each region and for the basin within each of the specified land uses. But we seek not to optimize for single, but for multiple compatible[1] land uses. Towards this end a matrix was developed with all prospective land uses on each coordinate. Each land use was then tested against all others to determine compatibility, incompatibility and two intervening degrees.

From this it was possible to reexamine the single optimum and determine the degree of compatibility with other prospective land uses. Thus, for example, an area that had been shown to have a high potential for forestry would also be compatible with recreation, including wildlife management. Within it there might well be opportunities for limited agriculture—pasture in particular—while the whole area could be managed for water objectives. Yet, in another example, an area that proffered an opportunity for agriculture as dominant land use could also support recreation, some urbanization and limited exploitation of minerals (Figure 11.9).

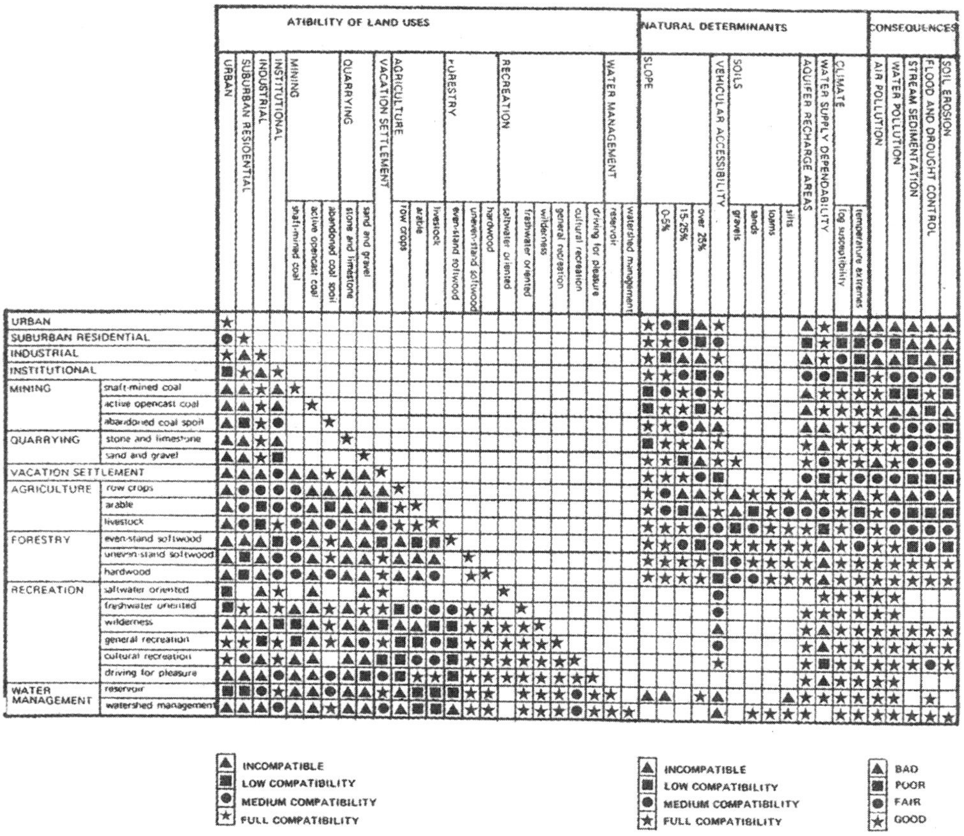

Figure 11.9 Degree of compatibility

Adjacent to the matrix[2] on intercompatibility is another that seeks to identify the resources necessary for prospective land uses-productive soils for agriculture, coal and limestone[3] for mining, flat land and water for urban locations, and so on. The final matrix is devoted to the consequences of the operation of these land uses. Where there is coal mining, there will be acid mine drainage;

---

[1] compatible: 兼容

[2] matrix: 矩阵

[3] limestone: 石灰石

agriculture is associated with sedimentation[1], urbanization with sewage[2], industry with atmospheric pollution, The sum of these, in principle, allows one to consider the intercompatibility of land uses, the natural determinants for their occurrence and the consequences of their operation.

When the results of the matrix are applied, the maximum potential conjunction of coexisting and compatible land uses for the basin is revealed. In every case the dominant or codominants are associated with minor compatible land uses.

When the results are examined, it is clear that mining, coal and water-based industry offer the maximum opportunity in the Allegheny Plateau, with forestry and recreation as subordinate uses. In the Ridge and Valley Province, the recreational potential is dominant, with forestry, agriculture and urbanization subordinate. In the Great Valley, agriculture is the overwhelming resource, with recreation and urbanization as lesser land use. The Blue Ridge exhibits only a recreational potential, but of the highest quality. The Piedmont is primarily suitable for urbanization with attendant agriculture and nondifferentiated recreation. The Coastal Plain exhibits the highest potential for water-based and related recreation and forestry, and a lesser prospect for urbanization and agriculture.

This is a method by which the nature of the place may be learned. It is because... and so, it varies. In its variety, it offers different resources. The place must be understood to be used and managed well. This is the ecological planning method.

Notes

1. Extracts from

Texts

(1) McHarg, Ian Lennox. *Design with Nature* [M]. New York: John Wiley & Sons, Inc. 1992: p5.

(2) McHarg, Ian Lennox. Ecological and Design [C]. George F. Thompson & Frederick R. Steiner. *Ecological Design and Planning*. New York: John Wiley & Sons, Inc. 1997: 321－330.

(3) McHarg, Ian Lennox. *Design with Nature* [M]. New York: John Wiley & Sons, Inc. 1992: 144.

Pictures

(1) Figure 11.8: Ian L. McHarg. *Design with Nature* [M]. New York: John Wiley & Sons, Inc. 1992: 1.

(2) Figure 11.9: Ian L. McHarg. *Design with Nature* [M]. New York: John Wiley & Sons, Inc. 1992: 144.

2. Sources of Additional Information

Bookrags

http://www.bookrags.com/Ian_McHarg

Ian McHarg Quotes

http://www.people.ubr.com/artists/by-first-name/i/ian-mcharg/ian-mcharg-quotes.aspx

---

[1] sedimentation：沉积作用

[2] sewage：污水

Unit 12

- Julius Gy. Fabos

- Peter Walker

- Martha Schwartz

- George Hargeaves

Julius Gy. Fabos

1. Introduction

朱利斯·Gy. 法布士（Julius Gy. Fabos）是一位学术严谨的国际著名的风景园林师、教育家，现为美国马萨诸塞大学风景园林与区域规划系风景规划荣誉教授，美国风景园林师协会（ASLA）常务理事。他是著名的"绿脉（Greenway）"的开创人和倡导者，美国"绿脉"论坛主任。1997年，美国风景园林师协会鉴于法布士对风景园林事业的巨大贡献，给他颁发了金质奖章。他也是一名著名的国际学者，讲学和研究项目遍及世界各地。

1961年，法布士得到了匈牙利布达佩斯园艺大学的荣誉学位和美国Rutgers大学的植物学学士学位，1964年获得了哈佛大学风景园林学硕士。1973年获得密执安大学风景规划与保护学博士学位。法布士是马萨诸塞大学大都市地区风景规划研究室（METLAND）的创办人。METLAND是一个典型的跨学科机构，由风景园林学家、生态学家、水文学家、土壤、社会、心理学家还有工程师和计算机学家等组成。他们完成了诸多的研究和实践项目并出版发表了许多重要学术论文和论著。他们的工作为拓宽风景园林学的边界做出了贡献。

法布士在20世纪80年代开始注重"绿脉"规划的研究，开创和领导"绿脉"运动，如今这一运动已经遍及整个美国，也正在成为一个全球性的运动。

2. Major Publications & Design Works

A. Major Publications
- *Land-Use Planning: From Global to Local Challenge*, New York, 1985.
- *Greenways: The Beginning of an International Movement*, 1995.

B. Major Design Works
- New England Greenway Vision Plan, 1999.

3. Ideas

A. Definition and Significance of Greenway

> 节选自美国绿脉论坛。本文阐述了法布士关于绿脉（Greenway）定义和重要性的基本观点。他认为绿脉是宽度不同，连接成网络的绿色走廊。绿脉可分三种类型：娱乐型、生态型和文化历史型。绿脉有三大价值：维持环境质量；提升场地经济价值；改善场地美学价值。

Julius Gy. Fabos defines greenways as "corridors of various widths, linked together in a network in much the same way as our networks of highways and railroads have been linked".

The major difference is that nature's super infrastructure—the greenway corridor networks is pre-existent. The river valleys have been carved out over many thousands of years. Our linear coastal system with thousands of miles of barrier beaches, rugged[1] cliffs, or extensive coastal wetland and

---

[1] rugged：崎岖的

floodplain systems have been formed by nature.

This "giant circulating system" identified by the US President's Commission (1987) is our greenway corridor network which needs to be treated with special care. (Greenways: The Beginning of an International Movement - Julius Gy. Fabos and Jack Ahern, 1995).

The intellectual framework for this project categorized greenways as:

- **Recreational Greenways** featuring paths and trails of various kinds, often of relatively long distances, based on natural corridors as well as canals, abandoned railbeds, and other public-rights ways. Trails and routes often have scenic quality as they pass through diverse and visually significant landscapes. Many successful recreational greenways and green spaces occur where networks of trails link with water-based recreational sites and areas.

- **Ecological Greenways** significant natural corridors and open spaces—usually along rivers and streams and ridge[1] lines, to provide for wildlife migration and biodiversity, nature study, and appropriate nature studies.

- **Cultural and Historic Greenways** places or trails with historic heritage and cultural values to attract tourists and to provide educational, scenic, recreational, and economic benefit. They are usually along a road or highway, the most representative of them making an effort to provide pedestrian access along the route or at least places to alight[2] from the car. They can also provide high quality housing environments at the edges of greenway (green space) for permanent and seasonal housing; accommodate water resources and flood prevention and sensitively located alternative infrastructure for communing (e. g. bike paths within urban areas, recycling of waste and storm water).

Greenways and green spaces are significant for at least three reasons:
(1) They maintain **environmental quality.**
(2) They can provide us with great **economic benefits .**
(3) They can increase **aesthetic values,** livability and quality of life.

- **Environmental Quality** derived from greenways and green spaces. Greenways and green spaces that are properly planned include from one third to two thirds of most landscapes of the United States and the entire globe. This is the magnitude of landscapes that planners need to identify or target for inclusion in greenway and green space networks. The greater the environmental sensitivity of a landscape (e. g. mountainous, like Vermont or one dominated with lakes and wetlands like Maine), the greater the percent of the landscape that needs to become part of greenway and green space networks. For example, the northern portion of New England is mountainous, defined by severe glaciation[3]. This results in steep and fragile wetland networks that need the highest level of protection to ensure the maintenance of acceptable environmental quality.

Scientists, planners and even legislators have recognized these needs. For example, legislators enacted legislation to fully protect all landscapes higher than 2,500 feet in Vermont. Legislation also

---

[1] ridge：山脊
[2] alight：落下
[3] glaciation：冻结成冰

protects 200 feet of land on both sides of rivers and streams in Massachusetts from development. This means that approximately 20% of the Massachusetts landscape falls within this zone of protection. However, this protection may be insufficient in places, for example, where floodplains or wetland systems are wider than the 200 foot-wide corridors. Such protected corridors along rivers and streams provide a natural greenway network, where water and land meet. These protected land corridors provide the most significant protection of water quality, hence environmental quality, while also providing logical linkages[1] for wildlife and recreational trails within these corridors. Interestingly, some 90% of the historical and cultural resources are also located along these rivers and streams as demonstrated by Lewis and Dawson. In summary, properly planned and maintained greenways and green spaces are essential to maintain environmental quality.

- **Economic Benefits** of greenways and green spaces. A recent study by the United States Department of Interior National Park Service lists three sets of significant economic benefits.

—Economic benefits of tourism, e.g. "in 1992 travel generated visitor expenditure in California (alone) reached approximately $52.8 billion".

—Residents' expenditures[2] on outdoor recreation, In Pennsylvania (alone) residents spent approximately $11.8 billion, or 12.6% of their total personal consumption dollars in leisure pursuit.

—Increased property values, e.g. the selling price of residential housing in Amherst, Massachusetts was $17,000 higher along public green spaces or around 10% higher than in conventional subdivisions in 1989.

—In addition, another study concluded that the average house value increased from 10% to 20% of the selling price when view/setting potential was maximized (see Fabos et al, Research Bulletin 653, UMass Experiment Station, 1978).

- **Increased Aesthetic Values**, livability and quality of life. As it is shown above, economists have attempted to place economic values on the visual or aesthetic quality of the landscape. In addition, planners and designers have recognized the intangible[3] aesthetic, livability and quality of life values provided by greenways and green spaces. Italian villas were located on hilltops around cities providing their residents with great views and ideal topoclimatic[4] benefits. Hill tops and proper orientations have been the places for the best urban neighborhoods around the world.

Similarly greenways and green spaces have enhanced livability and quality of life. The livability of parks has been recognized since the beginning of the park movement of the early 19th century. Today, many people seek out developments around golf courses. The majority of these people are non-golfers, who settle there primarily for the open space aesthetic. Many of these intrinsic[5] values, such as aesthetics, livability or quality of life issues are recognized by planners, but not yet proven like many economic values are proven through studies of market forces. In summary, greenways and green spaces do significantly increase aesthetic values and contribute greatly to

---

[1] linkage：连接

[2] expenditures：支出

[3] intangible：无形的

[4] topoclimatic：地形气候

[5] intrinsic：内在的，本质的

livability and the quality of life.

B. Greenway History

> 节选自美国绿脉论坛法布士撰写的绿脉发展史。本文描述了从20世纪开始的开放空间规划（open space planning）发展到20世纪末绿脉规划出现的历程。

The origin of greenways goes back to the 19th century park-planning era. The 20th century has been dominated by open space planning. Only the last decade of the 20th century has emerged to become the foundation of the greenway planning movement. It is anticipated that the greenway movement will dominate the 21st century and it will link together all major parks and open spaces into comprehensive networks of greenways and green spaces. It is anticipated that greenways and green spaces will be mapped and used by the general public as road maps are used today. Rhode Island was the first state to create a statewide greenway map in 1996, and it is distributing 100,000 copies of their statewide greenway map to residents and tourists yearly. Nationally, the New England states have been among the most advanced states for greenway planning in the United States. The following section is a brief review of the development of greenways in America.

...

Open Space Planning During the 20th Century

The planning and designing of urban parks and the establishment of more national parks, forests and reserves have continued during the 20th century. A new relevant planning concept was initiated, however in this century known as "open space planning". Open space planning has been done primarily at local and state levels and has been supported by federal agencies during most of this century (Figure 12.1).

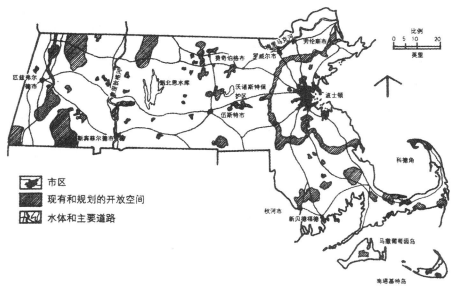

Figure 12.1 Open Space Plan for the Commonwealth of Massachusetts, 1928 (Fabos 1985).

The history of open space planning is especially relevant for this study. Once more, one of the New England states, Massachusetts was the location for the first comprehensive statewide open space planning. Once again, a landscape architect, Charles Eliot II, had a significant role in creating the "open space plan for the Commonwealth of Massachusetts in 1928". The Eliot plan was initiated by the governor, who appointed[1] an open space commission including Charles Eliot II, the nephew of Eliot[2], the planner of the Boston Park System three decades earlier. Eliot, as a member of the governor's Commission, drew up this statewide open space plan in 1928, at the start of the big depression. This vision plan was first put on the shelf. Later it was used as the framework for establishing state parks and protected areas in Massachusetts throughout the second half of the 20th century.

Phil Lewis[3], a landscape architect, initiated another relevant statewide vision plan in Wisconsin during the environmental movement of the 1960's (Figure12.2). The Lewis plan is significant for at least three reasons. First, the Lewis plan created a statewide network of green spaces and greenways that he called environmental corridors. Second, the great majority of his connections were along rivers, streams and wetland systems. Third, Lewis identified many cultural resources in addition to the natural resources used by planners concerned with recreation planning. Lewis mapped 220 natural and cultural resources with recreation values. Half of the 220 resources were natural, the other half were cultural resources. He found that the great majority of these resources (over 90%) co-occurred along river and stream corridors, hence he named these corridors the Wisconsin Heritage Trails.

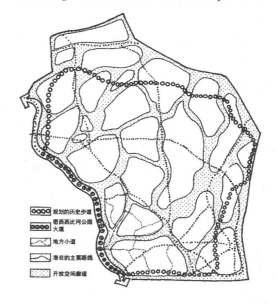

Figure 12.2 *Wisconsin's Heritage Trails Proposal*

The Beginning of Greenway Planning at the End of the 20th Century

The word, "greenway" appeared in the literature with increased frequency from the late 1970's on. Writers such as William White and agencies such as Housing and Urban Development started to use the word "greenways". The clearest statement about greenways, however, came from the President's Commission on American Outdoors for the United States in 1987. The commission's central recommendation advocated as a vision for the future: "A living network of greenways... to provide people with access to open spaces close to where they live, and to link together the rural and urban spaces in the American landscape... threading through cities and countryside's like a giant circulation system".

Three years after the President's Commission report, Charles Little, a well known environmental writer published a most influential book on greenways entitled Greenways for America (Little, 1990).

---

[1] appointed: 指定的，约定的
[2] Eliot: 查尔斯·艾略特（Charles Eliot, 1859～1897年），著名的风景园林师
[3] Phil Lewis: 菲利普·路易斯，第一位以环境廊道概念为核心进行景观规划的规划师

Since Charles Little's seminal[1] book, at least seven other books were published in the USA alone on trails and greenways.

Greenway information on planning and implementation has also been disseminated[2] through dozens of regional and state level conferences. The first national - international conference was held in January 1998, and organized by the Rails-to-Trails Conservancy[3]. The second international conference on trails and greenways was in June 1999, also organized by the Rails-to-Trails Conservancy (RTC). This non-governmental organization formed during the 1980's to facilitate the conversion of abandoned railroads to bicycle trails throughout the United States. The RTC has been credited with assisting and speeding up this conversion process. The first national trail and greenway conference celebrated the completion of 10,000 miles of rail trails. According to RTC, the United States has another 150,000 miles of abandoned railroad, the majority of which have the potential to become rail trails.

In addition to rail trails there are hundreds and perhaps, thousands of greenway segments planned and built yearly throughout the United States. The great majority of these greenways and trail segments are planned for hiking and other relevant recreation uses. According to Edward McMahon (the director of the American Greenway Program of the Conservation Fund), over half of the American states are involved in state wide greenway planning and implementation to a greater or lesser degree. At the national level, there are at least two significant developments. First, there is an ongoing national program to identify and protect bio-diversity. This national program has conducted a GAP analysis to identify critical areas that warrant[4] protection in each region of the United States.

The New England GAP analysis has been coordinated by two public universities. The University of Massachusetts has performed the GAP analysis for the three southern states, Massachusetts, Connecticut and Rhode Island. The University of Maine is primarily doing the analysis for the three northern New England states, according to Dr. Jack Finn at the University of Massachusetts. Professor Finn believes that southern New England's existing bio-diversity is sufficient, and approximately 15% of the critical resources could be protected if managed properly. Since the current definition of greenways and green spaces includes nature protection, this Federal program could help future greenway efforts.

The second Federal effort that enhances greenway planning and implementation is the American Heritage River Initiative (AHRI). Last fall, fourteen rivers across America were named as nationally significant Heritage Rivers. Three of the fourteen Heritage Rivers, the Connecticut, Blackstone and Woonasquatucket Rivers are in New England. Heritage Rivers are also being identified at the state level. New Hampshire, for example, has named the Merrimack River Corridor as a New Hampshire Heritage Trail. Please note that more detailed histories of greenways in New England are described under each state.

---

[1] seminal：种子的
[2] disseminated：散布
[3] conservancy：(自然物源的) 保护，管理
[4] warrant：使有正当理由

C. The New England Greenway Vision Plan

> 位于美国东北角的六个州被统称为新英格兰,这个地区的人口总密度超过美国全国的两倍多。新英格兰是美国差异性最大的区域,有吸引人的美丽风景,是美国重要的旅游区之一。新英格兰绿脉规划是奥姆斯特德等前人工作的延续,正如文中所说"做出连接"的时候到了。

In the tradition of landscape architecture, the members of the New England Greenway Vision Plan and the American Society of Landscape Architects believe that the time has arrived to "make the connections" among the thousands of parks and open spaces in New England. We can build on our amazing parks and open spaces to create the first multi-state, regional greenway network to serve as a model for America and the globe. We are poised[1] to create a cohesive[2] network for the 43 million acres of New England - making greenways as accessible to everyone as our roads are today (Figure 12.3).

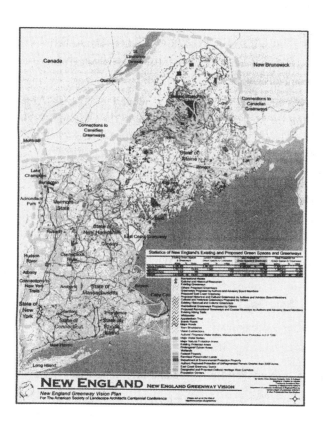

igure 12.3 The Greenway Vision Plan for New England

---

[1] poised:泰然自若的,平衡的
[2] cohesive:粘着的

Our Plan

The plan builds upon existing and proposed greenways in all six states and identifies the critical linkages that will make a New England system. The plan calls for:

- An additional 19,300 miles of greenways and trails to make all connections among green public open spaces, so that we have a cohesive green network for all New England.

- An additional 8 million acres of protected land to maintain a high quality environment for New England in perpetuity[1]. This will be done by protecting the most fragile and environmentally sensitive areas, where land meets water. This approach was first enacted into law by Massachusetts Legislators with the 1996 River Protection Act. Enacting similar laws in the other New England states will contribute greatly to this total.

- Action Plans for creating greenway legislation in all New England States to emulate[2] Rhode Island's successful legislation. (They now distribute over 100,000 greenway maps per year)

The fact that this conference is the centennial conference held at the birthplace of the American Society of Landscape Architects (ASLA), gives the impetus for a study that goes beyond a single site or a simple project. A brief review of the evolution of landscape architecture in Massachusetts shows how the profession here expanded the boundaries of landscape architecture from park design in the 19th century to open space planning in the 20th century.

It was Frederick Law Olmsted, Senior, the founder of Landscape Architecture in America who expanded his park planning for Massachusetts by planning the first significant park system for Boston and Brookline in 1867, commonly known as "The Emerald Necklace". This park system is known in the greenway literature as the first major greenway in America.

It was here in Massachusetts that Charles Eliot; a pupil of Olmsted expanded Olmsted's Park System by planning a Metropolitan wide park system. This was an open space/greenway network for the 250 square mile Boston Metropolitan Region at the turn of the twentieth century.

It was here in Massachusetts, where Charles Eliot II, the nephew of Charles Eliot coordinated the plan for the first statewide open space plan in America, for Massachusetts in 1928. This impressive history provides us with a challenge to expand further the boundaries of landscape architecture. Hence, it was decided to create a greenway vision plan for the entire New England region. The intent of this vision plan has been to "be the first step on the road to a national (greenway) system".

This project proposes for New England a bold vision, to make greenways and green spaces as accessible to everyone as our roads are today. This plan provides New England with a network of greenways that will serve the region as our state and interstate roads serve our cars and trucks today.

The literature on planning approaches used for greenways and green spaces is at the initial stages at this time. While the Conservation Fund listed thirty states that are involved in greenway planning, there have been no systematic studies about these statewide planning efforts. Our initial search of state level planning studies yielded only reports from twelve out of the fifty states. Analysis of these

---

[1] perpetuity: 永恒
[2] emulate: 仿效

twelve reports suggests that greenway planners have used at least four different approaches in planning greenways and green spaces. This analysis based on a previous study by the members of the Landscape Planning Studio directed by the principal author of this chapter at the Landscape Architecture and Regional Planning Department of the University of Massachusetts.

The four types of planning approaches are the resource assessment approach, the hubs[1] and corridors approach, the grassroots driven approach and balanced greenway planning.

- *Resource Assessment Approach*. The states of Rhode Island and Georgia appear to use this approach. Both states begin with a statewide resource assessment by identifying environmentally sensitive areas, e.g. wetland systems, for the green space protection and for logical river and other greenway corridors to link these green spaces.

- *Hubs and Corridors Approach*. Florida and New York provide a very broad framework for greenway planning. They identified dozens of major recreational and cultural hubs in Florida, and seven significant resource areas in New York, e.g. the Adirondacks, the Catskills, and the Fingerlake region. Then the planners proposed greenways to link the recreational and cultural hubs of their states.

- *Grassroots Driven Approach*. Maryland, Connecticut and Delaware have produced greenway vision maps that strongly rely on grassroots efforts. "The state of Maryland began their greenway efforts with a quick conceptual vision plan (including) existing and proposed greenways and green spaces. Connecticut began with the mapping of the locally initiated greenways and added other potential greenway corridors... The state of Delaware invited interested citizens to draw their greenway proposals on a state map which the state planners then refined".

- *Balanced Greenway Planning*. Vermont, New Jersey, North Carolina, South Carolina and Colorado are in various stages of developing greenway plans, balancing grassroots efforts with planning done by relevant state agencies.

Notes

1. Extracts from

Texts

(1) Fabos, Julius Gy. Definition and Significance of Greenway [EB/OL]. http://www.umass.edu/greenway/2GR-index.html, 2008-03-20.

(2) Fabos, Julius Gy. Greenway History [EB/OL]. http://www.umass.edu/greenway/2GR-index.html, 2008-03-20.

(3) Fabos, Julius Gy. The New England Green Vision Plan [EB/OL]. http://www.umass.edu/greenway/1PO-index.html, 2008-03-20.

Pictures

(1) Figure 12.1 朱利斯 GY. 朱利斯·法布士著,赵彩君等译. 土地利用规划 [M]. 北京：中国建筑工业出版社. 2007：117.

(2) Figure 12.2 朱利斯 GY. 朱利斯·法布士著,赵彩君等译. 土地利用规划 [M]. 北京：中国建筑工业出版社. 2007：118.

---

[1] hub: 中心

(3) Figure 12.3: Fabos, Julius Gy. The New England Green Vision Plan [EB/OL]. http://www.umass.edu/greenway/1PO-index.html, 2008-03-20.

2. **Sources of Additional Information**

Greenway Project Report

    http://www.railtrails.org.traillink2003

National Park Service

    http://www.nps.gov/grca/greenway

Peter Walker

1. Introduction

彼得·沃克（Peter Walker）1932年出生于美国加州的帕萨德纳（Pasadena），1955年毕业于美国加州大学伯克利分校，获风景园林学士学位，1957年获哈佛大学设计研究生院风景园林硕士学位，同年与哈佛大学设计研究生院佐佐木英夫教授（Hideo Sasaki，1919~2000年）共同创立了SWA（Sasaki & Walker Associates）景观设计公司。在20世纪60年代和70年代，作为SWA的总设计师，沃克成功地主持了许多区域规划、城市景观和风景园林设计项目，然而与此同时，他发现这些风景式的景观与他本人对极简主义艺术的兴趣相距甚远，最终于1976年离开了SWA而赴哈佛大学GSD任教并从事极简主义园林的研究工作。1978~1981年任哈佛设计研究生院风景园林系主任。1983年，沃克创办了自己的设计公司，从而得以把他对极简主义的探索付诸实践。1994年沃克与M.西蒙（M. Simo）合作出版了著作Invisible Gardens《不可见的园林》，阐述了1925年以来美国风景园林的发展和变革。1997年出版了作品集Minimalist Gardens《极简的园林》。

沃克特别推崇法国古典主义园林师勒·诺特尔，也欣赏日本禅宗园林简朴的风格，并受到了美国风景园林师丘奇、野口勇、埃克博和哈普林的影响，实际上，西方古典园林，特别是勒·诺特尔的园林、现代主义、极简主义和大地主义共同影响了沃克的设计。沃克的作品注重人与环境的交流，人类与地球、与宇宙神秘事物的联系，强调大自然的谜一般的特征，如水声、风声、岩石的沉重和稳定、飘渺神秘的雾以及令人难以琢磨的光。他对园林艺术的探索，达到了当代风景园林设计的一个新的高度。

沃克的极简主义园林在构图上强调几何和秩序，多用简单的几何母题如圆、椭圆、方、三角，或者这些母题的重复，以及不同几何系统之间的交叉和重叠。材料上除使用新的工业材料如钢、玻璃外，还挖掘传统材质的新的魅力。通常所有的自然材料都要纳入严谨的几何秩序之中，作为严谨的几何构图的一部分。沃克的极简主义景观并非是简单化的，相反，它使用的材料极其丰富，平面也非常复杂，但是极简主义的本质特征却得到体现。如无主题、客观性、表现景观的形式本身，而非它的背景；平面是复杂的，但基本组成单元却是简单几何形；用人工的秩序去统领自然的材料，用工业构造的方式去建造景观，体现机器大生产的现代社会的特质；作品冷峻、具有神秘感，与此并不矛盾的是他的作品具有良好的观赏性和使用功能。极简主义的手法使他的作品成为一个"美学的统一体"，而不是一种其他物体的背景或附属物。

沃克完成了大量的风景园林作品，不仅在美国，而且遍布世界上许多国家，如德国、法国、西班牙、日本、墨西哥甚至中国，成为当今世界最有影响力的风景园林师之一。

2. Major Publications & Design Works

A. Major Publications

* *Invisible Gardens: The Search for Modernism in the American Landscape*, New York: The MIT Press, 1994（与Melanie Simo合作出版）.
* *Minimalist Gardens*, New York: Spacemaker Press, 1997.
* *Peter Walker and Partners Landscape Architecture: Defining the Craft*, London: Thames & Hudson, 2005.

B. Major Design Works

* Tanner Fountain, Harvard University, Cambridge, Massachusetts, 1984.

- IBM Solana Westlake and Southlake, Dallas, Texas, 1984～1989.
- Ayala Triangle in the Makati district, Manila with Sdidmore, 1991～1993.
- Center for the Advanced Science & Technology, Hyogo Prefecture, Japan, 1993.
- World Trade Center Memorial: "Reflecting Absence", New York, New York, 2004.

3. Ideas

A. Minimalism in the Landscape

> 选自论文集 *Theory in Landscape Architecture* 中沃克同名论文。本文主要讨论极简主义（minimalism）的出现和主要特点，反映了沃克园林创作的主要观点。

Distinct from the specificity of its art world argument relating to a certain group of artists as a certain moment in the 1960s, minimalism in the landscape seems to me to represent a revival of the analytic[1] interests of the early modernists that parallel in many respects the spirit of classicism. It is the formal reinvention and the quest for primary purity and human meaning that dignify its spiritual strength: an interest in mystery and nonreferential content are thereby linked to the quest of classical thought.

As with the term and idea of classicism, minimalism has entered our fast-paced society and been further defined and redefined by varying artistic and cultural disciplines. In this greater context, minimalism continues to imply an approach that rejects an attempt to intellectually, technically, or industrially overcome the forces of nature. It suggests a conceptual order and the reality of changing natural systems with geometry, narrative, rhythm, gesture, and other devices that can imbue[2] space with a sense of unique place that lives in memory.

Despite the broader scope of my use of the term *minimalism*, a reference to the quintessential[3] minimalist artist in illuminating. Donald Judd insisted that minimalism is first and foremost an expression of the objective, a focus on the object in itself, rather than its surrounding context or interpretation. Minimalism is not referential or representative, though some viewers will inevitably make their own historical or iconic[4] projections. In correlation, though minimalist landscape exists in the larger context of the environment, and though it may employ strategies of interruption or interaction one can see beyond the designed "objects" to the larger landscape, the focus is still on the designed landscape itself, its own energy and space. Scale, both in context and internally experienced, remains primarily important. And as with minimal art, minimalist landscape is not necessarily or essentially reductivist, although these works often do have minimum components and a directness that implies simplicity.

With these parameters[5], minimalism in landscape architecture opens a line of inquiry that can illuminate and guide us through some of the difficult transitions of our time: the simplification or loss of craft, transitions from traditional natural materials to synthetics, and extensions of human scale to

---

[1] analytic: 分析的，用分析法的
[2] imbue: 使……浸染（颜色等），影响（感情等）
[3] quintessential: 精粹的，典范的
[4] iconic: 雕像的，圣像的，依照传统形式的，因袭的
[5] parameter: 参数，参量

the large scale, in both space and time, of our mechanically aided modern life. And minimalism in this context suggests an artistically successful approach to dealing with two of the most critical environmental problems we currently face: mounting waste and dwindling[1] resources.

An inquiry into minimalism in the landscape now seems to be especially timely. Recent developments in landscape architecture, architecture, and urban design during what has been termed our postmodern era have questioned the legitimacy[2] of modernist design, with some favoring a return to classicism. Much of the recent work and thought in this area has focused on formal and decorative issues on the one hand and sociological and functional issues on the other. Minimalism, one of the manifestations of the last moment of high modernism in the visual arts, has itself, of course many compelling affinities[3] with classicism. Rather than focusing on design and functional issues as mutually exclusive, minimalism leads to examination of the abstract and the essential, qualities of both classicist and modernist design.

It is interesting to recall that when the youthful Le Corbusier journeyed through the Middle Eastern and Mediterranean lands before World War I in 1911, he was drawn to Turkish mosques, Byzantine monasteries[4], and Bulgarian houses, because of certain qualities, particularly silence, light and simple, austere[5] form. On the Acropolis of Athens, however, he was overwhelmed and awed by the Parthenon, the "undeniable[6] master," which he later interpreted as a distillation of form, and unexcelled product of standardization. A moment had been reached, he concluded, when nothing more could be taken away. It was a moment of perfection, a defining of the classic. It so happens that I felt a similar response to one particular Le Nôtre garden when I visited it in the 1970s. Chantilly, a great garden of stone, water, space, and light, also represents a superb example of form reduced to its essential perfection. Chantilly seemed to me then and seems to me still to share in its essence an understanding and intent that is both classic and minimal.

These thoughts are a progress report of my personal journey as a landscape architect who came of age at the height of modernism in American environmental design. They are informed by the gardens, landscapes, designs, artists, and insights that have helped to shape my perceptions and to chart my particular course of inquiry to this point. They offer one personal approach to the making of environments that seems to be especially needed at this time in human history: environments that are serene[7] and uncluttered, yet still expressive and meaningful. More than ever, we need to incorporate in our built environment places for gathering and congregation, along with spaces for discovery, repose[8], and privacy in our increasingly bewildering, spiritually impoverished[9], overstuffed[10], and undermaintained garden Earth.

---

[1] dwindling: 渐渐变小，缩小，减少，衰落，失去重要性
[2] legitimacy: 正确（性），合理（性）
[3] affinity: 密切关系，构造相似，关系，姻亲关系
[4] monastery: 修道院，僧侣
[5] austere: （指人或其行为）严肃不苟的，严峻的，（指生活方式、地方、文体）质朴的，朴素无华的
[6] undeniable: 无可否认的，确实的
[7] serene: 晴朗的，宁静的
[8] repose: 休息，使休息或依靠，信赖
[9] impoverish: 使穷困，去除优点
[10] overstuffed: （指坐位等）用厚垫使柔软而舒适的，填塞极多东西的

B. Most Influential Landscape

> 选自美国 *Landscape Architecture* 杂志人物访谈类文章,该文沃克介绍了对自己影响最大的三个园林作品。

A difficult distillation, given the historic gardens of Japan, Italy, Spain, England, and the modern gardens of Marx, Church, Halprin, Barragán, Noguchi, et al. Still, three works do stand out, not in any particular order.

The first two are late Le Nôtre gardens[1], where the baroque[2] detail had given way to a more austere[3] minimalism of line and plane. Le Nôtre's "chantilly" (perhaps his last) is where an existing romantic sculptural chateau[4] and several separated levels were composed into a spatial collage of juxtaposed[5] planes and objects, seen first from a rising ramp, then from across, above, and below. These spaces are connected not by a single or cross axis, but rather by one of the greatest outdoor stairs of all time. Both processional and mysterious, urbane[6] and natural with extremely complex tensions, the garden is achieved by extraordinarily minimal means. Though relatively small, its surreal flatness and dimensional extension rivals Veaux-le-vicomte and even Versailles.

"Sceaux" is spatially a gothic cathedral of a garden. Almost without ornament, it is formed by a simple repetition of great linear spaces that build into a visually complex density. The grand verticality is achieved through exploitation of rhythm, reflection, and shadow. Directly and essentially interactive with the daily and seasonal change of light and atmosphere, "Sceaux" is a changing visual wonder at all times of the day and year. Unfortunately, it is also soon to be ruined by stupid contemporary plantsmen.

Kiley's "Miller House" is, to my knowledge, the first important non-painterly garden utilizing real constructivist spatial techniques (Figure 12.4). Seemingly classical (like the Barcelona Pavilion), this garden is radically modern in its multiple use of interlocking planes and its great variation of defined spaces. Kiley here composes a variety of plants and grounds into point and linear grids[7], allees[8] hedged rooms, and a dramatic change of grade. Though almost unphoto-graphable, it is in my view a giant

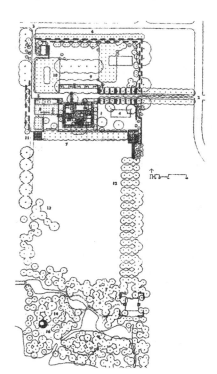

Figure 12.4 Miller Garden

[1] Le Nôtre gardens:勒·诺特园林
[2] baroque:(指建筑等)装饰得过分的,装饰得奇形怪状的
[3] austere:严厉的,简朴的
[4] chateau:城堡
[5] juxtapose:并列,并置
[6] urbane:文雅的
[7] grid:格子,栅格
[8] allee:(林荫)小径

advancement in modern landscape design. The garden is perhaps not understood sufficiently by the landscape and architecture professions, which have largely failed to take advantage of Kiley's genius.

My quest has always been to try to understand landscape as an art. These three gardens suggest to me the great range, simplicity of means, economy of scale, and formal distillation of nature performing in light that I feel most clearly reveal the modern artistic opportunity to bridge the industrial city, the agricultural plain, and the grandeur[1] of natural process.

C. IBM Solana Westlake and Southlake, Dallas, Texas (Peter walker, William Johnson and Partners)

> 得克萨斯州的索拉那（solana）IBM 研究中心园占地 850 英亩。建筑由墨西哥著名建筑师 R. 莱格雷塔（R. Legorreta）设计，强烈的色彩具有明显的墨西哥风格。沃克保护了尽可能多的现有环境的景观，对当地自然环境的保护贯穿整个计划始终。园区有优美的田园风光和略有起伏的山丘，当地原有景观的自然价值在设计中得以强调，保留了大片野生花草和树木，建筑和道路根据现存的植被的分布情况来布置。在建筑旁使用一些极端几何的要素，与周围环境形成强烈的视觉反差。这个项目获得了 1988 年美国风景园林协会城市设计与规划奖及 1990 年风景园林协会园林设计奖。

The decisive role of landscaping in linking architecture to its environment is one of the keys to understanding this fascinating creative discipline. This design by Peter Walker and his team shows how it is possible to reduce the environmental impact of a large-scale construction and preserve the site's landscape identity. This has also been helped by the contributions of the architects involved in planning this large complex which houses a wide range of functions, including recreational, commercial and administrative areas. The large site chosen for this ambitious project covers 850 acres in Texas, a few miles northwest of the Fort Worth-Dallas airport. The landscape is generally agricultural and rural with pastures and gently rolling hills. Walker's team had to integrate the architecture with its surroundings. And also plan the parking lots and the communication infrastructures. It was decided that the buildings should be no more than five stories high, and that, where possible, parking should be sited near the built modules[2] to reduce their impact.

The idea of creative responsibility is often rather disperse in a project that has lasted as long as the Solana project. Until 1990 planning was the responsibility of the Office of Peter Walker and Martha Schwartz, a partnership lasting from 1983 to 1989. Before that, Peter Walker had been a partner at Sasaki Walker Associates, consolidating[3] his career with works, mainly in the US. His work at Peter Walker, William Johnson and Partners has become well-known due to projects like this one, which won the 1988 Honor Award of the National Prairie Association and the 1990 Asia Design Award.

The masterplan for the Solana site consists of a sequential distribution, starting from the Highway 114 exit ramp, of the three main functional sectors: the residential and recreational area, by Ricardo Legorreta Arquitectos, the marketing centre, by the same company, and the administrative complex and IBM regional headquarters, by Mitchell/Giurgola Architects. Close collaboration

---

[1] grandeur：伟大，壮丽，华丽
[2] module：模数，模块
[3] consolidate：巩固

between architects, client and landscapers is the reason for the project's success, as it has sought a harmonic equilibrium[1] between aims and results, based on social, economic, aesthetic, cultural and environmental factors (Figure 12.5).

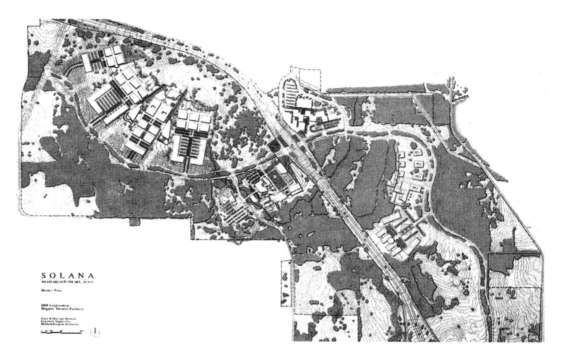

Figure 12.5 Master plan: IBM Solana Westlake and Southlake

Respect for the site's natural landscape has been decisive throughout the project, requiring a thorough analysis of the site's topographic[2], geological, climatic and ecological characteristics and a commitment to conserving the original flora[3]. The natural values of the local landscape were emphasized, maintaining the large wildflower prairies and trees, including post oaks and a few free-standing hickories[4]. The distribution of the buildings and the roads was determined by the existing distribution of vegetation.

Special emphasis is given to the act of entering the complex, showing the importance attached to movement as an essential component of the perception of landscape. Entrance is from the highway exit ramp, through the Arrivals Garden, a space that still has a rural feel, although it is highly stylised. Legorreta's architecture encourages this celebration of the concept of entrance, using red Cordovan stucco[5] walls and silent, monolithic[6], strongly vertical towers in bright colors. Rows of Indian hawthorn[7], planted like a vineyard, and a regular pool emphasise this sector's horizontal feel.

The recreational and marketing areas develop the idea of a perceptual sequence and the experience

---

[1] equilibrium: 平衡，均势
[2] topographic: 地形的，地形学的
[3] flora: (某地区或时代的) 植物的总称，植物区系
[4] hickory: (北美所产之) 山胡桃树，山胡桃木
[5] stucco: (粉饰墙壁用之) 灰泥
[6] monolithic: 独石的，似独石的
[7] hawthorn: 山楂

of entry. This was done using components like arcades, the vertical towers mentioned above, plazas (especially one acting as a focus for the recreational-residential area) and internal courtyards overlooking the landscape or sculpture gardens.

The radial distribution of modules in the corporate complex favours its integration with its environment. The transition between architecture and natural landscape is in the form of a 900-meter-long terraced garden. The coexistence[1] of geometric artificiality and biological forms creates an environment recalling the works of Kandinsky and miró.

Although water is present throughout the project, in the last sector it is especially important. There is a triangular pool (like those in the neighbouring ranches) flowing into a magnificent lake whose design is naturalistic. An important role is also played by water in the fountains, which are very suggestive landscape features. There are three fountains which differ in design, but their strategic use of mist means they all rely on a single cultural and symbolic concept, the contrast between the ancestral values of the native civilisations and new construction technologies. This is the same as the eternal contrast between tradition and the avant-garde[2] (Figure 12.6).

Figure 12.6 Water, free and controlled, played strategic role

Notes

1. Extracts from

Texts

(1) Walker, Peter. Minimalism in Landscape [C]. Simon Swaffield. *Theory in Landscape Architecture*. Philadelphia: University of Pennsylvania Press, 2002: 87-88.

(2) Walker, Peter. Most Influential Landscape—Peter Walker [J]. *Landscape Journal*, 1993, 12 (2): p173~174.

(3) Cerver, Francisco Asension. *The World of Environmental Design* (Vol.1) [M]. Spain: Arco Editorial, 1994: 14~21.

Pictures

(1) Figure 12.4 Walker, Peter. Minimalism in Landscape [C]. Simon Swaffield. *Theory in Landscape Architecture*. Philadelphia: University of Pennsylvania Press, 2002: 87.

(2) Figure 12.5 Francisco Asension Cerver. *The World of Environmental Design* (Vol.1) [M]. Spain: Arco Editorial, 1994: 15.

(3) Figure 12.6 Francisco Asension Cerver. *The World of Environmental Design* (Vol.1) [M]. Spain: Arco Editorial, 1994: 17.

2. Sources of Additional Information

(1) http://www.asla.org/awards/2004/medals/walker.htm.

---

[1] coexistence：和平共存

[2] avant-garde：先锋派，前卫

(2) Jamison Square, Portland, Oregon.
(3) The Nasher Foundation Sculpture Center.
(4) The Sony Center in Berlin.
(5) The McConnel Foundation.
(6) http://www.pwpla.com/frm_awards.php.

Martha Schwartz

1. Introduction

玛莎·施瓦茨（Martha Schwartz，1950年～）出生于美国费城，父母都是建筑师，他们的五个女儿都成长在艺术的氛围之中，后来成为教师、建筑师或艺术家。施瓦茨作为五个姐妹中的老大，从小学习艺术，会钢琴、长笛两种乐器，也学习芭蕾。她在费城艺术学院选修课程，1973年就读于密执安大学艺术系并获得学士学位。此时她开始熟悉大地艺术，并对风景园林设计产生了兴趣。然而大地艺术家主要是在远离城市的原始的自然环境中进行创作，而施瓦茨希望尝试将大地艺术运用到复杂的城市环境之中。1974年她进入密执安大学风景园林系学习，遇到了沃克。施瓦茨在密执安大学和哈佛大学设计研究生院学习并于1977年获得硕士学位，然后进入SWA公司，共同的兴趣与追求使她与沃克建立了家庭。1990年，为了各自事业的发展，施瓦茨建立了自己的设计事务所。

施瓦茨是一位在公共艺术领域开拓了风景园林新天地的设计师。她是风景园林师中的另类，是一位充满批判精神的创新家，一向以不走寻常路和挑战传统的设计手法而享誉国际风景园林界。她曾经学习了10年艺术，后来转向风景园林，其作品的魅力在于设计的多元性，深受极简主义、大地艺术、波普艺术及后现代主义等现代艺术流派的影响，作品极为大胆。她用现代主义理论武装自己，其作品的本质目的是面向大众的、平民化的，出发点或本质内涵是同于现代主义的，而在形式上保留了她作为一个艺术家的革命精神，即对原有专业原则的超越，对风景园林一词的自然化联想的批判。施瓦茨非常注重作品对生态系统所产生的社会影响力，喜欢在场景中采用技术手段而非自然标准或假定的自发性方案，酷爱鲜艳夺目的色彩和另类材料，而且对潮流非常敏感；她憎恨虚饰，主张设计应该诚实。其作品常常会与公众舆论相冲突，而招致同行的批评，但是，无论是赞同者还是反对者，都认为她是一位"始终孜孜不倦地探索景观设计新的表现形式，希望将景观设计上升到艺术的高度"而值得尊重的风景园林大师。作为风景园林师和艺术家双重身份的施瓦茨经常将西方古典园林的一些要素以现代的手法加以抽象和变形，体现在作品中。她的许多作品常常依据与基地相关的含义展开设计，使景观不仅可视，而且具有更深层的意义，反映了后现代主义注重历史文脉和地方特色的特点。2006年施瓦茨被授予第七届美国风景园林单元"国家设计奖"（National Design Award for Landscape Design）。

20世纪70年代以后，玛莎·施瓦茨一面在哈佛等几所著名高等学府任教，同时完成了从私家花园到城市园林的大量设计，引起了广泛关注，并且和一些著名的建筑师如菲利普·约翰逊、马克·麦克、矶崎新等合作过。自1992年她成为哈佛大学设计研究生院副教授以来，还在其他许多大学担任访问评论员，如罗得岛设计学院和加州大学，在世界范围都有较大的影响，成为20世纪中后期现代风景园林艺术的标志性人物。

2. Major Publications & Design Works

A. Major Publications

- *Martha Schwartz：Transfiguration of the Commonplace*，Spacemaker Press，Inc. 1997.
- *The Vangaurd Landscapes and Designs of Martha Schwartz*. United Kingdom：Thames and Hudson. 2004.

B. Major Design Works

- Bagel Garden，Massachusetts，1979.

- King County Jailhouse Garden, Seattle, Washington, 1987.
- Rio Shopping Center, Atlanta, Georgia, 1988.
- US Courthouse Plaza, Minneapolis, Minnesota, 1997.
- Denver Airport, Denver, Colorado, 2000.

3. Ideas

A. Landscape and Common Culture since Modernism

> 节选自 *Modern Landscape Architecture: A Critical Review* 中施瓦莎所写文章，内容有删节。在本文中，施瓦莎认为园林设计是与其他视觉艺术相当的艺术形式，也是一种表达当代文化并用现代材料制造的文化产品。她认为风景园林要进步，就必须以更开放的方式考虑材料，以增加设计语言。认为优秀的设计必须能为所有的阶层所享用，用混凝土、沥青、塑料等日常最普通、最廉价的东西代替那些昂贵的材料来设计景观，面向普通大众。认为风景园林师的思想应该具有文化活力。

Many ideas central to modernism are still attractive to me, and thus I distinguish my work from projects by historicist and neoclassicist[1] designers. If modernism's social agenda the basic optimism toward the future—where "good" design can be available to all classes—holds the most power. I view the manufacturing process not as a limitation but as an opportunity, and I see rationality in a positive light. Great landscapes can no longer be made in the tradition of carved stone and the fountains of Renaissance Europe. Instead they must be made today from concrete, asphalt[2], and plastic, the stuff with which we build our environment on a daily basis. Nonprecious materials and off-the-shelf items can be used artfully, and with this attitude we can build beautiful landscapes; not only for the rich, who today will no longer pay for fancy materials, but also for the middle class, who can't afford them. That we must embrace technology to find the opportunities inherent in mass production appears as valid today as it was to the early modernists. While these modernist sentiments are certainly not new attitudes in architecture, landscape architecture has been reticent[3] to deal with the aesthetics of technology, and has evolved a profession based on the romanticization of the past.

For example, "cheap" and ubiquitous landscape material such as asphalt and concrete are often regarded as lowly and are shunned by developers trying to sell an image of "quality". Developers often commit to budgets that can afford only lowly materials; the true (low) value of the project must be hidden from a prospective buyer by attempting to make the product look expensive. The decision to veneer[4] or stamp concrete into stone patterns, for example, ultimately fools no one and simultaneously expresses the lack of value and discomfort with this ruse[5]. It is possible, however, to appreciate asphalt and concrete for what they are—simple, cheap, and malleable[6]—and for their

---

[1] neoclassic：新古典主义的
[2] asphalt：沥青
[3] reticent：沉默寡言的
[4] veneer：薄板，单板，饰面，外表
[5] ruse：计策，策略，谋略
[6] malleable：（金属等）可锻的，可延展的，能适应的，顺应的

potential to be beautiful materials if used and maintained properly. This, I believe, is a more realistic and hopeful attitude than the reliance on "fine" materials applied only superficially (Figure 12.7).

...

If one wishes to work on the cutting edge in either fine art or design, one must be informed of developments in the world of painting and sculpture. Ideas surface more quickly in painting and sculpture than in architecture or landscape architecture, due to many factors including the immediacy[1] of the media and the relative low investment of money required to explore an idea.

Figure 12.7 City of Commerce, California. (Martha Schwartz, 1991)

Ideas must be challenged in order to prove their viability in a culture. Much art—such as that produced by Jeff Koons, Gordon Matta-Clark, Cindy Sherman, or Vito Acconci—may be important only in that it creates discussion and in the end critical self-reflection. Not every work of art or landscape need be a timeless masterpiece. More importantly, provocative[2] art and design foster an atmosphere of growth by questioning and by challenging the established standards.

In conclusion, the modernist architect's break from the Beaux-Arts[3] and classicism was an important event for landscape architecture. As the architects had to shed the old in order to develop an aesthetic and philosophical stance[4] to deal with the social needs of post-World War I Europe, we must now shed our romance with our wilderness heritage and the English landscape in order to deal effectively with our expanding urban—and suburbanization. The nostalgia[5] for the (imagined) English countryside (so idealized in English landscape and Hudson River School painting) has prevented us from seeing our landscape as it truly is and inhibited the evolution of an appropriate landscape approach to urbanization. We shake our heads in collective disgust at the ugliness of our man-made environments, and yet do little to fully consider the scope of the problem or its possible solution. To improve the visual blight, we place diminutive[6] mounds[7] in our median strips and at the bases of our buildings. Unthinkingly, we dredge[8] up the rolling English countryside like a universal balm[9], without questioning its appropriateness or viability in today's environments. Our profession's narrow and moralistic view of what constitutes a "correct" landscape has disallowed the questioning of this particular aesthetic and has hampered the exploration of other ideas and solutions that might address the problems of increased urbanization. While our culture

---

[1] immediacy：直接（性），目前，密切，即刻
[2] provocative：激怒的，引起议论，兴趣等的，激起的，刺激的
[3] Beaux-Arts：美术，艺术（包括绘画，雕塑，建筑，音乐，舞蹈，木刻等）
[4] stance：（高尔夫、板球）击球时所取的姿势
[5] nostalgia：思乡病，乡愁，留恋过去，怀旧
[6] diminutive：较通常为小的，小得多的
[7] mound：土墩，土堆，小山
[8] dredge：捞泥机，挖泥机，捞网（自水底捞泥，牡蛎，标本等所用者）
[9] balm：（取自某些种树中，用以止痛或疗伤之）香油，香膏，（喻）慰藉物，安慰

professes to be repelled by what and how we build, we still have been unable to conjure other formal vocabularies than those established by economic values, or to break from an ingrained[1] romantic attitude toward our landscape.

Landscape architecture, as a field, has barely touched upon the questions raised by modernism. To many practitioners, modernism and its attendant aspects of growth and embrace of technology are viewed as the cause of the degradation of our natural environment. A small but continuous stream of landscape designers, however, still works within, and perhaps beyond, the modernist tradition; designers who search for meaningful relationships between our natural and built environments, without romantic sentiment but with eyes opened to the world around them.

B. Color-coded Interior Spaces Define an Office Building

> 选自施瓦茨作品集 *The Vanguard Landscapes and Gardens of Martha Schwartz* 一书。该文介绍了德国慕尼黑瑞士再保险总部大楼设计方案。

The landscape surrounding a new office headquarters for a German insurance company divides the site into four quadrants[2], each assigned a different color: red, blue, yellow, green. Each quadrant is made up of strips of plants, hard materials and sculptural objects that radiate out from the building (Figure 12.8). A pool, also divided into colored quadrants, creates a tranquil[3] spot at the center of the building; one quadrant is filled with water lilies, and just beneath the still surface are vibrantly[4]

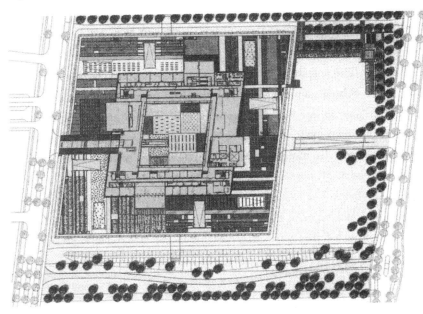

Figure 12.8 The plan is based on the division of the building into four quadrants

---

[1] ingrained: (指习惯, 癖性等) 深染的, 根深蒂固的, 彻底的
[2] quadrant: 象限, 一个圆或其圆周的 1/4
[3] tranquil: 安静的, 平静的, 宁静的
[4] vibrantly: 震动地, 震颤地

colored gazing globes, marbles, crushed glass, gravel and terracotta[1] pots. The whole building is encircled[2] by a vine-covered walkway, or floating hedge, at the level of the third story.

"This is a corporation where they have an interest in the employees' quality of life," Schwartz says. "They have a good chef who cooks everyone delicious and healthy meals. On this project, the landscape was not an afterthought. The building is sited in an office-park-no-man's-land near Munich airport. The company was one of the first kids on the block, and they wanted something quite inward-looking. The concept of the building was to create an internalized space with views to gardens."

There were serious practical problems to overcome, however. Every "landscape" surface of the building was on top of its structure, so lightweight materials had to be used. Many of the garden spaces were in permanent shade. And there was little soil to work with so smaller plants had to be used. "We never met earth," Schwartz says. "It was a little like doing gardens on a spaceship. Given the situation, we had to use materials other than plants to make these spaces interesting. The Germans love plants and are almost phobic about using inorganic[3] materials, so we used non-living organic materials." The idea of the strips came from the fact that the office park had been build on top of agricultural fields-the gardens were like miniature ploughed fields-and the color coding and distinctive garden themes were intended to aid orientation and create identity within the building (Figure 12.9).

Figure 12.9 "Technical materials," or non–plants were used in the many shaded areas where plants couldn't grow.

The graphic[4] nature of this design is a reflection of Schwartz's training as a printmaker and is the product of a working method that developed from that background. "I draw designs out on paper first", she says. "I start with a graphic depiction of space and work up from there, from 2-D to 3-D. I always start off in plan—that organizes my ideas, because when you go to visit these sites, their size can seem overwhelming. Often when I come back, I feel overwhelmed by the space; it seems impossible, chaotic[5]. The only way to deal with it is to reduce it to plan form and compose from that. To make something read in the landscape, which is inherently chaotic, you need something strong and simple at the core. But, typically, you have to get beyond the graphic and into the sculpture/spatial realm to make it work. You can't work too delicately in the landscape-you have to have something fundamentally stronger and more cohesive[6] than the context in order to hold it together. My intention is to create something that can be seen through contrast with a chaotic environment."

---

[1] terracotta：混合陶器，赤陶（用作花瓶、小雕像、建筑的饰物等）
[2] encircle：环绕，包围
[3] inorganic：无机的
[4] graphic：书写的，绘画的，（指描写）生动的
[5] chaotic：无秩序的，混乱，乱七八糟的
[6] cohesive：有附着力的，有凝聚力的

C. Introduction of Martha Schwartz

> 节选自施瓦茨作品集 *The Vanguard Landscapes and Gardens of Martha Schwartz* 一书序言，由 Tim Richardson 撰写。该文主要介绍了施瓦茨的设计哲学和思想。

Martha Schwartz's work is primarily a response to what she perceives as the visual chaos of the world outdoors, a way of superimposing[1] a human sensibility on the unwieldy[2], unpredictable and mutable[3] medium that is landscape. The exuberant[4] aspects of the Schwartz style-bright color, irreverent[5] humor, unbridled[6] imagination, unusual materials, a limited range of plants, surreal scaling of objects-are a continual source of delight and surprise, and mitigate[7] against the more conventional design values that she displays in her work: rigor[8], practicality and orderly methodology. But while a sense of fun is an essential facet of Schwartz's work, it should not mask the controlling seriousness of her artistic intention, and the practical basis for her designs. Unlike other practitioners[9], Schwartz does not try to manipulate the natural landscape in a subtle way, bending it to her ends by using nature's own palette[10] of trees, shrubs and flowers. For Schwartz, such an approach is lazy or even dishonest, since her argument is that even the notion that unsullied "nature" exists out there is patently[11] false. If that is the case, we can hardly be genuinely inspired by it. In this spirit, Schwartz views any tribute to the abstract majesty of "nature" as hopelessly misplaced, and outmoded[12] and irrelevant homage[13] to the romanticism of the 18th and 19th centuries, or the wilderness fantasies of our own time. Neither does Schwartz rely principally on the power of linear pattern to hint at cosmic awareness, as other garden formalists (from Bramante to Le Nôtre to the Modernists) have done.

Instead, Schwartz's unique contribution is to introduce a conceptual or psychic[14] element as the core of her design philosophy: a single idea based on the site's history (human and ecological), its context and its intended use, that is extrapolated[15] to inform every aspect of the design. The result is usually far more complex, symbolically and visually, than this first premise, but the concept nevertheless remains a constant presence-which is why every project in this book has been given a

---

[1] superimpose: 添加，附加
[2] unwieldy: 不易运用或控制的，庞大的，笨重的
[3] mutable: 可变的，易变的，不定的
[4] exuberant: 茂盛的，繁茂的，丰富的
[5] irreverent: 不虔敬的，不恭敬的
[6] unbridled: 不受约束的
[7] mitigate: 使缓和，使减轻，使镇静
[8] rigor: 严格，严厉，严厉执行
[9] practitioner: 从业者
[10] palette: 调色板
[11] patently: 明显的，显然的
[12] outmoded: 过时的，不流行的
[13] homage: 尊敬，敬意，尊崇
[14] psychic: 灵魂的，心灵的，（指力量）非自然的，非物质的
[15] extrapolate: 推断

conceptual surtitle. As a visionary[1] artist seeking an honest response to outdoor environments, Schwartz has devised a design vocabulary (literal, visual and symbolic) that is based on what she sees as the needs and aspirations of human beings rather than a vague concept of nature. As a practical landscape architect, this ethos is also a means of making places useful, meaningful and delightful to people. These artistic and utilitarian[2] aims have always run parallel in her work, but as Schwartz's career has developed they have become more or less indistinguishable.

Where does one place Martha Schwartz? In historical terms, the position she occupies as a formal landscape designer is not unique. Explicit[3] reference to historical design precedent is rare in Schwartz's work, but where she does honor older traditions-notably the parterres[4] of French 17th-centurey formal gardens, and the ancient Japanese garden tradition-the formalism referred to is founded on a profound relationship and dialogue with nature. Just as the formal tradition in Western garden design since the Renaissance is based on a neo-Platonic appreciation of the potential for pure form in nature (a purity recreated by the formal designer), so Schwartz's use of artificial materials and her conceptual methodology is intended to reveal some profound hidden truths of a place, rather than simply exist as an emulation[5] of nature. Schwartz arrived at her own pioneering version of conceptualism via Modernism—indeed, she still describes herself as a Modernist, and speaks of an abiding admiration for its social dimension, design discipline and functionalism. But only Schwartz's earliest solo work could be described as Modernist—as she gains confidence, the work becomes more exuberant, personal, witty and colorful; the emphasis on purity of line and the abstract passions produced by the arrangement of volumes is usurped[6] by a more humane conceptual narrative.

While Schwartz's rationale[7] is sited comfortably in a historical formalist tradition of landscape, her work is nevertheless out of kilter with the times. As such, she has suffered from the criticism and occasional scorn of her fellow professional landscape architects. If she can sometimes seem combative, a glance at some of the attacks leveled at her will explain why. To some, Schwartz's work is gimmicky[8], shallow and shoddy[9]. The humor that she includes in her designs is used against her, to demonstrate a supposed lack of integrity or intellectual rigor. She is accused of being aggressive and contrary, or willfully[10] obscure. But rather than conform to the comfortable lines of contemporary corporate Modernism, or to ally herself with a respectable architectural movement such as Post-Modernism, or to pay lip service to the now fashionable eco-revelatory of "process" design, Schwartz has chosen to remain at the periphery[11]. Schwartz's worst crime is that she is an original.

[1] visionary：（指人）有幻想的
[2] utilitarian：以实用为主的
[3] explicit：（指言语等）明白表示的，明确的
[4] parterre：（庭园之）花坛
[5] emulation：竞争，争胜
[6] usurp：篡夺，霸占
[7] rationale：（某事物的）基本理由，理论基础
[8] gimmicky：巧妙手法的
[9] shoddy：劣质的，冒充好货的
[10] willfully：（指人）刚愎地，任性地
[11] periphery：外围，表面

Notes

1. Extracts from

Texts

(1) Schwartz, Martha. *Landscape and Common Culture Since Modernism* [C]. Marc Treib. *Modern Landscape Architecture: A Critical Review*. London: The MIT Press, 1992: 260 – 265.

(2) Richardson, Tom. *The Vanguard Landscapes and Gardens of Martha Schwartz* [M]. London: Thames & Hudson, 2004: 199 – 205.

(3) Richardson, Tom. *The Vanguard Landscapes and Gardens of Martha Schwartz* [M]. London: Thames & Hudson, 2004: 6 – 18.

Pictures

(1) Figure 12.7 Tom Richardson. *The Vanguard Landscapes and Gardens of Martha Schwartz* [M]. London: Thames & Hudson, 2004: 166.

(2) Figure 12.8 Tom Richardson. *The Vanguard Landscapes and Gardens of Martha Schwartz* [M]. London: Thames & Hudson, 2004: 200.

(3) Figure 12.9 Tom Richardson. *The Vanguard Landscapes and Gardens of Martha Schwartz* [M]. London: Thames & Hudson, 2004: 203.

2. **Sources of Additional Information**

(1) http://www.marthaschwartz.com/

(2) http://www.wwnorton.com/thamesandhudson/new/spring04/551131.htm

(3) http://en.wikipedia.org/wiki/Martha_Schwartz

George Hargeaves

1. Introduction

乔治·哈格里夫斯（George Hargeaves）出生于 1952 年。幼年辗转生活于亚特兰大、休斯顿、纳什维尔和俄克拉荷马城和伊利诺斯州的弗农山区等地。他曾经在南伊利诺斯大学学习一年，在此期间的旅行中，他曾身临落基山国家公园的 Flattop 峰峰顶，四周的峰峦、初夏的积雪和沉静的湖面让他深深感动——"在和自然融为一体的过程中兴奋到恐惧的边缘。"哈格里夫斯后来说："我对于自人体和意识的融为一体的感受和经历，是从这里开始的。"

在一位任教于佐治亚州大学林业学院的叔父的指引下，哈格里夫斯随后进入了佐治亚大学环境设计学院，开始学习风景园林学。1977 年他获得风景园林学士学位，随后进入哈佛设计研究生院，于 1979 年获得风景园林硕士学位。在就学期间，他就以出众的才华引起了诸多风景园林大师广泛的注意。毕业后，他追随他的老师——Peter Walker，加入由其创办的著名的 SWA 事务所，两年后即担任主任设计师，并曾经在英格兰的 Cheshire 设计集团短暂工作。1983 年，他正式创办自己的事务所 Hargreaves Associates，开始了风景园林创作的新尝试。20 年来，他屡屡获得美国风景园林师协会大奖在内的各类国内外奖项，共计有数百项之多。1996～2008 年，他被哈佛大学设计研究生院聘为风景园林"彼德·路易斯·侯见克实践教授"，并于 1996～2003 年担任风景园林系主任一职。此外他还先后在宾夕法尼亚大学、哈佛大学、维吉尼亚大学、伊利诺斯大学等多处著名大学任客座教授，成为名副其实的世纪之交世界风景园林设计界教父级人物。

2. Major Publications & Design Works

A. Major Publications

(1) *Large Parks*：Princeton Architectural Press，2007.
(2) Prospect Green：*Garten+Landschaft*，October 1996.
(3) Most Influential Landscape：*Landscape Journal*. Fall 1993.
(4) Post Modernism Looks Beyond Itself：*Landscape Architecture*，July 1983.

B. Major Design Works

(1) Candlestick Point Cultural Park，San Francisco，California.
(2) Dayton Garden，Minneapolis，Minnesota.
(3) Bvxbee Park，Palo Alto，California.
(4) Guadalupe River Park.
(5) Sydney Olympics 2000 Master Plan，Sydney，Australia.

3. Ideas

A. Campus Green：University of Cincinatti

> 选自 *New World Landscape* 一书。本文介绍了哈格里夫斯的重要作品：辛辛那提大学校园规划。在辛辛那提大学设计与艺术中心，哈格里夫斯设计了一系列蜿蜒流动的草地土丘，创造出神秘的形状和变幻的影子。

George Hargeaves has been master planning the University of Cincinatti since 1989. He has

Figure 12.10 Campus Green in Cincinatti University

identified force fields-axial lines-which link the university's open spaces and he allowed for 185,000 square metres (2 million square feet) of new building development. He has reinforced the university's existing campus quadrangles[1] and introduced the character of the adjacent Burnet Woods. Finally, he has begun to establish links between the original West Campus (where the university was founded in 1819) and the more recent East Campus. The East and West campuses are like two squares touching at a corner, and Campus Green is the northeaster corner of the West Campus. In providing this link between the campuses, Hargreaves has re-established the pre-eminence[2] of the pedestrian.

Campus Green-a 2.4 hectare (6 acre) open space-was formerly a large parking lot. Now there are open lawns, gardens, a sculpture garden and arboretum[3]. The site was once crossed by a stream, and so to recall that origin the design lays new curving, streamlike paths like a braid over the straight geometry of the force fields (Figure 12.10). The former parking lot has become the social centre of the campus and now accommodates a bookshop, Alumni and Faculty centre, the College of Business and student housing.

In essence the design comprises a set of extensive lawns with three main features. First, there is an international arboretum and associated earthworks-international in the sense that it includes species that have a worldwide distribution such as black cherry (*Prunus serotina*)[4] and apple (*Malus sp.*)[5]. Second, there is the conical[6] mount with water stairs built of the local limestone, a major landmark. Emerging from the mount there is a curving wall marking the entrance to the campus from the north. To commemorate the old stream there is a new stream consisting of a stone runnel into which surface water runoff is fed. The third feature of the design comprises three raised triangular gardens with flowering tree to provide a place of peace and quiet within Campus Green. Hargreaves describes the development of the master plan in this way: "The first of the open spaces we created was McKicken Commons. It's a simple, green quad formed by the campus's most historic buildings. Then we went on to what we call object-spaces-Library Square and Sigma Sigma Amphitheater-where the open space functions like a building facade. Library Square, which is primarily paved, is about the unfolding of knowledge, embodied in a spiral Sigma Sigma Amphitheater is a convocation space for students, faculty and alumni and its message is expressed through a light tower. With Campus Green, we return to the theme of the quadrangle-only this time, the quad has been folded in a couple of times, so that other themes come into play. They give Campus Green more character than simply a green

[1] quadrangle：四角形，四边形，方院，
[2] pre-eminence：卓越
[3] arboretum：树园，植物园
[4] *Prunus serotina*：黑樱桃
[5] *Malus sp.*：苹果属
[6] conical：圆锥的，圆锥形的

space with some trees."

George Hargreaves is Professor and Chairman of Landscape Architecture at Harvard, and it shows. There is in his work the idea of a long-term commitment to place, an intellectual development, and an exploration of the ideas of land re-use and what he terms "connection". Those who work in universities know the ignorance and arrogance[1] exhibited by some academics and managers. At Cincinnati the university is fortunate to be led by those who have confidence in what a gifted landscape architecture can do: appreciate the site, realize its potential, and do so with great intellectual coherence and panache[2]. Of the transformation of parking lot into campus, Hargereaves says: "Our work acknowledges the simple truth that 'made' landscapes can never be natural. With increasing frequency our work deals with land which has been made and re-made."

B. Pure Design, Pure Image

> 选自美国 *Landscape Architecture* 杂志。本文介绍了哈格里夫斯的重要作品：德顿住宅花园（Dayton Garden）。哈格里夫斯创造的花园让主人与现代雕塑及开放空间自然形成相互对话交流的状态。

"I reveled in the scale of it," says George Hargreaves of the three-quarter-acre garden that he designed to accompany a contemporary house in a residential neighborhood close to downtown Minneapolis. "This is a scale that you can touch and feel. It wasn't fifty, a hundred, two hundred acres. It wasn't a ten-year project. It was a project that was all going to be done at one time, and I knew it was going to be done well. It was of such a scale that you could stake things out. We staked those mounds[3] out and tested the size of them. We staked out that lens shape. We staked out the curves. We were out there actually doing it" (Figure 12.11).

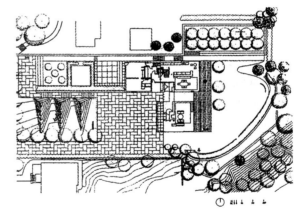

Figure 12.11 Dayton Garden

"When you think of this project vis-à-vis the work I've been doing in the last couple of years," says Hargreaves of the Minneapolis garden, "it's significantly different in two ways. The first way is that it's pure design, pure image. Although there are a lot of formal qualities in my work as well as process issues, there is nothing about natural processes here. The second thing is that it's very tied to architecture, and most of my work is not. I don't typically work with architecture of this quality."

The clients, Kenneth and Judy Dayton, decided to move into town from their thirteen-acre estate in Wayzata, where—not incidentally—Dan Kiley[4] had done a great deal of work over the years.

---

[1] arrogance：傲慢态度，自大
[2] panache：羽饰，华丽
[3] mound：土墩，护堤，垛
[4] Dan Kiley：丹·凯利，美国著名园林设计师

"We were getting older, and we were using the lake less. We love the orchestra[1] and the opera and the art museum, and we found ourselves on the highway more and more, coming into town," said Judy Dayton. An in-town house would provide a venue for many social occasions, including those associated with the Dayton's cultural interests.

The Dayton brought many of their favorite pieces into town with them, including works by Peter Shelton, Ellsworth Kelly, Richard Long, Richard Serra, Scott Burton, Martin Puryear, Joel Shapiro, David Nash, and Siah Armajani. In the country, explains Judy Dayton, "the sculpture was spread out all over. They weren't anywhere near each other. We had a path around that property, and you'd come to one and then the other and the other. But now they've all moved close by, and they all seem friendly with one another." "The Daytons told me," recalls Hargreaves, "that at night they'd have a drink and then go out and walk around and look at the sculpture. They had a circuit of their property, and the sculptures were around it. One of the things I tried to do was give them that seem circuit" on a smaller, urban scale.

Significantly, the landscape was not specifically designed to "hold" the sculpture. The two were not, however, unrelated. Says Hargreaves, "When you're working closely with two or three people over a year or two years, you naturally begin to say, 'Well, let's see if the Shelton can go there, if the Kelly can go in there.'" Hargreaves, James and the Daytons placed the sculpture using mockups[2] of canvas and wood—and by all accounts had a wonderful time in the process.

The result of all of these collective efforts is an exceptionally rich landscape that consists of two circuits, the circuit of landscape spaces and the circuit of sculptures. Or we might think of large party celebrating one of the Daytons' favorite cultural events, a party in which there are two intellectual group discussions going on—one among the landscape spaces, and the other among the sculptures—at the same time that some of the sculptures are enjoying little conversational tête-à-têtes, even flirtations[3], with the spaces in which they happen to find themselves standing.

C. Landscape Alchemist

> 选自哈佛大学在线学报，是一篇人物报道类的文章。本文介绍了哈格里夫斯的学习经历，主要学术观点和作品，以及在哈佛大学中的教学活动。

When George Hargreaves was 18, he went on a summer trip that changed his life.

While trekking[4] across the West with his friends, Hargreaves hiked to the top of Flat Top Mountain in the Rockies. The view was spectacular, but there was something else about the experience that moved him and that he kept struggling to define.

"It wasn't just the mountains or the trees or any of the individual elements. It was something about the sense of space itself. When I got back home I tried to explain this to my uncle who was dean

---

[1] orchestra：管弦乐队
[2] mockups：模式
[3] flirtation：调情，挑逗
[4] trek：旅行，艰苦跋涉

of forestry at the University of Georgia, and he said, 'Have you ever thought about going into landscape architecture?'"

Hargreaves followed his uncle's suggestion, earning a bachelor's degree in landscape architecture from the University of Georgia in 1977 and a master's in landscape architecture from the Harvard Graduate School of Design (GSD) in 1979.

As a practitioner, Hargreaves has had notable success in transforming his visions of outdoor spaces into reality. His San Francisco and Cambridge firm of Hargreaves Associates has designed a variety of innovative parks, plazas, and riverfront areas that have garnered numerous awards and wide critical acclaim.

Hargreaves began teaching at the GSD as an adjunct professor of practice in 1991. In July 1996, he was simultaneously promoted to full professor and appointed chairman of the Landscape Architecture Department.

In announcing the appointment, GSD Dean Peter Rowe called Hargreaves "one of the foremost practitioners of his generation. Over the past five years he has been one of the outstanding teachers at the GSD and has consistently exhibited a strong commitment to teaching students at all levels. We are indeed fortunate to have a person of George's talent and dedication heading the Landscape Architecture Department."

When Hargreaves talks about what inspires him, he is more apt to mention the impact of natural phenomena rather than the design conventions established by his predecessors. Like the Romantic poets of the early 19th century, he continually refers back to certain critical experiences in Nature as touchstones of his aesthetic development. One of these was his experience at the summit of Flat Top Mountain. A second was witnessing the fury of Hurricane Gloria from a Honolulu hotel in 1982.

"I had studied geology and I always thought of change as a slow process," he said. "But in Hawaii I was reminded of the power of immediate change, brought on by a 30-foot wall of water. It made me wonder, how do we imbue a project with the notion that the environment is not static, but is always changing?"

This quest to take account of Nature's dynamic qualities has led Hargreaves to a unique and intuitive method for approaching the process of landscape design.

"Convention is easy to slip into, to keep making the same pictures. What I try to find are those magic moments of clarity when you hear what the site is whispering to you."

What makes Hargreaves' approach even more distinctive is that typically the whispers he listens for come not from sites that embody the undisturbed beauty of Nature, but rather from those ruined and degraded landscapes created as an unfortunate by-product of the industrial world's activities."

"Our firm specializes in difficult sites", he said.

Design critics have taken note of Hargreaves' talent for rescuing abused sites. Writing in *Landscape Architecture*, critic John Beardsley wrote that "Hargreaves performs a kind of alchemy in which the dross[1] of post-industrial landscape is transformed into something approximating gold... combining a strong

---

[1] dross: (金属熔化时浮升至表面的) 渣滓, 无用之物

sculptural language with a sensitivity to both environmental process and social history..."

One of the difficult sites Hargreaves has transformed is Candlestick Point in San Francisco. Located at the edge of the city, on landfill facing San Francisco Bay, the site bears the brunt of strong winds and aggressive tides as well as the scars of its industrial heritage. Hargreaves collaborated with the firm of Mack Architects and artist Douglas Hollis to transform the site into an innovative cultural park.

Instead of trying to shut out the elements, Hargreaves decided to work with them. In some cases, this meant emphasizing the natural processes to which the site is exposed. The main entrance, for example, is positioned on the axis of the prevailing winds so that visitors are literally blown into the park. The experience is intensified by the presence of wind-activated organ pipes that announce the visitors' arrival.

Once inside, views of open fields become visible, sloping downward to the expanse of San Francisco Bay. This long, inclined plain is bounded on each side by water channels that amplify the experience of the tides by allowing the rising waters to penetrate into the park.

But there is also refuge from the elements. A series of dunes create wind-sheltered areas where visitors may picnic or sit. One of the dunes provides sanctuary[1] for an acoustically[2] efficient performance shed for outdoor concerts.

Candlestick Point Cultural Park is not the only site on which Hargreaves and his colleagues have transformed desolate wasteland into a useful and aesthetically pleasing facility without obscuring or falsifying its ignominious history.

Other examples include Byxbee Park in Palo Alto, Calif.; Guadalupe River Park in San Jose, California; Parque do Tejo e Trancao in Lisbon, Portugal; and many others, both in this country and abroad.

Recently, Hargreaves has undertaken two commissions which, because of their high visibility, have brought him attention beyond the design community. One is to produce a master design for Homebush Bay, the site of the 2000 Summer Olympics in Sydney, Australia (Figure 12.12). The second is another master plan to tie together a group of new buildings on the University of Cincinnati campus, each of which has been designed by a prominent contemporary architect.

As a teacher, Hargreaves believes that the GSD's Landscape Architecture Department has been doing an excellent job of providing professional education, but he would like to see an increase in the amount of research in the field. He believes that in the post-industrial era issues such as environmental systems management, flood control, highway design, and bioengineering belong at the heart of landscape architecture. He said there is also much work to be done in adopting computer-aided design systems (CAD) to the needs of landscape architects.

Hargreaves agrees that so far the work of landscape architects has not drawn the attention and notoriety that often accrues[3] to the efforts of architects. Part of this is due to the fact that buildings

---

[1] sanctuary：避难所
[2] acoustically：听觉上，声学上
[3] accrue：自然增加，产生

Figure 12.12 Olympics in Sydney

by their nature attract notice while landscapes are often taken for granted. But the problem, Hargreaves believes, is also that landscape architecture is a relatively young discipline and has not yet built up a significant body of work.

"The generation that's coming will realize that landscape and systems are intertwined and that they become an integral part of our lives. When this happens, I think we will begin to see works in landscape architecture that earn the same attention and care that we've given to great buildings".

Notes

1. Extracts from

Texts

(1) Laurence King Publishing. *New World Landscape* [M]. Laurence King Publishing, 2004: 30 - 31.

(2) Gillette, Jane Brown. *Pure Design, Pure Image* [J]. *Landscape Architecture*, 1998 (10): 93 - 95.

(3) Gewertz, Ken. Landscape Alchemist [EB/OL]. http://www.hno.harvard.edu/gazette/1997/02.06/LandscapeAlchem.html, 2007 - 11 - 28.

Pictures

(1) Figure 12.10 Laurence King Publishing. *New World Landscape* [M]. London: Laurence King Publishing Ltd., 2004: 31.

(2) Figure 12.11 Jane Brown Gillette. Pure Design, Pure Image [J]. *Landscape Architecture*, 1998 (10): 94.

(3) Figure 12.12 Michael Spens. *modern landscape* [M]. London: Phaidon Press Inc. 2003: 225.

2. Sources of Additional Information

(1) http://www.hno.harvard.edu/gazette/1997/02.06/LandscapeAlchem.html.

(2) http://www.hargreaves.com.

(3) http://www.gsd.harvard.edu/people/faculty/hargreaves/index.html.

附录 I Bibliography

Part 1

American society of landscape architects. *What is Landscape Architecture* [EB/OL]. http://www.asla.org/nonmembers/what_is_asla.cfm,2007-04-29. (Unit 1/ Further Reading B)

Booth, Norman K. *Basic Elements of Landscape Architectural Design* [M]. New York: Elsevier Science Publishing Co. Inc, 1983: 283-304. (Unit 9/ Text)

Bureau of Labor of U.S. Department of Labor. Occupational Outlook Handbook (2006-07 Edition): *Landscape Architects* [EB/OL]. http://www.bls.gov/oco/ocos039.htm, 2007-04-28. (Unit 1/ Further Reading C)

Carson, Rachel. *Silent Spring* [M]. Boston: Houghton Mifflin Company, 1962: 5-8. (Unit 7/ Text)

Conway, Hazel. *People's Parks—The Design and Development of Victorian Parks in Britain* [M]. New York: Cambridge University Press, 1991: 76-81. (Unit 5/ Further Reading A)

Dines, Nicholas T. et al. *Landscape Architect's Portable Handbook* [M]. New York: McGraw-Hill Publishing Company, 2003: 402-407. (Unit 3/ Further Reading C)

Eckbo, Garrett. et al. *Landscape Design in the Urban Environment* [C]. Marc Treib. *A Critical Review*. London: The MIT Press: 1993: 78-82. (Unit 4/Further Reading B)

Eckbo, Garrett. *Landscape for Living* [M]. New York: Architectural Record with Duell, Sloan, & Pearce, 1950: 57-60. (Unit 2/Further Reading C)

Elizabeth K. Meyer. The Post-Earth Day Conundrum: Translating Environmental Values into Landscape Design [C]. Michel Conan. *Environmentalism in Landscape Architecture*. Washington D.C.: Dumbarton Qaks Research Library and Collection, 2000: 187-191 (Unit 7/Further Reading B)

Fordney, Chris. New Birth for Gettysburg [J]. *Landscape Architecture*, 2002 (8): 46-49. (Unit 6/Further Reading B)

Freeman, Raymond L. & Wayne D. Iverson. *National Parks & National Forests* [C]. 南京林业大学园林学院. 园林专业英语. 南京: 南京林业大学自编讲义: 2005: 7-13. (Unit 5/ Further Reading C)

Goulty, Sheena Mackellar. *Heritage Gardens: Care, Conservation and Management* [M]. London and New York: Routledge Inc., 1993: 54-59. (Unit 6/ Text)

Hackett, Brian. *Planting Design* [M]. London: E &F. N. Spon Ltd., 1979: 2-11. (Unit 8/ Text)

Hannebaum, Leroy G.. *Landscape Design: A Practical Approach* [M]. New Jersey: Prentice

Hall Career & Technology, 1994: 23 - 26. (Unit 9/ Further Reading B)

Hargreaves, George. *Post Modernism Looks Beyond Itself* [J]. *Landscape Architecture*, 1983 (7): 60~65. (Unit 2/ Further Reading A)

International Federation of Landscape Architecture. Definition of the Profession of Landscape Architect for the International Standard Classification of Occupations [EB/OL]. http://www.iflaonline.org/resources/policy/pdf/ifla_definition.pdf, 2007 - 04 - 29. (Unit 1/ Further Reading A)

James B. Root. *Fundamentals of Landscaping and Site Planning* [M]. Westport: The Avi Publishing Company, Inc., 1985: 66 - 68. (Unit 9/ Further Reading A)

Jellicoe, Geoffrey. *Contemporary Meanings in the Landscape* [J]. *Landscape Architecture*, 1980 (1): 51 - 54. (Unit 2/ Text)

Jellicoe, Geoffrey. *The landscape of Civilisation—Created for the Moody Historical Gardens* [M]. East Sussex: Garden Art Press Ltd., 1989: 17 - 19. (Unit 6/Further Reading C)

Laurie, Michael. *An Introduction to Landscape Architecture* [M]. New York: Elsevier Science Publishing Co., Inc. 1986: 74 - 85. (Unit 5/ Text)

Laurie, Michael. *An Introduction to Landscape Architecture* [M]. New York: Elsevier Science Publishing Co. Inc, 1986: 74 - 85. (Unit 8/ Further Reading B)

Motloch, John L. *Introduction to Landscape Design* [M]. New York: Van Nostrand Reinhold, 1991: 82 - 86. (Unit 8/ Further Reading C)

Newton, Norman T. *Design on the Land: The Development of Landscape Architecture* [M]. Cambridge: Harvard University Press, 1991: 596 - 615. (Unit 5/ Further Reading B)

Ogrin, Dušan. *The world Heritage of Gardens* [M]. London: Thames and Hudson Ltd, 1993: P14 - 22 (Unit 6/Further Reading A)

Perry, Kevin Robert. Sustainable Stormwater Management Program [EB/OL]. http://asla.org/awards/2006/06winners/341.html, 2007 - 05 - 28 (Unit 7/ Further Reading C)

Reiniger, Clair. Bioregional Planning and Ecosystem Protection [C]. Frederick Steiner, William Thompson. *Ecological Design and Planning*. New York: Wiley & Sons, 1997: 185 - 199. (Unit 7/ Further Reading A)

Rose, James C.. Plants Dictate Garden Forms [C]. Marc Treib. *A Critical Review*. London: The MIT Press: 1993: 73 - 75. (Unit 8/ Further Reading A)

Simonds, John Ormsbee. *Landscape Architecture: A Manual of Site Planning and Design* [M]. New York: McGraw - Hill Publishing Company, 1997: 4 - 9. (Unit 1/ Text)

Sudjic, Deyan. *The 100 mile City* [M]. New York: Harvest/HBJ Book, 1992: 305 - 309 (Unit 4/ Further Reading C)

Tuan, Yi - Fu. *Topophilia: a study of Environment Perception, Attitudes, and Values* [M]. New York: Columbia University Press, 1990: 230 - 234. (Unit 4/ Further Reading A)

Turner, Tom. *City as Landscape: A Post - postmodern View of Design and Planning* [M]. London: E &F. N. Spon Ltd., 1996: 8 - 10. (Unit 2/ Further Reading B)

Turner, Tom. *City as Landscape: A Post - postmodern View of Design and Planning* [M]. London: E &F. N. Spon Ltd., 1996: 169 - 172. (Unit 9/ Further Reading C)

United Nations Conference on Environment and Development. *The Covention on Biological Diversity* [EB/OL]. http://www.cbd.int/convention/convention.shtml, 2008 - 03 - 20 (Unit 3/ Further Reading A)

Whiston Spirn, Anne. *The Granite Garden: Urban Nature and Human Design* [M]. New York: Basic Books, Inc. Publishers, 1984: 3 - 5. (Unit 4/Text)

IInd International Congress of Architects and Technicians of Historic Monument. The Venice Charter 1964 [EB/OL]. http://www.iflaonline.org/resources/policy/pdf/charter/venice _ charter.pdf, 2007 - 09 - 22. (Unit 3/Text)

诺曼 K. 布思. 风景园林设计要素 [M]. 北京: 中国林业出版社, 1989: 267 - 287. (Unit 9/ Text)

吴良镛. 国际建协《北京宪章》——建筑学的未来（中英文版）[M]. 北京: 清华大学出版社, 2002: 3 - 14, 177 - 184. (Unit 3/ Further Reading B)

张京祥. 西方城市规划思想史纲 [M]. 南京: 东南大学出版社, 2005: 272 - 275. (Unit 3/ Text)

Part 2

Beveridge, Charles E. Olmsted—His Essential Theory [J]. *The Journal of the Victoran Society in America*, 2000, 20 (2): 32 - 37. (Unit 10/ Lesson 1 - C)

Brackenbury, Martin. Sir Geoffrey Jellicoe [EB/OL]. http://users.eggconnect.net/mandvbrackenbury/Geoffrey%20Jellicoe.html, 2008 - 03 - 20. (Unit 10/ Lesson 3 - C)

Cerver, Francisco Asension. *The World of Environmental Design* (Vol. 1) [M]. Spain: Arco Editorial, 1994: 14~21. (Unit12/ Lesson 2 - C)

Church, John Thomas. *Gardens are for People* [M]. Berkeley: University of California Press, 1995: 1~9. (Unit 10/ Lesson 2 - C)

Church, John Thomas. *Gardens are for People* [M]. Berkeley: University of California Press, 1995: 29 - 34. (Unit 10/ Lesson 2 - A)

Church, John Thomas. *Gardens are for People* [M]. Berkeley: University of California Press, 1995: 182 - 186. (Unit 10/ Lesson 2 - B)

Elivson, Sima. *The Gardens of Roberto Burle Marx* [M]. Portland: Sagapress Inc./Timber Press, Inc., 1991: 7. (Unit 10/ Lesson 4 - A)

Elivson, Sima. *The Gardens of Roberto Burle Marx* [M]. Portland: Sagapress Inc./Timber Press, Inc., 1991: 119 - 122. (Unit 10/ Lesson 4 - B)

Elivson, Sima. *The Gardens of Roberto Burle Marx* [M]. Portland: Sagapress Inc./Timber Press, Inc., 1991: 179 - 180. (Unit 10/ Lesson 4 - B)

附录 I Bibliography

Fabos, Julius Gy. *Definition and Significance of Greenway* [EB/OL]. http://www.umass.edu/greenway/2GR-index.html, 2008-03-20. (Unit 12/ Lesson 1-A)

Fabos, Julius Gy. *Greenway History* [EB/OL]. http://www.umass.edu/greenway/2GR-index.html, 2008-03-20. (Unit 12/ Lesson 1-B)

Fabos, Julius Gy. *The New England Green Vision Plan* [EB/OL]. http://www.umass.edu/greenway/1PO-index.html, 2008-03-20. (Unit 12/ Lesson 1-C)

Gewertz, Ken. *Landscape Alchemist* [EB/OL]. http://www.hno.harvard.edu/gazette/1997/02.06/LandscapeAlchem.html, 2007-11-28. (Unit 12/ Lesson 4-C)

Gillette, Jane Brown. *Pure Design, Pure Image* [J]. Landscape Architecture, 1998 (10): 93-95. (Unit 12/ Lesson 4-B)

Halprin, Lawrence. *The RSVP Cycles: Creative Processes in the Human Environment* [M]. New York: George Braziller, 1969: 1-5. (Unit 11/ Lesson 2-A)

Jellicoe, Geoffrey. & Susan. *The Landscape of Man: Shaping the environment from prehistory to present day* [M]. London: Thames & Hudson, 1988: 155-157. (Unit 10/ Lesson 3-A)

Kiley, Dan. & Jane Amidon. *Dan Kiley - The Complete Works of America's Master Landscape Architecture* [M]. Boston: Bulfinch Press, 1999: 8-11. (Unit 11/ Lesson 3-A)

Kiley, Dan. & Jane Amidon. *Dan Kiley - The Complete Works of America's Master Landscape Architecture* [M]. Boston: Bulfinch Press, 1999: 42-43. (Unit 11/ Lesson 3-C)

Kiley, Dan. & Jane Amidon. *Dan Kiley - The Complete Works of America's Master Landscape Architecture* [M]. Boston: Bulfinch Press, 1999: 98-103. (Unit 11/ Lesson 3-B)

Laurence King Publishing. *New World Landscape* [M]. Laurence King Publishing, 2004: 30-31. (Unit 12/ Lesson 4-A)

Leccese, Michael. *Mystical Pragmatist* [J]. Landscape Architecture, 1990 (3): 79-83. (Unit 11/ Lesson 1-C)

Maryman, Brice. & Liz Birkholz. *Freeway Park: Past, Present and Future* [EB/OL]. http://www.tclf.org/features/freeway/index.htm, 2008-03-20. (Unit 11/ Lesson 2-B)

McHarg, Ian Lennox. *Design with Nature* [M]. New York: John Wiley & Sons, Inc. 1992: p5. (Unit 11/ Lesson 4-A)

McHarg, Ian Lennox. *Design with Nature* [M]. New York: John Wiley & Sons, Inc. 1992: 144. (Unit 11/ Lesson 4-C)

McHarg, Ian Lennox. *Ecological and Design* [C]. George F. Thompson & Frederick R. Steiner. *Ecological Design and Planning*. New York: John Wiley & Sons, Inc. 1997: 321-330. (Unit 11/ Lesson 4-B)

Montero, Marta Iris. *Roberto Burle Marx: the Lyrical Landscape* [M]. London: Thames & Hudson, 2001: 1. (Unit 10/ Lesson 4-C)

Richardson, Tom. *The Vanguard Landscapes and Gardens of Martha Schwartz* [M]. London: Thames & Hudson, 2004: 6-18. (Unit 12/ Lesson 3-C)

Richardson, Tom. *The Vanguard Landscapes and Gardens of Martha Schwartz* [M]. London: Thames & Hudson, 2004: 199-205. (Unit12/ Lesson 3-B)

Schuyler, David. & Jane Turner Censer. *The Papers of Frederick Law Olmsted* 1865~1874 [M]. Maryland: The Johns Hopkins University Press, 1992: 234-236. (Unit 10/ Lesson 1-A)

Schuyler, David. &Jane Turner Censer. *The Papers of Frederick Law Olmsted* 1865~1874 [M]. Maryland: The Johns Hopkins University Press, 1992: 664~671. (Unit 10/ Lesson 1-B)

Schwartz, Martha. *Landscape and Common Culture Since Modernism* [C]. Marc Treib. *Modern Landscape Architecture: A Critical Review*. London: The MIT Press, 1992: 260-265. (Unit 12/ Lesson 3-A)

Simonds, John Ormsbee. *Lessons* [M]. Washington, D. C.: ASLA Press, 1999: 37-39. (Unit 11/ Lesson 1-A)

Simonds, John Ormsbee. Simonds, John Ormsbee. *Modern Architects* [M]. 编者王欣从西蒙兹处获得的复印资料. (Unit 11/ Lesson 1-B)

Thompson, J. William. *The Power of Place* [J]. *Landscape Architecture*, 1997 (7): 63-70. (Unit 11/ Lesson 2-C)

Walker, Peter. *Minimalism in Landscape* [C]. Simon Swaffield. *Theory in Landscape Architecture*. Philadelphia: University of Pennsylvania Press, 2002: 87-88. (Unit 12/ Lesson 2-A)

Walker, Peter. *Most Influential Landscape——Peter Walker* [J]. *Landscape Journal*, 1993, 12 (2): p173~174. (Unit 12/ Lesson 2-B)

Waymark, Janet. *Modern Garden Design* [M]. London: Thames & Hudson, 2003: 185-189. (Unit 10/ Lesson 3-B)

注 因本书所精选的英文原文的作者众多，我们无法事先与原作者一一取得联系，在此一并致谢。如有版权事宜，请与出版社联系。

附录 Ⅱ Illustration Credits

Part 1

Figure 1.1：John Ormsbee Simonds. *Landscape Architecture：A Manual of Site Planning and Design* [M]. New York：McGraw-Hill Publishing Company, 1997：6.

Figure 1.2：John Ormsbee Simonds. *Landscape Architecture：A Manual of Site Planning and Design* [M]. New York：McGraw-Hill Publishing Company, 1997：7.

Figure 1.3：Bureau of Labor of U.S. Department of Labor. *Occupational Outlook Handbook*（2006-07）：*Landscape Architects* [EB/OL]. http://www.bls.gov/oco/ocos039.htm, 2007-04-28.

Figure 2.1：大卫·路德林，尼古拉斯·福克 著. 王健，单燕华 译. 营造21世纪的家园 [M]. 北京：中国建筑工业出版社，2005：39.

Figure 2.2：Tom Richardson. *The Vanguard Landscapes and Gardens of Martha Schwartz* [M]. London：Thames & Hudson, 2004：47.

Figure 2.3：Hargreaves Associates. *Harlequin Plaza* [EB/OL]. http://www.hargreaves.com/projects/harlequin/index.html, 2007-04-28.

Figure 2.4：Tom Turner. *City as Landscape：A Post-postmodern View of Design and Planning* [M]. London：E &F. N. Spon Ltd., 1996：9.

Figure 2.5：Tom Turner. *City as Landscape：A Post-postmodern View of Design and Planning* [M]. London：E &F. N. Spon Ltd., 1996：9.

Figure 2.6：Tom Turner. *City as Landscape：A Post-postmodern View of Design and Planning* [M]. London：E &F. N. Spon Ltd., 1996：10.

Figure 3.1：吴良镛. 国际建协《北京宪章》——建筑学的未来（中英文版）[M]. 北京：清华大学出版社，2002：81.

Figure 5.1：Laurie, Michael. *An Introduction to Landscape Architecture* [M]. New York：Elsevier Science Publishing Co., Inc. 1986：85.

Figure 5.2：Hazel Conway. *People's Parks—The Design and Development of Victorian Parks in Britain* [M]. New York：Cambridge University Press, 1991：78.

Figure 5.3：Hazel Conway. *People's Parks—The Design and Development of Victorian Parks in Britain* [M]. New York：Cambridge University Press, 1991：79.

Figure 5.4：Hazel Conway. *People's Parks—The Design and Development of Victorian Parks in Britain* [M]. New York：Cambridge University Press, 1991：80.

Figure 5.5: Norman T. Newton. *Design on the Land: The Development of Landscape Architecture* [M]. Cambridge: Harvard University Press, 1991: 613.

Figure 5.6: Norman T. Newton. *Design on the Land: The Development of Landscape Architecture* [M]. Cambridge: Harvard University Press, 1991: 614.

Figure 6.1: Chris Fordney. *New Birth for Gettysburg* [J]. *Landscape Architecture*, 2002 (8): 49.

Figure 6.2: Chris Fordney. *New Birth for Gettysburg* [J]. *Landscape Architecture*, 2002 (8): 49.

Figure 6.3: Geoffrey Jellicoe. *The landscape of Civilisation—Created for the Moody Historical Gardens* [M]. East Sussex: Garden Art Press Ltd., 1989: 18.

Figure 7.1: Rachel Carson. *Silent Spring* [M]. Boston: Houghton Mifflin Company, 1962: 1.

Figure 7.2: Kevin Robert Perry. *Sustainable Stormwater Management Program* [EB/OL]. http://asla.org/awards/2006/06winners/341.html, 2007-05-28.

Figure 7.3: Kevin Robert Perry. *Sustainable Stormwater Management Program* [EB/OL]. http://asla.org/awards/2006/06winners/341.html, 2007-05-28.

Figure 8.1: 郦芷若，朱建宁. 西方园林 [M]. 郑州: 河南科学技术出版社, 2002: 325.

Figure 8.2: John L. Motloch. *Introduction to Landscape Design* [M]. New York: Van Nostrand Reinhold, 1991: 82.

Figure 8.3: John L. Motloch. *Introduction to Landscape Design* [M]. New York: Van Nostrand Reinhold, 1991: 83.

Figure 8.4: John L. Motloch. *Introduction to Landscape Design* [M]. New York: Van Nostrand Reinhold, 1991: 84.

Figure 8.5: John L. Motloch. *Introduction to Landscape Design* [M]. New York: Van Nostrand Reinhold, 1991: 85.

Figure 8.6: John L. Motloch. *Introduction to Landscape Design* [M]. New York: Van Nostrand Reinhold, 1991: 86.

Figure 9.1: Norman K. Booth. *Basic Elements of Landscape Architectural Design* [M]. New York: Elsevier Science Publishing Co. Inc, 1983: 302, 303.

Figure 9.2: Hannebaum, Leroy G.. *Landscape Design: A Practical Approach* [M]. New Jersey: Prentice Hall Career & Technology, 1994: 25.

Part 2

Figure 10.1: David Schuyler, Jane Turner Censer. *The Papers of Frederick Law Olmsted 1865~1874* [M]. Maryland: The Johns Hopkins University Press, 1992: 664.

Figure 10.2: David Schuyler, Jane Turner Censer. *The Papers of Frederick Law Olmsted 1865~1874* [M]. Maryland: The Johns Hopkins University Press, 1992: 669.

Figure 10.3: John Thomas Church. *Gardens are for People* [M]. Berkeley: University of California Press, 1995: 182.

Figure 10.4: John Thomas Church. *Gardens are for People* [M]. Berkeley: University of California Press, 1995: 186.

Figure 10.5: John Thomas Church. *Gardens are for People* [M]. Berkeley: University of California Press, 1995: 49.

Figure 10.6: Geoffrey and Susan Jellicoe 著,刘滨谊 主译. 图解人类景观——环境塑造史论 [M]. 上海:同济大学出版社. 2006: 330.

Figure 10.7 Geoffrey and Susan Jellicoe 著,刘滨谊 主译. 图解人类景观——环境塑造史论 [M]. 上海:同济大学出版社. 2006: 343.

Figure 10.8 Janet Waymark. *Modern Garden Design* [M]. London: Thames & Hudson, 2003: 189.

Figure 10.9 Marta Iris Montero. Roberto Burle Marx: *the Lyrical Landscape* [M]. London: Thames & Hudson, 2001: 58.

Figure 10.10 Elivson, Sima. *The Gardens of Roberto Burle Marx* [M]. Portland: Sagapress Inc. / Timber Press, Inc., 1991: 121.

Figure 10.11 Elivson, Sima. *The Gardens of Roberto Burle Marx* [M]. Portland: Sagapress Inc. / Timber Press, Inc., 1991: 180.

Figure 10.12 王向荣,林箐. 西方现代景观设计的理论与实践 [M]. 北京:中国建筑工业出版社,2002: 113.

Figure 11.1: Michael Leccese. *Mystical Pragmatist* [J]. Landscape Architecture, 1990 (3): 83.

Figure 11.2: John O. Simonds. *Miami Lakes New Town* [J]. Parks & Recreation, 1970 (10): 17.

Figure 11.3: Lawrence Halprin. *The RSVP Cycles: Creative Processes in the Human Environment* [M]. New York: George Braziller, 1969: 4.

Figure 11.4: Maryman, Brice. & Liz Birkholz. *Freeway Park: Past, Present and Future* [EB/OL]. http://www.tclf.org/features/freeway/index.htm, 2008-03-20.

Figure 11.5: Thompson, J. William. *The Power of Place* [J]. Landscape Architecture, 1997 (7): 67.

Figure 11.6: Dan Kiley. *Dan Kiley: The Complete Works of America's Master Landscape Architecture* [M]. Boston: Bulfinch Press, 1999: 98.

Figure 11.7: Dan Kiley. *Dan Kiley: The Complete Works of America's Master Landscape Architecture*. Boston, 1999: 42.

Figure 11.8: Ian L. McHarg. *Design with Nature* [M]. New York: John Wiley & Sons, Inc. 1992: 1.

Figure 11.9: Ian L. McHarg. *Design with Nature* [M]. New York: John Wiley & Sons, Inc.

1992: 144.

Figure 12.1 朱利斯 GY. 朱利斯·法布士 著，赵彩君等译. 土地利用规划 [M]. 北京：中国建筑工业出版社. 2007: 117.

Figure 12.2 朱利斯 GY. 朱利斯·法布士 著，赵彩君等译. 土地利用规划 [M]. 北京：中国建筑工业出版社. 2007: 118.

Figure 12.3: Fabos, Julius Gy. *The New England Green Vision Plan* [EB/OL]. http://www.umass.edu/greenway/1PO-index.html, 2008-03-20.

Figure 12.4 Walker, Peter. *Minimalism in Landscape* [C]. Simon Swaffield. *Theory in Landscape Architecture*. Philadelphia: University of Pennsylvania Press, 2002: 87.

Figure 12.5 Francisco Asension Cerver. *The World of Environmental Design* (Vol.1) [M]. Spain: Arco Editorial, 1994: 15.

Figure 12.6 Francisco Asension Cerver. *The World of Environmental Design* (Vol.1) [M]. Spain: Arco Editorial, 1994: 17.

Figure 12.7 Tom Richardson. *The Vanguard Landscapes and Gardens of Martha Schwartz* [M]. London: Thames & Hudson, 2004: 166.

Figure 12.8 Tom Richardson. *The Vanguard Landscapes and Gardens of Martha Schwartz* [M]. London: Thames & Hudson, 2004: 200.

Figure 12.9 Tom Richardson. *The Vanguard Landscapes and Gardens of Martha Schwartz* [M]. London: Thames & Hudson, 2004: 203.

Figure 12.10 Laurence King Publishing. *New World Landscape* [M]. London: Laurence King Publishing Ltd., 2004: 31.

Figure 12.11 Jane Brown Gillette. *Pure Design, Pure Image* [J]. *Landscape Architecture*, 1998 (10): 94.

Figure 12.12 Michael Spens. *modern landscape* [M]. London: Phaidon Press Inc. 2003: 225.

附录 Ⅲ Glossary

abet：教唆；煽动；帮助；支持 Unit 2 - B
abstract expressionism：抽象表现主义
　　　　　　　　　　　　　　　Unit 2 - A
abuse：辱骂 Unit 9 - C
accent：强调 Unit 8 - C
acclimatize：使服水土；（喻）使适应新
　环境 Unit 6 - C
accomplish：完成 Unit 7 - A
accrue：自然增加；产生 U12 - D
acknowledged：公认的 U10 - A
acoustically：听觉上；声学上 U12 - D
acquiescence：默许 Unit 4 - B
acre：英亩，1英亩＝0.405公顷 Unit 5
acreage：以英亩计算之土地面积；
　英亩数 Unit 5
acumen：敏锐 Unit 1 - C
adjoining：邻接的；隔壁的 Unit 8 - B
adversely：逆地；反对地 Unit 3 - C
aerosol：浮尘 Unit 7
aesthetic：审美的；有审美能力的；
　（对于艺术等）有高尚趣味的 Unit 5
aesthetics：美学 Unit 1 - C
aesthetics：美学；美术理论；审美学；
　美的哲学 Unit 8 - A
affiliation：联系 Unit 7 - B
affinity：密切关系；构造相似；关系；
　姻亲关系 U12 - B
affluent：富有 Unit 4 - C
aftermath：结果；后果；余波 Unit 6 - B
aggrandize：增加；夸大 U10 - A
aggregate：聚集；聚集成团 Unit 6
aggregation：总和 U10 - A
alchemy：魔术；炼金术 Unit 7
alight：落下 U12 - A
allee：（林荫）小径 U12 - B
allegorical：寓言的 U10 - C
alley：（花园或公园中的）小径 Unit 5 - B

allotment：所配得的一份；（尤指英国）
　租来作为菜园用的一小块公地 Unit 5
almond：杏树 Unit 8
Alpine：高山的；阿尔卑斯山的 Unit 8
ambiguity：含糊；不明确 Unit 2 - A
amble：（指人）骑马或走路缓缓而行
　　　　　　　　　　　　　　Unit 6 - B
ameliorate：改良 U11 - B
amoebic：阿米巴性的 U11 - D
amorphous：无定形的 U11 - D
analytic：分析的；用分析法的 U12 - D
anathema：极令人讨厌之事物 Unit 6 - B
anglophile：亲英派的人 U10 - C
animated：活生生的；活泼的；动的；
　愉快的 Unit 8 - A
animation：尤指活泼；有生气 Unit 5 - A
antagonism：对抗；敌对；对立 Unit 5 - B
anticipate：预期；期望 Unit 3 - C
antidote：解毒剂；抗毒药 Unit 5
antiquated：陈旧的 Unit 4 - B
antique：古老的 Unit 2 - B
antiquity：古代；古老；古代的习俗；
　古事；古迹；古物 Unit 6 - A
antiquity：上古 U11 - B
apex：顶点 U10 - D
apparatus：器械；设备；仪器 Unit 4 - B
appointed：指定的；约定的 U12 - A
approximately：近似地；大约 U10 - A
aquifer：蓄水层 Unit 7 - A
arbitrarily：武断的 U11 - C
arboretum：树园；植物园 U12 - D
arboretum：植物园 Unit 5 - A
arcade：[建]拱廊；有拱廊的街道 U10 - A
archaeological：考古学 Unit 4 - A
archaic：陈旧的 Unit 2 - A
archaic：古老的；陈旧的 Unit 4 - B
archery：射箭 Unit 4 - B

· 333 ·

archetype：原型	U11 - A	bar：沙洲；沙滩	Unit 6 - C
arid：干旱的；贫瘠的	U11 - C	bark：树皮	Unit 8 - B
arrogance：傲慢态度；自大	U12 - D	Baroque：巴洛克式	Unit 2 - C
Art Deco：艺术装饰派	Unit 2 - B	baroque：（指建筑等）装饰得过分的，	
artefacts：人工物	Unit 8	装饰得奇形怪状的	U12 - B
artifact：人工制品	Unit 6 - B	barren：单调的	Unit 9
artifacts：史前古器物	U10 - D	battering：用坏，损坏	U10 - C
artisans：工匠	Unit 4 - A	be apt to：易于	Unit 9
artistry：艺术之性质	U10 - C	be called away：被召到别的地方	Unit 5 - B
asbestos：石棉	U10 - B	Beaux Arts：巴黎美术学院派艺术	
ascertain：确定	Unit 7 - B		Unit 4 - B；Unit 8 - A
ash：岑树	Unit 8	beaux - arts：美术，艺术（包括绘画，	
asphalt：沥青	U12 - C	雕塑，建筑，音乐，舞蹈，	
aspiration：热望；渴望	U10 - C	木刻等）	U12 - C
assessments：评估	Unit 1 - A	beech：欧洲山毛榉	Unit 8
assimilate：吸收	Unit 2 - C	beholden：对……表示感谢	Unit 7 - B
assimilation：同化	Unit 2 - C	beleaguered：受到侵害	U11 - D
assimilation：同化；同化作用	U11 - C	belly：肚子；腹部	Unit 6 - C
assumption：假定；设想	U10 - C	belvedere：望景楼	U10 - C
astonishing：令人惊讶的；惊人的	Unit 4	bewitched：迷惑的	U11 - A
asymmetrically geometric：不对称几何形		bilaterally：双边	U11 - C
	Unit 2 - A	biomass：生物量	Unit 7 - A
asymmetry：不对称	U10 - C	biosphere：生物圈	U10 - C
attune：使调和；使一致；使适合	Unit 6 - A	bisected：平分	U11 - C
audacious：勇敢的	U11 - B	bistros：小酒馆	Unit 2 - B
auditory：耳的；听觉的	U10 - C	blocking：堵塞	Unit 8 - C
austere：（指人或其行为）严肃不苟的；		bombardment：袭击	Unit 7
严峻的；质朴的；朴素无华的	U12 - B	boredom：厌烦；厌倦	Unit 5
austere：严峻的，严厉的；操行上一丝		botanic：植物的；植物学的	Unit 5 - A
不苟的；简朴的	U10 - B	botanic：植物学的	Unit 5
austere：严厉的，简朴的	U12 - B	boulevard：大马路；大道（两旁常植	
authentic：真实的；可信的；可靠的	Unit 6	有树木）；林荫大道	Unit 5 - B；Unit 6 - C
authenticity：真实；真确；可靠	Unit 6 - A	bounty：恩赐	U11 - D
avant - garde：先锋派；前卫	U12 - B	Brazil：巴西	U10 - D
aviary - conservatory：大型鸟舍温室	U11 - A	brew：酿造	Unit 7
avish：丰富的	U11 - A	brickmason：砌砖工人，泥水匠	Unit 2 - C
awning：雨篷	Unit 9	broad - leaved evergreen：阔叶常绿树	Unit 8
awry：歪曲的；错误的	Unit 2 - A	bromide：套话	Unit 9 - C
axial：轴的；形成轴的	Unit 5 - A	brutalistic：野兽派	Unit 2 - A
backdrop：背景	Unit 8 - C	brutally：野蛮地；残忍地	Unit 6 - B
balm：香油；香膏；（喻）慰藉物；		bulldozers：推土机	Unit 1
安慰	U12 - C	bulrush：芦苇	Unit 6
balustrade：栏杆	U10 - C	bureaucratic：官僚	U11 - B

cacti：仙人掌（复数）	Unit 7 - A	compatible：兼容的	U11 - D
camouflage：伪装	Unit 8 - A	compatriot：同胞	Unit 7 - A
campground：野营地；露营场所	Unit 5 - C	compliance：依从；和……一致	Unit 1 - A
canopy：天篷；遮篷；这里指上层乔木		conceal：隐藏；隐匿；隐瞒	Unit 5 - A
	Unit 8 - C；Unit 9	concede：承认；让与；容许	Unit 6 - B
canvass：讨论，游说	U10 - C	conceit：想法	Unit 7 - B
capriciousness：多变性	U11 - B	concentric：同心	Unit 4 - A
caption：加上标题；加上说明	Unit 9 - A	concrete：建筑物	Unit 8 - A
carnage：大屠杀；残杀（人类）	Unit 6 - B	confederacy：州联盟；邦联	Unit 6 - B
carnations：荷兰石竹；康乃馨	Unit 8	conical：圆锥的；圆锥形的	Unit 8 - B
carnelian：玛瑙	U11 - B	conical：圆锥的；圆锥形的	U12 - D
carnivore：食肉动物	Unit 2	conifer：针叶树	Unit 8
carriageway：车道	Unit 5 - B	consciousness：意识；觉悟	Unit 7 - B
cascade：小瀑布；喷流	U10 - C	consequence：结果；影响	Unit 4
catalysts：催化剂	Unit 7 - B	conservancy：（自然物源的）保护；管理	
categorize：加以类别；分类	Unit 8 - B		U12 - A
cemetery：墓地；公墓	Unit 6 - B	conservation：保护；保存（以免损失、	
chaotic：无秩序的；混乱的；乱七八糟的		浪费、损坏等）	Unit 6
	U12 - C	conservative：保守的	U10 - B
chateau：城堡	U12 - B	console：安慰	Unit 9 - C
chateaux：城堡	Unit 1 - B	consolidate：巩固	U12 - B
cherry laurel：桂樱树	Unit 8	constellation：星座；星群	Unit 4
chic：时髦	U11 - A	constituent：要素	Unit 2 - B
chickadees：黑头山雀	U11 - D	contamination：污染	Unit 7
chronicle：编入编年史	Unit 7 - B	contemplative：冥想的	Unit 7 - B
chrysalis：蛹；茧	U11 - C	contextual：文脉上的；前后关系的	Unit 2 - B
circulation pattern：交通模式	Unit 9 - C	continuum：连续	Unit 4
civic：民事	Unit 4 - C	continuum：连续统一体	U10 - C
clamour：喧闹	U11 - C	contour：（海岸等的）轮廓、周线、围线；	
clarify：使清晰或易懂	Unit 9 - C	（地图、图案等）各着色区域间	
claustrophobia：幽闭恐怖症	U10 - B	的区分线	Unit 6
client：客户	Unit 9 - C	contour：等高线	Unit 9
clientele：客户	Unit 2 - C	contractor：承包人	Unit 2 - C
clutter：乱塞；乱堆	Unit 6 - B	contradictory：矛盾	Unit 2 - C
coexistence：和平共存	U12 - B	contrived：人为的	Unit 8
cognizant：认知的	Unit 7 - B	controversial：争议的	Unit 3 - C
cohesive：有附着力的；有凝聚力的	U12 - C	conventionally：常规地	Unit 4 - C
cohesive：粘着的	U12 - A	convivial：欢乐	Unit 4 - C
collide：互撞；碰撞	Unit 6 - B	cordon：用警戒线围住	Unit 2 - B
commercialism：商业主义；重商主义		cornerstone：基础	Unit 5 - C
	Unit 2 - C	cornucopia：希腊神话里象征丰富的	
commitment：承诺	Unit 7 - C	富饶羊角	U11 - D
compartmentalize：划分	Unit 7 - A	correlate：和……相关	Unit 9 - A

cosmic：宇宙的	Unit 7
cost-benefit：成本—收益	Unit 7-A
co-tenants：共同的居住者	U11-D
crackling：爆炸	Unit 4-C
cramped：局促	Unit 4-C
creed：信条	Unit 2-B
crescent：新月状的	Unit 2-A
crudity：生硬	Unit 2-A
crux：症结	U11-A
cubism：立体派	U10-C
cubist：立体主义	Unit 2-A
culmination：顶点；极点	Unit 5-B
cultivated：耕种的；耕植的	Unit 4
cumulative：累积的	Unit 2-C
curb cut：路缘缺口	Unit 7-C
curb：路缘	Unit 7-C；Unit 9
curio：希奇古怪而其价值即在此的艺术品；古董；古玩；珍品	Unit 6-B
curiosity：好奇心	U10-D
curiosity：好奇心	Unit 5-A
curriculums：课程	Unit 1-C
cypress：落羽杉	U11-C
cypresses：柏木属植物	Unit 8
dappled：有斑点的	Unit 8-C
Dark Ages：欧洲中世纪早期，大约在公元476～1000年	Unit 8
dearly：极；非常	Unit 6
debar：阻止；禁止	U10-C
debris：残骸	U11-B
debris：碎片	Unit 2-A；Unit 7-C
decadence：颓废	Unit 2-C
deciduous tree：落叶树	Unit 8-B
deciduousness：落叶的	Unit 4-B
decode：解译	Unit 7-B
deforestation：砍伐树木	Unit 7-A
deftly：巧妙地	Unit 2-B
degraded：退化的	Unit 1-C
deliberate：深思熟虑的	Unit 7
deliberately：故意地	Unit 7-C
delineate：描画；描绘；描写	Unit 3-C；Unit 6
demarcation：划分	Unit 7-A
demolition：拆除；推翻；毁坏；破坏	Unit 6-B
demolition：破坏	U11-B
demonstrate：阐明	Unit 7-C
depict：描述；描写	U10-D
depict：描述；描写	U10-D
deplore：悔恨	U11-D
deposition：沉积	Unit 7-B
deprave：败坏其道德；使腐败	Unit 5
depression：美国经济大萧条时期（1929～1933年）	Unit 5-C
derelicts：流浪汉	Unit 4-C
designation：名称；称号	Unit 5-B
despoliation：掠夺性	U11-D
deStijl：荷兰风格派（建筑）	Unit 2-A
desultory：杂乱的	U10-A
deterioration：退化	U11-D
detriment：损害；伤害	Unit 6
devastate：毁坏；破坏；使荒凉；使成废墟	Unit 6-B
devoid：缺乏的	Unit 7-B
dialectic：辩证的	Unit 7-B
dichotomy：二分法	Unit 2-C
dictate：指示	Unit 7-A
dilation：扩张	U11-C
dim：暗淡的；悲观的	U10-D
diminutive：较通常为小的；小得多的	U12-C
disconnect：使分离	Unit 7-C
discourse：论题	Unit 7-B
discretionary：任意的；自由决定的	Unit 3-C
disdain：蔑视	Unit 1
disillusion：醒悟	Unit 2-B
disjunctive：造成分离的	U11-C
disorientation：方向知觉的丧失；迷惑	U10-C
disparate：全异的	Unit 7-B
disseminated：散布	U12-A
distinguishable：可区分的	Unit 7-A
disturbing：烦人的	Unit 7
diversify：使多样化	Unit 7
dogma：教条	Unit 2-C
dogwood：山茱萸	Unit 8

donor：赠与者；捐赠者	Unit 5 - A
drab：（喻）乏味的；单调的	Unit 5
drainage pattern：排水模式	Unit 9 - A
drainage：排水设施	Unit 4 - A
dramatics：业余演出	Unit 4 - B
drastic：（指行动、方法、药品等）激烈的；猛烈的	Unit 6
dreary：沉闷的	Unit 4 - B
dredge：捞泥机；挖泥机；捞网（自水底捞泥，牡蛎，标本等所用者）	U12 - C
dredge：挖掘；疏浚	Unit 3 - C
dross：（金属熔化时浮升至表面的）渣滓；无用之物	U12 - D
drudgery：辛苦而令人讨厌的工作	Unit 5
Druid：德鲁伊教团员	U11 - A
dubious：（指事物、动作等）可疑的；其价值、真实性等有问题的	Unit 6
dwindling：渐渐变小；缩小；减少；衰落；失去重要性	U12 - B
dyke：堤	Unit 6 - C
dynamism：活力	Unit 8 - C
easel：画架	Unit 2 - C
eccentricity：古怪	Unit 2 - B
eclectic：折衷的，折衷学派的	Unit 8 - A
eclectic：折衷的；折衷学派的	U10 - C
eclecticism：折衷主义	Unit 2 - C
edaphic：土壤的	Unit 9 - C
efficacious：（不用以指人）有效的	Unit 6 - A
egocentrism：中间路线	Unit 2 - C
egress：出口	Unit 7 - C
elaborate：精心制作；详细阐述	Unit 9
elevation：起伏；小山；高处；高地	Unit 5 - B
eligible：符合条件的；合格的	Unit 3 - C
eligible：合适的	U11 - D
elite：精英，杰出人物	Unit 2 - C
elm：榆树	Unit 5 - B；Unit 8
embellishment：装饰；修饰；润色	Unit 8 - A
embodiment：能具体表现他物者；化身；被具体表现者	Unit 6 - A
embody：体现	Unit 7 - B
embosom：围绕；围护	U10 - A
empathy：移情作用	U10 - C
emulate：仿效	U12 - A
emulation：竞争；争胜	U12 - C
enamored：使人迷恋的	U11 - D
encircle：环绕；包围	U12 - C
enclave：被包围的领土	Unit 2 - B
encroach 超出正当范围；侵入；侵害	Unit 6 - B
endeavour：尽力；竭力	Unit 2 - B
engender：造成	Unit 7 - B
English School of Landscape：英国自然风景园林流派	Unit 8
enliven：使活泼；使有生气	Unit 5 - A
enmesh：陷入	Unit 7 - B
entity：实体；本质	Unit 4
environed：环绕	Unit 4 - B
environs：郊外；近郊	Unit 6 - B
envisage：正视	U10 - C
envisioned：构想	Unit 1
eon：无限长的时代	Unit 7
ephemerality：短暂；瞬息	Unit 6
epic：描写英雄事迹的诗；史诗；叙事诗	Unit 6 - C
episodical：插曲式的	Unit 5 - A
equilibrium：平衡，均衡	Unit 2
equilibrium：平衡；均势	U12 - B
equipoise：平衡，均衡	Unit 2
erratically：不规律地	Unit 2 - C
erupts：喷发	Unit 4 - C
escapism：逃避现实主义	Unit 2 - C
Esperanto：世界语	Unit 2 - B
eternity：永远；不朽	U10 - C
evaluation：评价	Unit 9
evapotranspiration：水分蒸发	Unit 7 - A
eventually：最终	Unit 7 - C
excavation：挖掘；发掘	Unit 3 - C
excavations：挖掘	U11 - A
exhaustive：彻底的；详尽的	Unit 3 - C
exodus：许多人离开；很多人离去	Unit 5
exotic：外来的	Unit 5 - B
exotic：异国情调的；外来的	Unit 8
expenditures：支出	U12 - A
explicit：（指言语等）明白表示的；	

明确的	U12-C	fortitude：坚韧	U11-C
exterior：外部；表面	U10-C	fossiliferous：含有化石的	Unit 4
extracted：选录；摘取	U11-A	fragile：虚弱的	Unit 7-B
extramural：围墙外	Unit 4-A	frank：率直的	Unit 4-B
extrapolate：推断	U12-C	freight trucks：货车	Unit 1
exuberant：茂盛的；繁茂的；丰富的	U12-C	fringe：边缘	Unit 4-C；Unit 7-B
fabric：结构；构造；构造物	Unit 6	frivolity：无聊的举动	Unit 2-C
facade：正面	Unit 8-A	fundamental：基本原则；基本原理	U10-B
fallacious：不合理的	Unit 4-B	furnace：火炉；熔炉	Unit 5
Far East：远东地区	Unit 8	furrow：犁沟；畦	Unit 6-A
fauna：动物群	Unit 1-A	fusion：融合	U11-D
feasibility studies：可行性研究	Unit 9-A	futile：徒劳	Unit 4-C
feasibility：可行性	Unit 9-A	gargoyle：怪兽状滴水嘴，（突出的）	
felt-tip pen：毛毡笔	Unit 9-C	怪兽饰	U10-C
fern：蕨类植物	Unit 8	gauge：测量	U11-A
fervently：热情的	Unit 2-A	generalist：通才	U11-A
fervently：热情的	Unit 2-A	genial：亲切的	Unit 2
feudalism：灭亡	Unit 2-C	geological substructure：地质基础	Unit 9-A
filthy：不洁的；污秽的	Unit 5-B	geranium：天竺葵	Unit 1
flag：枯萎，衰退	Unit 1	germinal：初期的	Unit 4-B
flexible：灵活的	Unit 4-B	gimmicky：巧妙手法的	U12-C
flint：燧石	Unit 2-C	glaciation：冻结成冰	U12-A
flirtation：调情；挑逗	U12-D	gopher：囊地鼠	Unit 1
flogging：鞭打	Unit 2-B	gophers：土拨鼠	U11-A
flora：（某地区或时代的）植物的总称；		gorge：障碍物	U11-B
植物区系	U12-B	Gothic：哥特式的	Unit 2-C；Unit 5
flora：植物群	Unit 1-A	grandeur：伟大；壮丽；华丽	U12-B
fluctuating：上下波动	Unit 2-A	granite：花岗岩	U11-B
fluid：可以改变的	U11-C	granite：花岗岩；坚硬的	Unit 4
fluidly amorphous：自由曲线形	Unit 2-A	graphic：书写的；绘画的；（指描写）	
fluorescent light：明亮的荧光色；		生动的	U12-C
荧光灯	Unit 4-B	grate：笆子	Unit 7-C
foetus：胎儿	U10-C	gratifying：欣喜	Unit 6
foliage：树叶	Unit 1	grazing：放牧	Unit 5-C
foliage：树叶；植物	Unit 8-B	greensward：草皮	Unit 6
foot：英尺，1英尺=0.305米	Unit 5-B	grid：格子；栅格	U12-B
footboard：踏板	Unit 5-A	gridiron：网格状	Unit 5-B
foray：袭击	Unit 7-B	gridlock：高压封锁	Unit 4-C
foreground：前景；最显著的位置	U10-B	grizzly bear：灰熊	Unit 5-C
formal patterns：规则式图案	Unit 8	grooved rush：（植）灯芯草	Unit 7-C
formidable：难以克服的；艰难的	Unit 6-C	grope：探索；摸索	Unit 2
forms：[计]窗体	Unit 8-B	grotesquery：怪诞；古怪	Unit 2-C
forsaken：被抛弃的	Unit 2-A		

grotto：洞穴（尤指人造的花园中的洞室）		iconic：雕像的；圣像的；依照传统	
	Unit 6 - A	形式的，因袭的	U12 - B
grotto：洞穴；岩穴；人工洞室	U10 - C	ideology：意识形态	Unit 7 - B
grove：丛树；小树林	Unit 6 - A	idyll：田园景色	U11 - D
guarantee：保证	Unit 4 - B	imbue：使……浸染（颜色等）；影响	
gusto：爱好；趣味	U10 - D	（感情等）	U12 - B
hackneyed：陈腐的；常见的	U10 - C	immediacy：直接（性）；目前；密切；	
hag：女巫；老丑婆	Unit 6	即刻	U12 - C
hanker for：渴望；追求	Unit 2	immerse：沉浸	Unit 7 - B
haphazardly：无规则地	Unit 1	immune：免疫的	Unit 7 - A
harbour：港	Unit 6 - C	impervious：不可渗透的	Unit 7 - C
hawthorn：山楂	Unit 8；U12 - B	impetuous：冲动的；猛烈的	Unit 7
hazardous：危险的；冒险的；碰运气的		implementation：执行	Unit 1 - C
	U10 - A	impound：关在栏中	Unit 2 - C
headwaters：（河川的）源流	Unit 5 - B	impoverish：使穷困；去除优点	U12 - B
hectare：公顷	Unit 5 - A	impracticable：不可行的	Unit 5 - B
heed：注意	U11 - C	imprecise：不严密的	Unit 2 - B
heedless：不注意的	Unit 7	imprecision：不严密；不精确	Unit 5 - B
herbaceous：草本的	Unit 9 - A	Impressionism：印象派	Unit 2 - B
hickories：山胡桃树	Unit 8	impressionistic：印象主义的	Unit 2 - A
hickory：（北美所产之）山胡桃树；		impulse：推动	Unit 7 - B
山胡桃木	U12 - B	inadequacy：不充分	Unit 4 - B
high-tech：高技派	Unit 2 - A	inasmuch：由于；因为	Unit 5
hodgepodge：杂乱的一堆东西；		incentive：动机	Unit 7 - A
杂混在一起的东西	Unit 6 - B	inconsequential：不合逻辑的；	
Holm oak：圣栎（冬青栎）	Unit 8	不合理的	Unit 2 - C
homage：尊敬；敬意；尊崇	U12 - C	inculcate：劝导	Unit 7 - B
homo erectus：直立人	Unit 2	indicate：显示	Unit 7 - C
homogeneity：同质性	U11 - C	indigenous vegetation：乡土植物	Unit 9 - A
horticultural：园艺的	Unit 8 - B	indigenous：（尤指动植物等）当地	
horticulturist：园艺家	Unit 2 - C	出产的；土产的	Unit 5 - B
hostility：敌意；恶意；对抗；反对	Unit 2	indigenous：本土的	Unit 2 - C
hub：中心	U12 - A	inexhaustible：无穷无尽的	Unit 8 - A
hue：色调；样子；颜色；色彩	Unit 8 - B	infiltrate：渗透	Unit 7 - C
humanism：人文主义	Unit 8	infinitesimal：微小的	U11 - C
hurricane：飓风	Unit 6 - C	influx：流入；注入	Unit 6
hyacinth：风信子属	Unit 8	informal：不规则的	Unit 2 - C
hybridization：杂交；杂种培植	Unit 8	infuse：鼓舞	Unit 7 - B
hydrologic：水文的	Unit 4；Unit 7 - C	ingenious：设计精巧的	U11 - B
hydrological：水文学的	Unit 9 - C	ingenuity：独创性；精巧；灵活性	U10 - C
hydrologist：水文学者	Unit 1 - C	ingrained：（指习惯，癖性等）深染的；	
hydrology：水文学	U11 - A	根深蒂固的；彻底的	U12 - C
hypersensitive：非常敏感的	Unit 1	ingredient：成分	Unit 7

英文词条	位置
initial：开始的；最初的	Unit 9 - C
inlet：湾；插入物；镶入物	Unit 6 - C
inorganic：无机的	U12 - C
inquiry：调查	Unit 7 - B
instigate：鼓动	Unit 7 - B
instinct：本能	Unit 2
insurance：保险	Unit 3 - C
intangible：无法明了的	Unit 7 - A
intangible：无形的	U12 - A
intangible：无形的	Unit 4 - B
integral：不可或缺的；作为组成部分的	Unit 5
intended：有意的	Unit 7
intense：剧烈的	Unit 7 - C
interdisciplinary：各学科间的	Unit 3 - C
intermediary：中间物；媒介	U10 - C
intermingle：混合	Unit 7 - B
interpretation：阐述	Unit 7 - B
interpretative：解释的	Unit 7 - C
interspersed：点缀，散布	U11 - A
intertwine：（使）交织	U11 - C
interventions：干涉	Unit 7 - B
intimate：亲密的，私密的	Unit 8 - C
intrigue：谋划；考量	Unit 7 - C
intriguing：迷人的；有迷惑力的；引起兴趣（或好奇心）的	U10 - B
intrinsic：（指价值或性质）固有的；内在的	Unit 6
intrinsic：本质的	U11 - C
intrinsic：内在的；本质的	U12 - A
intrinsic：实在的，本质的	U11 - C
intrusion：闯入；侵挠	Unit 6
intuit：凭直觉知道	U11 - C
inundation：泛滥；淹没；洪水	Unit 6 - C
inventory：清查；对能力、资产或资源的评估或调查	Unit 9
invisibility：看不清	Unit 7 - B
ironic：讽刺的	U11 - D
irrational：无理性的	Unit 2
irretrievably：不可挽回地；不能补救地	Unit 6 - A
irreverent：不虔敬的；不恭敬的	U12 - C
irreversible：不能逆转的	Unit 7
issue：问题	Unit 4
juniper：[植] 刺柏属丛木或树木	U10 - B
juxtapose：并列；并置	U12 - B
juxtaposition：毗邻；并置	U10 - D
kettle：水壶	Unit 2 - B
laboriously：费劲地	Unit 1
laburnum：金链花属	Unit 8
lacquered：漆器	U11 - A
landfill：垃圾填埋场	Unit 1 - C
landform：地形	Unit 9 - C
larval：幼虫	U11 - C
latent：潜在的	U11 - 2；C
layoff：失业	Unit 1 - C
layout：设计	Unit 1 - A；Unit 8
leap：跳跃	Unit 7
legibility：（指字迹、印刷物）易读的；清楚的	Unit 6 - B
legislation：立法；法律的制定（或通过）	Unit 3 - C
legitimacy：正确（性）；合理（性）	U12 - B
lethal：致命的	Unit 7
levee：防洪堤	Unit 7 - A
liberalism：自由主义	Unit 8
lilacs：丁香属	Unit 8
limestone：石灰石	U11 - D
linear：线形	Unit 4 - C
linkage：联接	U12 - A
litany：枯燥重复的连续不断的叙述	Unit 2 - A
livability：宜居性	Unit 7 - C
loom：隐约地威胁性地出现	Unit 6 - B
luxuriant：丰产的；丰富的；肥沃的；奢华的	U10 - D
luxuriant：富饶的	U11 - C
luxurious：奢侈的；豪华的	Unit 4 - B
M.I.T：麻省理工学院	Unit 4 - B
macrocosm：宏观世界	U10 - C
magnitude：巨大	Unit 7
mainland：大陆	Unit 6 - C
malleable：（金属等）可锻的；可延展的；能适应的，顺应的	U12 - C
mammals：哺乳动物	Unit 2
manifestation：显示；表现	U11 - C

manifesto：声明	U11－D
mannerism：（言语、写作中的）特殊习惯；怪癖	U11－C
mantle：覆盖物	Unit 1
marginal：边际土地	Unit 5－A
marginalize：忽视、排斥	Unit 7－B
marionettes：牵线木偶	Unit 4－B
marvelous：绝妙的；了不起的	Unit 2
masonry：石工	Unit 4－C
matrix：矩阵	Unit 8－B；U11－D
matte：不光滑的	Unit 8－B
maturity：成熟	Unit 4－B
mechanization：机械化	Unit 8
mediate：仲裁	Unit 7－B
Medieval：中世纪	Unit 4－A
memorable：难忘的	Unit 7－B
menace：威胁	Unit 4－C
mentality：心智	Unit 7－A
merit：优点	Unit 7－B
mess：散乱；杂乱	Unit 5－B
metaphysical：形而上学的；纯粹哲学的；超自然的	U10－C
metropolis：大城市	Unit 4－B
microcosm：被认作人类或宇宙之缩影的某事物（尤指人）；小天地	Unit 6－A
microcosm：微观世界	U10－C
mile：英里，1英里＝1.609千米	Unit 5
milieu：环境	Unit 1
millennium：千年	Unit 4－A
mimic：模仿	Unit 2－A
mineral extraction：矿物开采	Unit 9－C
minnow：鲦鱼（一种小淡水鱼）	U11－A
misinterpret：误译；误解；误以为	Unit 5－B
misinterpretation：误译；误解；误以为	Unit 6
mitigate：减轻	Unit 8－C
mitigate：使缓和；使减轻；使镇静	U12－C
moat：护城河；城壕	U10－C
mockup：模式	U12－D
modulation：韵律的和谐的运用	Unit 8
module：模数；模块	U12－B
monarchs：君主政体；君主政治；君主国	Unit 2
monastery：修道院；僧侣	U12－B
monoculture：单一栽植	Unit 7－A
monoculture：单作	U11－C
monolithic：单路	Unit 7－B
monolithic：独石的；似独石的	U12－B
monotonous：单调的；无变化的	U10－A
monotony：单调；无聊	Unit 5－B；Unit 8
morphological：形态学的	U11－B
motif：（艺术作品的）主题；主旨	Unit 6－A
motivate：推动	Unit 7－B
mound：土墩；护堤；垛	U12－C；D
municipal：市的；市政的；自治城市的	Unit 5－A
murderous：危险的	Unit 1
musing：冥想的	U11－A
mutable：可变的；易变的；不定的	U12－C
mutation：变化；突变	U10－D
myriad：无数	Unit 4
mysteriously：神秘地	Unit 7
mystic：神秘主义者	U11－A
mysticism：神秘主义	Unit 2－A
myth：神话	U10－C
mythical：神话的；仅存在于神话中的	Unit 6－A
mythology：神话学；神话的总称	Unit 6－A
Neoclassic：新古典主义的	U12－C
Neoclassicism：新古典主义	Unit 2－B
Neo－Moderns：新现代主义	Unit 2－B
neutralize：抑制	Unit 7－A
nondescript：无特征的	Unit 9
nonsupervisory：非管理的	Unit 1－C
northerly orientation：指北针	Unit 9－A
nostalgia：思乡病；乡愁；留恋过去；怀旧	U12－C
novelty：新颖	Unit 5－A
nuance：细微差别	U10－C
nudged：推动	U11－A
nuthatches：五子雀	U11－D
obfuscating：模糊	U11－B
objecthood：客观事物	Unit 7－B
objectification：对象化；客观化	Unit 7－B
obligation：义务；债务	Unit 7
obliterate：涂去；使湮没	Unit 9－C

英文：中文	位置
oblivion：被完全遗忘的状态；淹没	Unit 4 - B；Unit 5
obsolete：过时	Unit 4 - C
occupational：职业的	Unit 4 - B
octagon：八边形；八角形	Unit 2 - C
offensive：讨厌的；无礼的；攻击性的	Unit 8 - A
offices：事务所	Unit 9 - C
off-site：外部场地	Unit 2 - A
oleander：夹竹桃	Unit 6 - C
opportunistic：机会主义的	Unit 7 - B
opportunity：机会；有利环境	Unit 4
optimum：最适宜的	Unit 4 - B
oracle：神谕	U11 - C
orator：演讲者；雄辩家	Unit 1
orchestra：管弦乐队	U12 - D
organisms：有机体；生物体	Unit 4
ornament：装饰物；装饰	U10 - D
orthogonal：直角的	U11 - D
osprey：鱼鹰	U11 - D
outcrop：（矿脉）露出地面的部分；露头	Unit 4
outmoded：过时的；不流行的	U12 - C
outworn：用旧的；废弃的	Unit 8 - A
overarch：（在……上面）造成拱形	Unit 5 - B
overgraze：过度放牧	Unit 7 - A
overhaul：追上；赶上	Unit 6 - B
overlays：叠层	Unit 9 - A
overriding：最重要的；高于一切的	Unit 2 - C
overstuffed：（指坐位等）用厚垫使柔软而舒适的；填塞极多东西的	U12 - B
palette：调色板	U12 - C
palpable：明显的	Unit 7 - B
pamphlet：小册子	Unit 5 - B
panache：羽饰；华丽	U12 - D
panoramic：全景的	Unit 9 - A
parakeet：[鸟] 长尾小鹦鹉	Unit 1
parallels：轨道	Unit 2 - A
parameter：参数；参量	U12 - B
parameters：参数	U11 - C
paramount：最重要的	Unit 8
parkway：公园路；风景大道；休闲路	Unit 5 - B
parterre：（庭园之）花坛	U12 - C
parterre：花坛；花圃	Unit 8
partial：部分的；不完全的	Unit 6
participate：参加	Unit 4 - B
partition：隔离物	Unit 4 - B
passer-by：过路人；行人	Unit 5 - A
pastel：柔和的	Unit 9 - C
pastoral：牧歌，田园诗	Unit 7 - B
pastoral：田园诗；牧歌	Unit 6 - B
patently：明显的；显然的	U12 - C
patron：资助人	Unit 7 - B
pavement：人行道；铺过的道路	Unit 4
paver：砌块	Unit 7 - C
pavilion：（尖顶）大帐蓬；（公园、庭园等的）亭子	Unit 5 - A
pedestal：柱基；塑像或艺术品的座	Unit 5 - A
pedestrian：步行的	Unit 9 - C
pedestrian：步行者	U11 - B
pedestrian：步行者；行人	Unit 1 - B；Unit 4 - C；Unit 5
pelican：塘鹅	U11 - A
penetrate：渗透	Unit 7 - C
perceptions：有理解的	Unit 2
perennial：四季不断的；终年的；长期的；永久的；（植物）多年生的	U10 - D
perforated：有孔的或多孔的	Unit 4 - B
pergola：（藤本植物的）棚架；藤架；绿廊；凉棚	Unit 4 - B
periphery：外围；表面	U12 - C
periphery：外围；表面	Unit 5 - A
permeate：渗透；渗入	Unit 4
perpendicular：垂直的；正交的	Unit 2 - A；Unit 7 - C
perpetuate：使不朽	Unit 7 - B
perpetuity：永恒	U12 - A
persistence：坚持	Unit 7 - A
pertain：（与 to 连用）属于；关于；适合于	Unit 6
pertinent：有关的	Unit 9 - A
petri：培养皿（实验室用于培养细菌等的有盖小玻璃盆）	Unit 1

英文	中文	位置
physiography	地貌学	U11 - A
physiological	生理学的	Unit 1
physiology	生理学	Unit 2
piazza	广场，走廊，露天市场	Unit 2
picnic grills	野餐烧烤屋	Unit 4 - B
picturesque	独特的	Unit 8 - A
piecemeal	一个个地	Unit 1
pilgrimage	朝圣者的旅程	Unit 6 - B
plage	海滩（尤指在时髦的海滨游乐地）	Unit 6 - C
plague	困扰	Unit 4
plasticity	可塑性	Unit 4 - B
plural	复数的；多于一个的	Unit 2 - B
pluralism	多重性	U11 - B
pluralist	兼职者	Unit 2 - B
poachy	（土地）易被牲口踩得泥泞的，湿而松软的	U10 - A
poised	泰然自若的；平衡的	U12 - A
polyanthus	西洋樱草	Unit 8
porous	能渗透的；多孔的	Unit 8 - C
potentiality	潜力；可能性	Unit 8 - A
pottery	陶器	U11 - C
practitioner	从业者	U12 - C
pragmatist	实践主义者	U11 - A
prairie	大草原；牧场	Unit 8
prairie	牧场	U11 - A
precipitates	沉淀物	Unit 4 - C
precipitation	降水	Unit 7 - A
precisely	正好	U10 - D
preconceptual	预概念的；概念前的	Unit 7 - B
preeminence	卓越	U12 - D
preliminary	初步的	Unit 1 - C
preoccupation	当务之急	Unit 2 - A
preparatory	初步的	Unit 6 - C
preponderance	优势；占优势	Unit 4 - B
preposterous	荒谬的	Unit 2 - B
preservation	保护；贮藏；维持；留存	Unit 6
prettification	美化，装饰	Unit 9 - C
prevailing winds	盛行风	Unit 9 - A
primeval	原始的	Unit 2 - C
primordial	原生的；原始的；最初的	Unit 6 - A
primroses	报春花属	Unit 8
privilege	给与特权	Unit 7 - B
probe	探查	Unit 7 - B
proclamation	宣言；公布	Unit 6 - C
prod	以尖物推或刺	Unit 6 - B
prodigious	巨大的	Unit 1
proficient	精通	Unit 1 - C
promenade	（为运动或散心所做的）散步；骑马；散步或骑马的地方	Unit 5 - B
prompt card	提示卡片	Unit 9 - C
proportion	比例；均衡；面积；部分	Unit 8 - B
protestation	声明；主张	Unit 2
province	研究范围；管辖范围	Unit 2
provision	准备；防备（尤指为未来的需要者）	Unit 6
provisions	规定	Unit 3 - C
provocative	激怒的；引起议论，兴趣等的；激起的；刺激的	U12 - C
proximity	接近	Unit 3 - C
proximity	接近	Unit 5
psychic	灵魂的；心灵的；（指力量）非自然的；非物质的	U12 - C
purple finches	紫雀	U11 - D
purple	紫色的	Unit 8 - B
purports	意谓；声称；声言	Unit 6
quadrangle	四角形；四边形；方院	U12 - D
quadrant	象限；一个圆或其圆周的四分之一	U12 - C
qualitative	性质上的；定性的；质量上的	Unit 4 - B
quantifiable	可以计量的	Unit 8 - B
quarry	采石场	Unit 5 - A
quay	码头；横码头	Unit 6 - C
quintessential	精粹的；典范的	U12 - B
radical	激进的	Unit 2 - C
ramification	结果	Unit 9 - C
ramp	坡道	Unit 9
ramparts	垒	Unit 4 - A
ranger	（美）森林看守人员	Unit 6 - B
ransack	搜索	Unit 2 - A

rasping：锉磨声的；令人焦躁的	Unit 8 - A
rational：推理的	Unit 2 - B
rationale：（某事物的）基本理由；	
理论基础	U12 - C
rationality：理性观点；理性实践	U11 - C
rationalization：合理化；合于经济原则	
	Unit 2 - C
ravines：沟壑；峡谷	Unit 1
raze：彻底破坏（城市，建筑物）；	
尤指夷为平地	Unit 6 - B
real‑estate：不动产	Unit 5 - B
recess：（墙壁的）凹室；壁龛；	
隐密地方；难进入的地方	Unit 5 - A
recession：经济不景气	Unit 1 - C
recognition：共识	Unit 7 - A
reconsideration：再考虑	Unit 7 - B
reconstruction：重建；再建	Unit 6
rectify：矫正；调整	U10 - D
recuperation：复元；休养；恢复	Unit 5
red‑tailed hawk：红尾鹰	U11 - D
redundant：多余的	Unit 9
reef：礁；暗礁	Unit 6 - C
refinement：精致；（言谈；	
举止等的）文雅；精巧	Unit 8 - B；Unit 9
refurbishment：整修	Unit 1 - C
rehabilitation：恢复；修复；恢复原有	
地位、正常生活等	Unit 1 - B；Unit 6 - B
reinforce：加强；增援；补充；增加……	
的数量；修补；加固	Unit 8 - B
reinstatement：使复原位或原状；	
恢复	Unit 6
reinstitute：重新建立	Unit 2 - A
reiteration：重复；反复	Unit 9
relegate：转移；转入	Unit 7 - B
relics：遗迹	Unit 2 - A
reluctant：不顾的；勉强的；	
难得到的；难处理的	U10 - A
rely：依靠	Unit 5 - C
remediation：纠正	Unit 1 - C
reminiscent：回忆往事的	U10 - C
remote：很久以前的	Unit 4
renaissance：文艺复兴时期	U11 - C
Renaissance：文艺复兴时期，大约	
在公元14～16世纪	Unit 1 - B；Unit 8
renewable：可更新的	Unit 7 - A
renewal：更新；再始；换新	Unit 5；Unit 6
reoccur：重复出现	U10 - C
repertoire：节目；全部技能	U10 - C
repose：休息；使休息或依靠；信赖	U12 - B
reproduction：繁衍	Unit 7 - A
requisite：需要的；必不可少的；	
必备的	Unit 3 - C
resembling：类似	U11 - A
reservoirs：水库	U11 - A
residential subdivision：居住小区	Unit 9 - A
resonance：共鸣；反响；中介	U10 - C
restrictive：限制性的	Unit 7 - B
resurface：铺（路等）之新表面；	
换装新面	Unit 6
resurgence：复兴	Unit 2 - A
reticent：沉默寡言的	U12 - C
retrofit：式样翻新	Unit 7 - A
reversible：可逆的	Unit 6
rhythm：节奏；韵律	U10 - B
rhythmically：有节奏地；有韵律地	Unit 6 - A
ridge：山脊	U12 - A
rigor：严格；严厉；严厉执行	U12 - C
rigorously：残酷地	Unit 7
ritual：仪式；典礼	U11 - A
ritual：仪式；典礼；宗教仪式	
	Unit 6 - A；Unit 7 - B
rivalry：竞争；敌对	Unit 5 - C
rodent：啮齿动物	Unit 7
rostrum：讲台；讲坛	Unit 2 - B；Unit 6 - B
rugged：崎岖的	U12 - A
runoff：径流	Unit 7 - C
ruse：计策；策略；谋略	U12 - C
rustic：有乡村居民特色的；朴素的	Unit 5
rustic‑style：乡村风格	Unit 5 - C
sage：贤人；圣人	Unit 1
salesmanship：推销术，说服力；	
销售，推销	Unit 8 - A
salubrious：有益健康的	Unit 1
sanctuary：避难所	U12 - D
sanctuary：动物保护区	U11 - A
sand dune：沙丘	Unit 9 - C

sanitary：清洁的，卫生的	Unit 5	spatial：空间	Unit 4 - C
sanitation：卫生	Unit 4 - A	specification：说明书	Unit 9 - A
sassafras：檫木	Unit 8	spectrum：光；光谱；型谱	Unit 8 - B
savannah：大草原	Unit 2	spillway：溢洪道；泄洪道	Unit 2 - C
scavenge：提取	Unit 2 - A	spine：脊椎；脊柱	Unit 5 - A
scrub：矮树	Unit 6 - C	splash：一种强烈但是短暂的印象	U11 - C
scrutiny：详细审查	Unit 8 - B	spontaneity：自发性	U11 - C
secluded：（尤指地方）安静的；		squirrels：松鼠	U11 - D
幽僻的	Unit 5	stadia：（复数）露天大型运动场	Unit 4 - B
sediment：沉积物	Unit 7 - C	stagnant：停滞的；迟钝的	Unit 2 - C
sedimentation：沉积作用	U11 - D	stance：（高尔夫、板球）击球时所取的	
seedling：秧苗；树苗	U10 - C	姿势	U12 - C
segregate：隔离	Unit 8 - A	stature：身高；身材；（精神、道德等的）	
self - mutilation：自取毁灭的		高度	Unit 8 - B
（生活习惯）	U11 - D	statute：法令；条例	Unit 3 - C
seminal：种子的	U12 - A	stereoscopic：有立体感的	Unit 2
sensual：感觉的	Unit 8 - C	sterile：缺乏想象力的；缺乏新意的	Unit 2 - C
sentiment：情感；情绪	Unit 2 - B	sterility：思想贫乏	Unit 2 - C
sequoia：巨杉；红杉	Unit 5 - C	steward：管理员	U11 - D
serene：晴朗的；宁静的	U12 - B	stewardship：工作	Unit 7 - C
severity：严肃；严格；严重；激烈	Unit 3 - C	stilt：高跷	Unit 6 - C
sewage：污水	U11 - D	stockpile：存储	Unit 7 - A
shedding：蜕落	Unit 2 - B	straddle：跨骑	Unit 2 - A
sheer：全然的；纯粹的	Unit 4 - B	strata：stratum 的复数，[生]层	
shimmer：微光	Unit 4		Unit 4；Unit 8 - C
shoddy：劣质的；冒充好货的	U12 - C	stressful：紧迫的	Unit 7 - A
signage：标识	Unit 7 - C	strontium：锶	Unit 7
silhouette：侧面影象；轮廓		stucco：（粉饰墙壁用之）灰泥	U12 - B
	Unit 8；Unit 8 - B	stumble：绊倒；使困惑	Unit 2 - C
silviculture：造林术；森林学	Unit 4 - B	subconscious：下意识的	U10 - C
simulate：模拟	Unit 7 - C	subdivisions：细分	Unit 4 - A
simultaneous：同时的	Unit 7 - B	subjugate：使屈服；征服；使服从；	
simultaneously：同时地	Unit 9	克制；抑制	Unit 8 - B
sinister：险恶的	Unit 7	subordinate：次要的；从属的；	
site - dominated：受控场地	Unit 2 - A	下级的	U10 - A
site - specific：特定场地	Unit 2 - A	subordination：下级的；次要的；	
slight：轻微的	Unit 7	附属的	Unit 6 - A
solicit：恳求	Unit 3 - C	subservient：有用的	Unit 7 - B
solitude：独居；单独；孤独	Unit 5 - A	subsidiary：次要的；附属的；辅助的	
soothing：抚慰的；使人宽心的	Unit 8 - B		Unit 5 - A；Unit 9 - C
sophisticated：诡辩的；久经世故的	Unit 8 - B	substantially：相当大地	Unit 8 - C
sordid：肮脏的	Unit 8 - A	suggestively：暗示地；提醒地；	
sparse：稀少的；稀疏的	Unit 8 - B	引起联想地	Unit 6 - C

sumptuous：华丽的	U11 - A
superb：庄重的；华丽的	U10 - C
superbly：雄伟地；壮丽地	U10 - C
superimpose：添加；附加	U12 - C
superintendent：监督（者）；	
管理（者）	Unit 5 - B
supersede：代替；取代	U10 - C；U11 - D
supersede：代替；取代；接替	Unit 2
surrealism：超写实主义	Unit 6 - C
swamp：湿地	U11 - C
sycamor：小无花果树	Unit 8
symbolic：象征的；符号的	U10 - C
symbolism：象征主义	Unit 2 - A
symmetrical：对称的；均匀的	U10 - B
symmetry：对称；匀称；调和；	
对称美	Unit 5 - A
syntax：句法；语法	Unit 7 - B
synthesis：综合	Unit 2 - B
synthetic：合成的	Unit 7
tactics：策略	Unit 7 - B
tandem：串联	Unit 7 - C
tapestry：交织	Unit 7 - B
technics：工艺	Unit 4 - B
technocratic：技术统治论者	Unit 2 - B
tenet：原则；主义；信条；教理；教条	
Unit 6 - A；Unit 7 - B	
tentatively：试验性地；暂时地	Unit 9 - A
terrace：台地	Unit 9
terraced：露台	Unit 4 - C
terracotta：混合陶器；赤陶（用作花瓶、	
小雕像、建筑的饰物等）	U12 - C
terrain：地域；地带；（尤指从军事	
观点而言的）地势；地形	Unit 6 - B
territory：用地	10 - A
test boring：测试钻探	Unit 9 - A
thoroughfare：大街	Unit 1 - B
three - dimensionally：三维的	Unit 4 - B
threshold：开始；开端；极限	Unit 2
thrive：繁茂；蔓延	Unit 4
timber：木材	Unit 4
titmouse：长尾山雀	U11 - D
topiary：灌木修剪法	Unit 8；Unit 8 - B
topoclimatic：地形气候	U12 - A

topographic：地形的；地形学的	U12 - B
topographical undulations：地形起伏	
	Unit 9 - A
topographical：地形学的	Unit 1
topography：地形学	
Unit 1 - C；Unit 4 - B；Unit 7 - A；U11 - C	
tortuous：弯曲的；多扭曲的	Unit 5 - B
tot：儿童	Unit 5
trajectory：轨道	Unit 7 - B
tranquil：安静的；平静的；宁静的	U12 - C
tranquility：安静；平静；宁静	Unit 6
transposition：转换；换置；换位；	
移调	Unit 6 - A
trek：旅行；艰苦跋涉	U12 - D
trellis：（支架蔓生植物的）棚架；	
格子架；格子棚	Unit 9
trench：沟渠	Unit 7 - C
triangular：三角形的	U10 - D
triggering：触发；控制	Unit 3 - C
trivial：微不足道的	U11 - D
Tudor：英国都铎式建筑式样的	U10 - C
tupelo：篮果树	Unit 7 - C
turf：草地	U10 - A
ubiquitous：普遍存在；无所不在的	
	Unit 4；Unit 4 - A
unbridled：不受约束的	U12 - C
unconscious：不省人事的；	
无意识的	Unit 4 - B
uncontrollable variables：不可控	
变量	Unit 8 - B
uncoordinated：不协调的	Unit 4 - B
undeniable：无可否认的；确实的	U12 - B
undercurrent：潜流	Unit 7 - B
underpinnings：基础	U11 - A
underplanting：下木栽植	Unit 8
underrate：低估；看轻	U10 - C
understory：林下叶层；这里指中层乔木	
	Unit 8 - C
underutilized：未充分使用的	Unit 7 - C
undulate：（表面）波动；起伏	Unit 5 - A
unencumbered：不受妨碍的	Unit 7 - C
unity and variety：统一和变化	
（形式美原则之一）	Unit 2 - C

unobtrusive：不唐突的；不多嘴的；	
客气的；谦虚的	U10 - A
unprecedented：空前的	Unit 2 - C
unpredictable：不可预知的	Unit 8 - B
unwary：不注意的；粗心的；不警惕的	
	Unit 2 - C
unwieldy：不易运用或控制的；	
庞大的；笨重的	U12 - C
upkeep：保养	Unit 6
urban fabric：城市肌理	Unit 7 - C
urbane：文雅的	U12 - B
urbanization：城市化	Unit 4
urn：瓮；缸	U10 - C
usurp：篡夺；霸占	U12 - C
utilitarian：功利的	Unit 7 - A
utilitarian：以实用为主的	U12 - C
utopian：乌托邦的；理想化的	Unit 2 - B
valuation：评价	U11 - B
vandalism：故意破坏艺术的行为	U10 - C
vantage：优势；有利的地位	Unit 5 - B
variegated：杂色的；斑驳的；多样化的	
	Unit 8 - B
vehement：强烈的；猛烈的	Unit 5 - B
vehicular：车行的	Unit 9 - C
vehicular - pedestrian access：车行—	
人行入口	Unit 9 - A
veneer：薄板；单板；饰面；外表	U12 - C
venerate：崇敬	Unit 2；Unit 6 - A
Venetian：威尼斯式	Unit 4 - A
ventilation：通风；流通空气	Unit 8 - C
venues：地点	Unit 7 - B
verdant：青翠的	U11 - A
vernacular：本国的	Unit 7
veronica：婆婆纳属	Unit 8
verse：诗歌	Unit 4 - A
vertical：垂直面；竖向	Unit 8 - C
vibrant：明快的	Unit 9 - C
vibrantly：震动地；震颤地	U12 - C
vicious：坏的	Unit 7 - A
virtuosity：艺术鉴别力	Unit 8 - A
virtuoso：艺术品鉴赏家	U10 - A
visionary：（指人）有幻想的	U12 - C
vista：深景，尤指人透过如两排建筑	
或树木之间空隙看到的远景或	
视觉感受	Unit 2 - C
vitalize：激发	Unit 8 - B
vitalizer：活力激发者	Unit 2 - C
vogue：流行	U11 - B
void：空隙	Unit 4
warehouse：仓库	U11 - A
warrant：使有正当理由	U12 - A
wavy：有波状卷曲的	Unit 6 - A
weir：堰，拦河坝	Unit 2 - A
whet：（喻）促进；刺激（胃口，欲望）	
	Unit 6 - C
whimsical：古怪的	Unit 2 - A
whimsy：奇想	Unit 2 - A
whirlpool：旋涡；涡流	Unit 2
willfully：（指人）刚愎地；任性地	U12 - C
workhorse：驮马	Unit 7 - C
wreck：破坏	Unit 9 - C
Yosemite National Park：（美国加利	
福尼亚州中部）约塞米蒂国家	
公园	Unit 5 - C